FINITE MATHEMATICS APPLIED

CLEMENT E. FALBO

California State College, Sonoma

Wadsworth Publishing Company, Inc.
Belmont, California

TO AUDREY

Production Editor: Joanne Cuthbertson
Designer: Ann Wilkinson
Copy Editor: Carol S. Reitz
Technical Illustrator: Mark Schroeder

©1977 by Wadsworth Publishing Company, Inc., Belmont, California 94002. All rights reserved. No part of this book may be reproduced, stored in a retrieval system or transcribed, in any form or by any means, electronic, mechanical, photocopying, recording or otherwise, without the prior written permission of the publisher.

Printed in the United States of America
2 3 4 5 6 7 8 9 10—81 80 79 78

Library of Congress Cataloging in Publication Data

Falbo, Clement E.
  Finite mathematics applied.

  Includes index.
  1. Mathematics—1961-   I. Title.
QA39.2.F34   510′.2′43   76-22650
ISBN 0-534-00481-4

CONTENTS

PREFACE ix

1

SETS 1

1.1 Set Diagrams and Counting 1
1.2 Set Combinations 14
1.3 Subsets 27

2

INEQUALITIES AND SUMS 35

2.1 Inequalities—Basic Concepts 35
2.2 Inequalities as Constraints in the Management of Resources 44
2.3 Summation—An Introduction 54

3
GRAPHS AND COORDINATE GEOMETRY   61

3.1	Reading and Writing Graphs	61
3.2	Relations and Functions	73
3.3	Function Notation	86

4
LINES   97

4.1	Linear Equations	97
4.2	Point-Slope Form of the Line Equation	106
4.3	Simultaneous Linear Equations	109

5
INTRODUCTION TO LINEAR PROGRAMMING   119

5.1	Inequalities in the X-Y Plane	119
5.2	Geometric Linear Programming	130
5.3	The Use of Slack Variables to Find Corner Points	138
5.4	The Slack Variable Linear Programming Method	151

6
DETERMINANTS AND VECTORS   159

6.1	Introduction to Determinants	159
6.2	Third-Order Determinants	167
6.3	Vectors	176

7
MATRICES 185

7.1	Introduction to Matrices	185
7.2	Matrix Multiplication	199
7.3	Inverse of a Matrix	209
7.4	Leontief's Prize-Winning Matrix	219

8
GAUSS-JORDAN ELIMINATION 225

8.1	Introduction	225
8.2	Solving for Systems of Basic Variables	238

9
THE SIMPLEX METHOD FOR LINEAR PROGRAMMING 249

9.1	The Incoming Variable	249
9.2	The Outgoing Variable	264
9.3	Obtaining an Initial Feasible Solution	274
9.4	The Simplex Algorithm—History and Summary	286

10
INTRODUCTION TO PROBABILITY 291

10.1	Random Events and Simulation	291
10.2	Finite Probability Model	299
10.3	Sample Spaces with Equiprobable Events	306
10.4	Arranging and Counting Events—Permutations and Combinations	312

11
CONDITIONAL PROBABILITIES — 323

11.1	Conditional Probabilities and Independent Events	323
11.2	Total Probability	332
11.3	Partitions and Cross Partitions	341

12
PROBABILITY ANALYSIS — 351

12.1	Discrete Random Variable	351
12.2	Expected Values	361
12.3	Bayes' Theorem	368

13
THE MARKOV PROCESS — 379

13.1	Probability Vectors and Matrices	379
13.2	States and Transitions	387
13.3	The Markov Process—Theoretical Foundations	398

14
INTRODUCTION TO STATISTICS — 409

14.1	Elementary Concepts	409
14.2	Data Variability	419
14.3	Binomial Probability Distribution	430
14.4	Normal Distribution	438

15
CORRELATION — 445

| 15.1 | Linear Regression and Correlation | 445 |
| 15.2 | Multiple Regression and Path Models | 458 |

16
INTRODUCTION TO GAME THEORY — 471

16.1	Matrix Games	471
16.2	Mixed Strategies	485
16.3	Combined Strategies	495
16.4	Linear Programming Applied to Zero-Sum Games	502

ANSWERS TO ODD-NUMBERED PROBLEMS — 513

INDEX — 570

INDEX TO APPLIED PROBLEMS AND EXAMPLES — 574

PREFACE

The use of mathematics in the social and biological sciences is increasing at a hectic pace. Fundamental concepts that had been (awkwardly) described in ordinary language a few years ago are now being (elegantly) described by systems of equations. Advanced research is more and more frequently being based upon mathematical models, and the published results contain pages of mathematical expressions. This is happening in many fields from geography to medicine.

Conversely, the social and life sciences have started to provide mathematicians with interesting problems and methods. Mathematics journals publish an increasing number of articles concerned with new mathematical models applied to a variety of subjects, such as nerve impulses, transportation networks, tissue formation, and prison riots.

Finite mathematics courses are becoming part of the required curriculum for students majoring in such fields as management, sociology, and biology. This is due, in part, to the recognition that we are now more able to handle (through the use of computers) the large number of variables required to describe sociological and biological systems in mathematical terms. This situation presents mathematics departments with a new audience—students with little mathematics background and an urgent need to learn mathematics applied to their own fields of interest.

Many of the new finite mathematics textbooks, including this one, are attempting to meet these new requirements by providing modern applications and by assuming fewer prerequisites. In this text, I have used *applied problems* in a wide range of fields (triangles mark their beginnings and ends) to introduce most of the mathematical topics. I have also included relevant background material. Students who have used the trial versions of this text have often expressed their appreciation for this approach.

Although the text is written at a level that assumes approximately one year of high-school algebra or one semester of intermediate college algebra, it does contain some elementary material such as inequalities, coordinate geometry, and linear functions. From my experience, these are the most useful subjects to review in order to help those students whose backgrounds are weak. On the other hand, stronger students also enjoyed this treatment, especially the applied problems. A large number of my applied problems are inspired by primary sources. They are interesting enough to be assigned as independent study in an advanced class that could skip the explanatory material.

In addition to the elementary topics, some difficult ones are also included (Markov processes, multiple regression, and path analysis, for example). The difficult problems are starred by an asterisk *. An instructor will find that by picking and choosing topics, this text is suitable for a variety of levels. (A diagram of chapter interdependence is given at the end of the preface.)

The following brief assessment of each chapter should help instructors to determine which ones are most suitable for their own students' needs.

Chapter 1   **Sets** The primary objective of the material on sets is to provide an immediate method for solving real problems involving counting and classifying. A secondary objective is to provide enough set theory for the mathematics in subsequent chapters, especially the ones on probability.

Chapter 2   **Inequalities and Sums** Sections 2.1 and 2.3 are standard. Section 2.2, "Inequalities as Constraints in the Management of Resources," is very useful and shows how powerful a simple idea in mathematics can be when applied to the real world. Section 2.2 should not be skipped, even by the most advanced classes.

Chapter 3   **Graphs and Coordinate Geometry** The material is fairly standard and can be skipped by classes with the appropriate prerequisites (analytic geometry or one of its variants). This chapter introduces practical uses of graphs, rates, areas, and functional notation.

Chapter 4   **Lines** This is a standard treatment of linear functions; it does provide, however, more realistic problems than those usually associated with this topic.

Chapter 5   **Introduction to Linear Programming** This chapter goes far enough into linear programming (including minimizing problems and problems with mixed constraints as well as more than two variables) for a first-semester course. The explanation of slack variables, in terms of both their geometric significance and their real-life meaning, seems to be helpful to all students. If linear programming is to be covered at all in a course, then this chapter must be included; it contains the foundation of the simplex method.

Chapter 6	**Determinants and Vectors,** and
Chapter 7	**Matrices** Both chapters represent fairly standard treatments. The applications are drawn from a wide variety of fields. My students were particularly interested in Section 7.4 concerning the Leontief input-output model of economics.
Chapter 8	**Gauss-Jordan Elimination** Chapter 8 prepares the student for the final assault on linear programming problems. It provides the mechanical technique for the simplex method.
Chapter 9	**The Simplex Method for Linear Programming** This has been the most popular and most successful topic as the text evolved through several mimeographed versions. Two simplex methods are devised—one for maximizing and one for minimizing. They are practically the same (a simple matter of reversing the order of subtraction of two numbers distinguishes one from the other). These methods are easier to work with, easier to explain, and even more efficient than the usual *dual simplex method*. (For example, there is no need to write an entirely new system of inequalities.) Duality theory is unnecessary and is not even mentioned in this chapter.
Chapter 10	**Introduction to Probability,**
Chapter 11	**Conditional Probabilities,** and
Chapter 12	**Probability Analysis** These topics are applied to real problems, not just drawing colored balls out of containers. The unusual features include: an introduction to most of the mathematical ideas through the use of real-life examples, a discussion in Chapter 10 on random versus deterministic events, and the table in Figure 11.4 displaying several possible relationships between two events. The aim of this approach is to help students develop some intuition about probability theory.
Chapter 13	**The Markov Process** and
Chapter 15	**Correlation** These chapters, especially Sections 13.2, 13.3, and 15.2 present topics of more mathematical sophistication than those in the rest of the book. These sections should probably not be selected except in those courses with more advanced students.
Chapter 14	**Introduction to Statistics** This is a standard treatment of the elementary topics in statistics.
Chapter 16	**Introduction to Game Theory** I have tried to show that game theory can be used in everyday life. This introduction has stimulated some of my students to pursue the topic further.

Chapter Interdependence

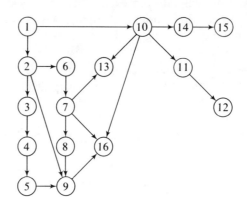

Possible Courses

Course Title	Chapters
Linear Programming	2, 4, 5, 6, 7, 8, 9, 16
Applied Elementary Probability	1, 2, 6, 10, 11, 12, 13
Probability and Statistics	1, 10, 11, 12, 14, 15
Math for Social and Life Sciences (two semesters)	
First semester	1, 2, 3, 4, 5, 6, 7
Second semester	8, 9, 10, 11, 12, 14, 16

ACKNOWLEDGMENTS

A number of people were involved in the development of this book. Hundreds of students at Sonoma State College were taught from the various mimeographed versions. Almost every faculty member of the mathematics department has participated in testing this material at some time or another. In particular, I would like to thank William J. Barnier and Norman Feldman for their valuable comments and suggestions. I am indebted to Charles J. Phillips for reading and criticizing an early version of the chapters on probability. A special thanks goes to Sally Cochran, who not only typed large parts of the manuscript, but also worked some of the problems.

I would also like to express my appreciation to Wadsworth's mathematics editor Don Dellen and to the many reviewers for their constructive criticism. I am pleased with the careful editing and the imaginative format provided by the editorial and design staff members who worked on my book.

1
SETS

1.1
SET DIAGRAMS AND COUNTING

The classification of such things as blood samples, jobs, and student attitudes into various types or categories is an application of the *mathematics of sets*. The set concept is so fundamental that the word *set* will be a **primitive term** in this text. That means that we will not try to define it in terms of anything more simple. We will, however, establish certain basic agreements about sets in this introduction.

The first common notion needed in any discussion of set theory is that *a set must be well defined*. There must be a precise description of the set, and this can be in almost any form: prose, poetry, equations, legal jargon, scientific jargon, tables, graphs, or even sign language. For a collection of objects to be a set, it must be defined well enough so that it will be easy to determine whether or not any given item or number or person belongs to that set.

For example, it is inaccurate to speak of the "set of stockholders of AT&T common shares," since the stockholders may change from hour to hour. On the other hand, one may speak of the (well-defined) *set of people or institutions who are AT&T stockholders of record at the close of trading on Friday, August 22, 1975*. It is important to know the exact membership of this set because only these stockholders can receive the third-quarter dividends on October 1, 1975.

The set of AT&T stockholders (of record August 22, 1975), or any other set, can be geometrically represented by a set diagram called a **Venn diagram.** Such a diagram is often a circle, or sometimes an ellipse, or a rectangle. The area *inside* the figure represents elements *in* the set, and the area *outside* the figure represents elements *not in* the set. See Figure 1.1.

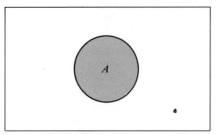
Shaded region is a set A;
A = AT&T stockholders of record.

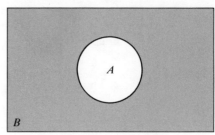
Shaded region is the set B;
B = the people who are not AT&T stockholders of record.

Figure 1.1
Venn diagrams for sets A and B.

Two or more sets may *overlap* (have elements in common), and this can also be represented by Venn diagrams. See Figure 1.2.

Diagrams are useful in certain counting problems involving combinations of sets, as shown in the following example.

Example 1.1* A brokerage house surveyed its customers and found that

25,000 of them held shares in AT&T.
30,000 of them held shares in Greyhound Corp.
10,000 held both AT&T and Greyhound shares.
40,000 held neither Greyhound nor AT&T shares.
How many customers does this brokerage house have? How many customers do not hold *any* AT&T stock?

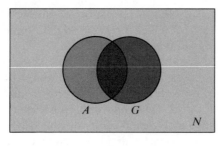

● A = AT&T stockholders of record.

● G = Greyhound stockholders of record.

● N = people who are neither.

● Overlapping area = stockholders of record in both companies.

Figure 1.2
Combination of two sets.

*You will find a triangle at the beginning and end of applied material.

1.1 Set Diagrams and Counting

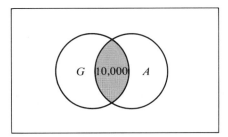

Figure 1.3

To show that this is not just a simple problem in addition, we point out that 105,000 is the *wrong answer* to the first question and 40,000 is the *wrong answer* to the second question.

Solution While there may be several ways to work this problem, we will illustrate the method using Venn diagrams. Let A be the set of AT&T stockholders, and let G be the set of Greyhound stockholders (of record on a certain date). Since 10,000 people are in both sets G and A, a diagram can be drawn showing 10,000 in the overlapping part of these two sets. See Figure 1.3.

This 10,000 is also counted among the 30,000 Greyhound stockholders and the 25,000 AT&T stockholders; therefore, the rest of G contains 20,000 members and the rest of A contains 15,000 members. See Figure 1.4.

Notice that Figure 1.4 shows four distinct nonoverlapping sets:

1. Those stockholders who are in both sets A and G

2. Those who are in G, but not in A (they own Greyhound stock, but no AT&T stock)

3. Those who are in set A but not in set G

4. Those who are in neither set

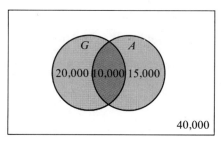

Figure 1.4

To see how many stockholders there are altogether, we simply add the numbers in the four nonoverlapping sets:

> 10,000 who are in both sets
> 20,000 who are in G, but not in A
> 15,000 who are in A, but not in G
> <u>40,000</u> who are in neither A nor G
>
> 85,000 Total number of stockholders

The diagram also tells how many people do not have any AT&T stock; they are all those outside of set A (the 40,000 who hold neither stock, plus the 20,000 who hold Greyhound, but not AT&T, for a total of 60,000). ▲

The above example illustrates an important logical principle for counting the number of elements in sets. The different types of elements in the sets must be separated into nonoverlapping sets before the total can be computed. If a set contains one type of element, we can't necessarily assume that it doesn't contain another type as well.

A problem involving only two types of elements (such as AT&T and Greyhound stockholders) is fairly simple. We get into more complicated problems involving double and triple counting of elements when we work with three or more sets, such as in the next example with three overlapping sets.

▲ **Example 1.2** Suppose you decide to start a rental service store, but first you want to determine: 1. the total number of such stores in your city, 2. the number that concentrate on one type of rental service, and 3. the number that offer various combinations of rental services. Such a survey of the current competition is essential to your decision to enter this business. Your analysis of the present situation reveals that the stores offer a variety of rental equipment and some have a wider range than others, but generally they rent equipment in three categories: garden, camping, and household.

The first step in solving this problem is to introduce some set notation, one set for each category of equipment. Let

> X = the set of all stores that offer garden equipment for rent.
>
> Y = the set of all stores that rent camping equipment.
>
> Z = the set of all stores that rent household equipment.

These sets overlap because some stores belong to all three sets, others belong to one or two of the sets. Figure 1.5 shows the general situation.

1.1 Set Diagrams and Counting

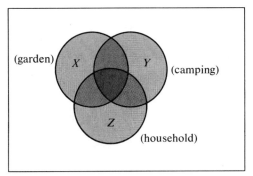

Figure 1.5
Rental stores.

In your survey you obtain the following data:

A total of seven stores rent all three types of equipment: garden, camping, and household.

A total of twelve stores rent garden and camping equipment. (These twelve stores overlap with the seven already counted as renting all three types of equipment.)

A total of nine stores rent camping and household equipment (including the seven renting all three types).

A total of twenty stores rent garden and household equipment (including the seven renting all three types).

A total of twenty-six stores rent garden equipment (including those that rent garden plus other equipment).

A total of fourteen stores rent camping equipment (including those that rent other equipment).

A total of thirty-eight stores rent household equipment (including those that rent other equipment).

Here are your questions:

a. How many stores are in this survey?

b. How many rent only camping equipment? (That is, they rent no garden or household equipment.)

Solution We will use Venn diagrams working from the inside out. We start with the number in all three sets as shown in Figure 1.6a. Seven stores are in the overlap of all three sets. Next, in Figure 1.6b the twelve elements in the overlap of X and Y are shown. Seven of them have already been counted as stores in which all three types of equipment are rented; this means that there are five remaining

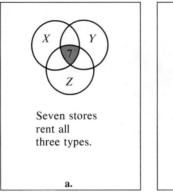
a.
Seven stores rent all three types.

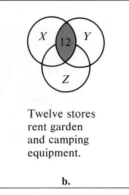
b.
Twelve stores rent garden and camping equipment.

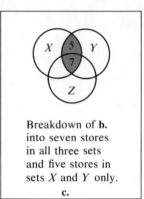
c.
Breakdown of **b.** into seven stores in all three sets and five stores in sets X and Y only.

Figure 1.6

stores renting garden and camping equipment, but no household equipment. These five plus the seven renting all three types give us a total of twelve stores renting both garden and camping equipment.

Similarly, Figure 1.7 illustrates the breakdown for the combination of stores renting camping and household equipment.

Figure 1.8 illustrates the breakdown for the combination of stores renting garden and household equipment.

We summarize Figures 1.6c, 1.7c and 1.8c in Figure 1.9.

In this diagram, we see that twenty-five (5 + 7 + 13) garden equipment stores also rent other types of equipment. The original data said that twenty-six stores rent garden equipment; therefore this leaves *one* store that rents only garden equipment. This result is shown in Figure 1.10.

This figure shows that there four different types of stores that rent garden equipment. The description and number of these types are

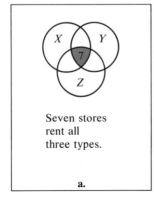
a.
Seven stores rent all three types.

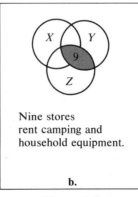
b.
Nine stores rent camping and household equipment.

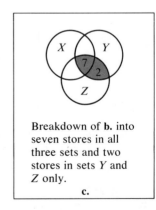
c.
Breakdown of **b.** into seven stores in all three sets and two stores in sets Y and Z only.

Figure 1.7

1.1 Set Diagrams and Counting

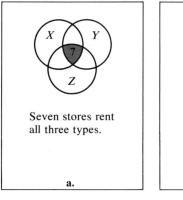
Seven stores rent all three types.

a.

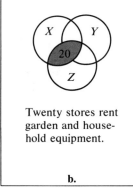
Twenty stores rent garden and household equipment.

b.

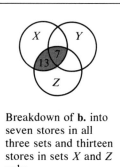
Breakdown of **b.** into seven stores in all three sets and thirteen stores in sets X and Z only.

c.

Figure 1.8

Stores that rent *garden,* camping, and household equipment	7
Stores that rent *garden* and camping equipment only	5
Stores that rent *garden* and household equipment only	13
Stores that rent *garden* equipment only	1
Total stores that rent garden equipment	26

Similarly, Figure 1.9 shows that fourteen (5 + 7 + 2) stores rent camping equipment. The original data said a *total* of fourteen stores rented camping equipment, so no stores rent *only* camping equipment. See Figure 1.11.

Finally, of the thirty-eight stores renting household equipment, twenty-two (13 + 7 + 2) of them have already been counted among those stores also renting other types of equipment, as shown in Figure 1.9. This leaves sixteen stores that rent only household equipment. See Figure 1.12.

We can now summarize all of the data in one diagram. Figure 1.13 completely separates all the sets into nonoverlapping parts.

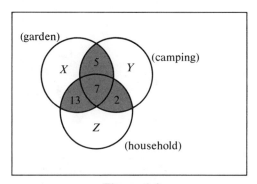

Figure 1.9

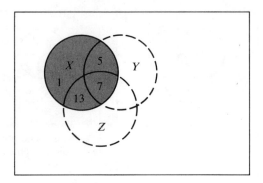

Figure 1.10
The twenty-six stores that rent garden equipment (set X).

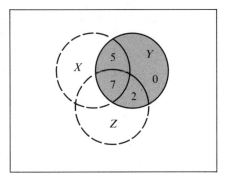

Figure 1.11
The fourteen stores that rent camping equipment (set Y).

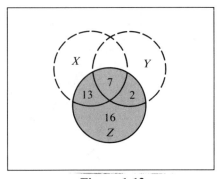

Figure 1.12
The thirty-eight stores that rent household equipment (set Z).

1.1 Set Diagrams and Counting

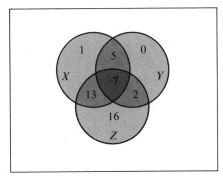

Figure 1.13
The complete breakdown for the rental stores.

The solutions to the two parts of the problem are

a. The number of stores in the survey:

   1 garden only
13 garden and household only
   5 garden and camping only
   7 garden, camping, and household
   0 camping only
   2 camping and household only
<u>16 household only</u>
44 total number of rental stores

b. The number of stores that rent only camping equipment:

none

▲

EXERCISE 1.1

1. From the diagram, find the number of elements

   a. in set A

   b. in set B, but not in set A

   c. in neither set A nor set B

   d. in both sets A and B

   e. in either set A or set B (Note: An element in both sets is also in either set.)

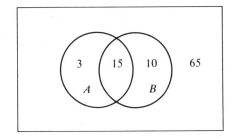

2. From the diagram, find the number of elements
   a. in both sets
   b. in set Y
   c. in either set X or set Y
   d. in set X only or in set Y only, but not in both

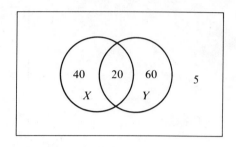

3. From the diagram, find the number of elements
   a. in set B
   b. in both sets A and C, but not in B
   c. in set C only
   d. in either set A or C
   e. in set B and at least one other set

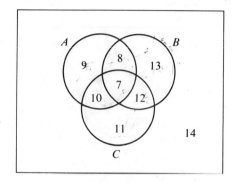

4. From the diagram, find the number of elements
   a. in set Z
   b. in both sets Y and X, but not set Z
   c. in at least two sets

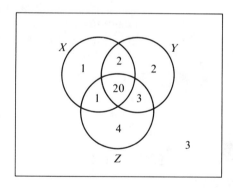

5. Suppose A and B are sets as shown with the overlapping part containing 100 elements. If set A has a *total* of 247 elements and set B has a *total* of 194 elements:
   a. How many elements are in set A but not in set B?
   b. How many elements are there altogether?

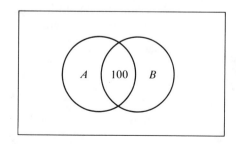

1.1 Set Diagrams and Counting

6. Suppose sets X and Y are as shown with 37 elements in the overlapping part. If set X has a *total* of 52 elements and set Y has a *total* of 75 elements, what is the total number of elements?

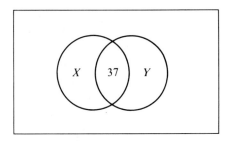

▲ 7. In preparation for starting your own business, you make a study of the fast-food stores in your city. You find that

   4 sell hamburgers, chicken, and ice cream
   14 sell hamburgers and ice cream
   22 sell chicken and ice cream
   5 sell chicken and hamburgers
   16 sell hamburgers
   25 sell chicken
   35 sell ice cream

   a. How many stores are in the city?

   b. How many sell only ice cream? (The answer is *not* 35; see the discussion on rental service stores in Example 1.2.)

   c. How many sell ice cream and chicken but no hamburgers?

8. In XYZ county in California, a survey reveals that fruit farmers grow apples, plums, grapes, or none of these fruits. The data are:

   8 grow grapes, apples, and plums
   22 grow grapes and apples
   16 grow apples and plums
   11 grow grapes and plums
   38 grow grapes
   54 grow apples
   68 grow plums
   129 grow none of these fruits

   a. How many farmers are there in XYZ county?

   b. How many grow *only* apples?

   c. How many grow *only* plums?

9. Of the people purchasing a certain brand of portable tape recorder, 950 completed and mailed their consumer survey cards. The following information was gathered:

232 already owned an FM radio, a TV set, and a stereo set
 50 owned FM and TV, but no stereo
400 owned only TV (no stereo and no FM)
332 owned both TV and stereo
243 owned both FM and stereo
 25 owned only stereo (no TV and no FM)
 10 owned none of these

a. How many owned FM only (no TV and no stereo)?

b. How many TV owners were there?

c. How many owned either TV or FM?

d. How many owned either a TV set or both FM and stereo?

e. What percentage of the entire group owned both TV and FM?

*10. In the article, "Student Attitudes Regarding the Temporary Closing of a Major University" by K. E. Rudestam and B. J. Morrison in *American Psychologist,* May 1971, a study was made of student attitudes toward the closing of Miami (Ohio) University in May 1970—a period of student unrest following the U.S. invasion of Cambodia. In the questionnaire submitted to the students, three questions reflecting attitudes toward campus closing were:

1. "Miami University will have to change its educational policies in the future."
2. "Trouble will never end at Miami until the whole university administrative structure is overturned."
3. "Violence seems to be the only way to get the administration to listen to us."

There was a choice of four answers ranging from "definitely false" to "definitely true." Students giving affirmative answers to question 1 are said to be in set Y_1. Similarly, an affirmative answer to question 2 puts that student in set Y_2, and yes on question 3 puts the student in set Y_3. Of the 432 students returning the questionnaires, the numbers in the various sets were:

193 students were in set Y_1 (drastic educational change needed)

---

*The asterisked problems throughout the text indicate exceptionally difficult or lengthy problems.

1.1 Set Diagrams and Counting

82 students were in set Y_2 (administration overturn)
112 students were in set Y_3 (violence is the only way)
62 students were in both sets Y_1 and Y_2
69 students were in both sets Y_1 and Y_3
36 students were in both sets Y_2 and Y_3
30 students were in all three sets, Y_1, Y_2, and Y_3

  a. Draw a Venn diagram for these results and put the appropriate numbers in the nonoverlapping areas.

  b. How many of the 432 students were not in any of the sets, Y_1, Y_2, or Y_3?

  c. What percentage of the 432 students thought that either there would have to be a drastic educational change or the administration would have to be overturned?

  d. What percentage of the 432 students felt that violence was the only way and the administration must be overturned, but that there was no need for a drastic educational change? (Hint: They answered no to question 1, and yes to questions 2 and 3).

*11. Human blood is classified according to whether or not it contains three materials: the Rh antigen, the agglutinogen A, or the agglutinogen B. People with the Rh antigen are said to have positive blood, or Rh-positive blood (denoted Rh^+). Those without this antigen are said to be Rh-negative (Rh^-). People with both agglutinogens A and B have type AB blood. If this type also contains the Rh antigen, it is AB Rh^+ (or simply, AB positive) blood. People whose blood contains neither A nor B are said to have type O blood, which can be positive or negative according to the Rh factor.

  a. Draw a Venn diagram of these three blood materials and make a list of all eight blood types represented by the eight areas in the diagram.

  b. In a sample of 1000 human beings

500 have the A agglutinogen
180 have the B agglutinogen
850 have the Rh antigen
 80 have both the A and B agglutinogens
425 have the Rh antigen and the A agglutinogen
153 have the B agglutinogen and the Rh antigen
 68 have all three factors

How many are AB-negative? O-positive? O-negative?

1.2

SET COMBINATIONS

Real-life applications in the previous section illustrated such processes as 1. describing when an element belongs or doesn't belong to a set, 2. studying elements in two or more overlapping sets and the separation of these elements into nonoverlapping pieces, and 3. counting the total number of elements in two or more intricately related sets.

These and other concepts can be expressed in a mathematical shorthand introduced in this section. Mathematical notation and terminology help to streamline discussions of various ideas because brief symbols can be used in place of longer awkward phrases. You will be able to see this better after we look at the following examples. We begin with the symbol for set membership.

SET MEMBERSHIP, SET COMPLEMENT

The symbol $\in$ means *is an element of*, or *belongs to*. The symbol $\notin$ means *is not an element of*, or *does not belong to*. Thus, if A is a set and x is an element of A, we can write $x \in A$. If B is a set, the statement that y is not a member of B is written $y \notin B$.

Example 1.3 Let Z be the set of all *whole numbers* (also known as *integers*). Then $1 \in Z$, $2 \in Z$, $3 \in Z$; also $0 \in Z$, $-1 \in Z$, $-2 \in Z$, because $1, 2, 3, \ldots$, etc., are integers. Some examples of numbers that are not elements of Z are $1/2$, $35/17$, etc. (Since $1/2$ is not an integer, $1/2$ is not an element of Z.) Similarly, $35/17 \notin Z$ because $35/17$ is not an integer. See Figure 1.14.

In the above example, $1/2$ is not an element of Z, but it is still an element of some set. It belongs to a set called **complement** of Z, consisting of all the numbers that are not in Z.

In general, if A is any set, the complement of A is the set of *all elements under consideration that are not in A*. Thus, if $y \notin A$, then y is an element of the complement of A. One notation for the complement of A is $-A$. Hence, we write: If $y \notin A$, then $y \in -A$. Also, if x is not a member of the complement of A, then x is a member of A. In symbols: If $x \notin -A$, then $x \in A$.

Example 1.4 Sometimes junior high-school students play a game called In-Out. A group of people standing in a circle pretend to be the Ins and everybody else is a member

1.2 Set Combinations

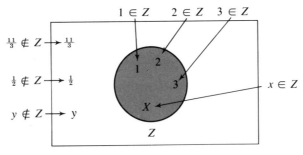

Z = shaded area = set of integers.
Figure 1.14

of the Outs. The Ins joke, point at people who are not with them, and in general act as if they are sharing some great secret. This game can be described in set notation for easy study. Let the Ins be denoted by I and the Outs by $-I$.

Suppose that Mike is a student in a certain junior high school; let the letter m stand for Mike. Express the following sentences in set notation:

a. Mike is a member of the In group.

b. Mike is a member of the Out group.

c. Mike is not a member of the Out group.

Also show that statements **a** and **c** mean the same thing.

Solution **a.** $m \in I$. **b.** $m \in -I$. **c.** $m \notin -I$.

Now, since every student in the junior high school must be either in I or in $-I$, then from part **c** $m \notin -I$, we can conclude that $m \in I$.

As the above example illustrates, one (informal) description of the complement of a given set would be the set of all elements *outside* that set. This description suggests a way to depict set complements by Venn diagrams.

We can emphasize the elements in a set by shading the *interior* of the circle for that set in a Venn diagram. The complement of the set is emphasized in another diagram by shading the *exterior* area. Look back at Figure 1.1. The shaded region inside the circle in the first diagram is the set A of AT&T stockholders; the shaded region exterior to set A in the other diagram is the set B of people who are not AT&T stockholders. That is, $B = -A$.

This interior-exterior shading scheme for sets and their complements is analogous to a photographic print and its negative. See Figure 1.15.

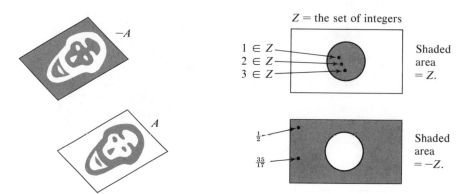

Figure 1.15
Sets and their complements.

SET-BUILDER NOTATION

Braces—{ and }—are used to contain the elements of a set; they are called **set builders.** If a list of items (separated by commas) appears between braces, then those items are elements of the set built by the braces. For example, if a, b, and c are elements of a set X, then we write $X = \{a,b,c\}$, which reads "X is the set containing the elements a, b, and c." Let $Y = \{n\}$, then Y is the set whose only element is n. (The *set* Y and the *element* n in Y are not the same thing.) A set that contains only one element is called a **singleton set.**

Example 1.5 In the 400-year history of world chess championships, only three chess players from the Western Hemisphere have ever been recognized as world champions. Let W = the set of all people of the Western Hemisphere who have been the world champion chess player at some time, then W = {Morphy, Capablanca, Fischer}.

Often braces are used to enclose a *general description* of the elements in the set rather than a list of all the elements; thus, for example, we can write

$$W = \{m : m \text{ is a person from the Western Hemisphere who has been recognized as the world chess champion at some time}\}.$$

This reads "W is the set of all elements m, such that m is a person from the Western Hemisphere who has been recognized as the world chess champion at some time." Notice that the colon reads *such that*. Another notation for this phrase (such that) is a vertical bar $|$. Example: Let

$$A = \{m | m \text{ was world chess champion in 1575}\},$$

then A is the singleton set: A = {Lopez}.

1.2 Set Combinations

The value of set-builder notation with a general description is most evident when the elements are so numerous that it would be difficult, if not impossible, to list them all.

Example 1.6 Let E be the set of all even positive integers from 2 through 200. We can write

$$E = \{x: x \text{ is an even number between 2 and 200, inclusive}\}.$$

A partial list of the elements may be given assuming that the pattern for continuing the list is obvious. For example, If $S = \{2,4,6, ..., 200\}$, one may assume that set S is the same as the set E. Such an assumption is practical, but it may be logically invalid, since we haven't actually described how the terms represented by the ... are to be filled in.

INTERSECTION OF SETS

If A and B are two sets, then the **intersection** of A and B is the set of all elements that belong, simultaneously, to both sets. For example, in Figure 1.3, the 10,000 AT&T stockholders (set A) who also hold Greyhound stock (set G) are in the intersection of A and G. In Figure 1.6a, the stores that rent all three types of equipment are members of the intersection of all three sets: X (garden equipment rental stores), Y (camping equipment), and Z (household equipment).

The notation for intersection is ∩. That is, the intersection of two sets, X and Y, is written as $X \cap Y$. The intersection of three sets, X, Y, and Z, is denoted by $X \cap Y \cap Z$.

Example 1.7 The Venn diagrams in Figure 1.16 show the intersections of various collections of sets.

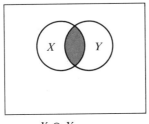
$X \cap Y$
Intersection of
X and Y

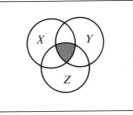
$X \cap Y \cap Z$
Intersection of
X, Y, and Z

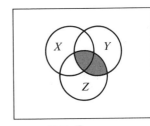
$Y \cap Z$
Intersection of
Y and Z

Figure 1.16
The shaded areas represent intersections.

Example 1.8 Find $\{-1,0,3,4,5,6,15,21,22\} \cap \{-3,-2,0,5,6,7,11,15,18\}$.

Solution $\{0,5,6,15\}$

The intersection of a set A with the complement of B is the set of all elements in set A that are not in set B. It is sometimes denoted by $A - B$, read "A minus B." See Figure 1.17.

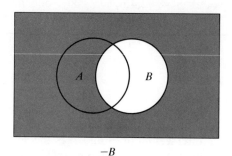
$-B$
Shaded area is the complement of B (part of set A is in $-B$).

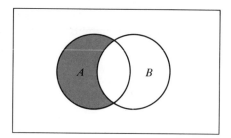
$A - B = A \cap (-B)$
Shaded area is the elements of A that are also in the complement of B.

Figure 1.17
The set $-B$ and the set $A - B$, that is, $A \cap (-B)$.

Example 1.9 Let A be the set of all positive integers from 1 through 15. In the notation of the set-builder braces

$$A = \{1,2,3,4,5,6,7,8,9,10,11,12,13,14,15\}.$$

Let B be the set of all multiples of 3 from 3 through 15; i.e., $B = \{3,6,9,12,15\}$. Let C be the set of all even numbers between 1 and 15.

a. Express C as a set of elements in the notation of the set-builder braces.

b. Express $B \cap C$ in this same brace notation.

c. List the elements of $A - B$. Describe this set in words.

d. List the elements of $C - B$.

e. List the elements of $B - C$.

Solution
a. $C = \{2,4,6,8,10,12,14\}$.
b. $B \cap C = \{6,12\}$.

1.2 Set Combinations

c. $A - B = \{1,2,4,5,7,8,10,11,13,14\}$. $A - B$ is the set of elements in A that are not in B. Therefore, $A - B$ is the set of integers from 1 through 15 that are not multiples of 3.

d. $C - B = \{2,4,8,10,14\}$.

e. $B - C = \{3,9,15\}$.

UNION OF SETS

If A and B are sets, then the **union** of sets A and B is the sum total of all the elements that belong to at least one of the sets. If an element is *in either one set or the other* (or in both sets), then it is in the union of the two sets. Any element in the union of A and B must be in at least one of the sets. The union of sets A and B is denoted by $A \cup B$. The union of two sets is also called the *join* of the sets, since it joins together all of the elements in either set. In symbols, $x \in A \cup B$ if and only if $x \in A$ or $x \in B$.

Example 1.10 In Example 1.1 (page 2), the 45,000 stockholders who hold shares in Greyhound (set G), in AT&T (set A), or in both are members of the union of the sets A and G. See Figure 1.18a.

In Example 1.2 (page 4), any store that rents *at least one* of the three types of equipment belongs to the union $X \cup Y \cup Z$. See Figure 1.18b. The other two parts of Figure 1.18 represent various combinations of unions and intersections.

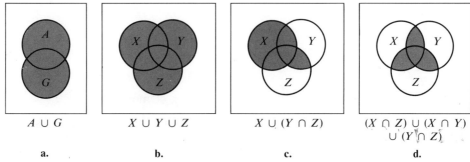

Figure 1.18
The shaded areas represent the indicated unions.

Figure 1.18c, for example, represents the set of stores that rent either garden equipment or both camping and household equipment.

The construction of new sets by combining other sets provides some important practical applications of set theory. In statistical applications, for example, set descriptions (in words, tables, or equations) can really define a portion of the population. The set of all men whose height is between $5\frac{1}{2}$ feet and 6 feet is a precise *sample* of the whole human population. A *characteristic* of the sample is the property of being a man whose height is between $5\frac{1}{2}$ feet and 6 feet.

If sets A and B represent different population samples, then the union of A and B is the sample consisting of those people who have properties characteristic of either one or the other of the two samples. The intersection of A and B is a population sample consisting of those people who have properties characteristic of both groups.

Example 1.11 If A = set of all men whose height is between $5\frac{1}{2}$ feet and 6 feet, and B = set of all men who have completed one year of college, then the union of the two sets is

$A \cup B$ = the set of all men who either are between $5\frac{1}{2}$ feet and 6 feet tall *or* have completed one year of college, or both.

The intersection of the two sets is

$A \cap B$ = the set of all men who are between $5\frac{1}{2}$ feet and 6 feet tall *and* have completed one year of college.

In general, the union of two sets contains more elements than the intersection of the same two sets. There are even cases in which the intersection contains no elements at all!

Example 1.12 If $A = \{1,2,3\}$ and $B = \{4,5,6\}$,

    a. How many elements are in $A \cup B$?

    b. How many elements are in $A \cap B$?

Solution   **a.** $A \cup B = \{1,2,3,4,5,6\}$, which has six elements.

    b. A and B are nonoverlapping; therefore, they have no elements in common and their intersection, like old Mother Hubbard's cupboard, is bare. That is, $A \cap B$ is empty, null, and void.

1.2 Set Combinations

We conclude this section with a discussion of the set containing no elements.

THE EMPTY SET

If a set contains no elements, it is called the **empty set.** In mathematics, we assume all empty sets are the same; that is, there is one and only one empty set. The set of live elephants on the sun is exactly the same empty set as $A \cap B$ in Example 1.12.

Riddle How is the set of Nazguls like the set of Hobbits?

Answer They are both the empty set (because there are no such things as Nazguls or Hobbits).

The empty set may arise naturally in an experimental situation.

Example 1.13 Suppose an experimenter is planning to study the relationship between certain personality traits and heights in men. One of the steps in designing the experiment would be to divide the *height variable* into several categories. The experimenter may choose the following height categories:

> Set H_1: men below $4\frac{1}{2}$ feet tall
>
> Set H_2: men between $4\frac{1}{2}$ feet and 5 feet tall
>
> Set H_3: men between 5 feet and $5\frac{1}{2}$ feet tall
>
> Set H_4: men between $5\frac{1}{2}$ feet and 6 feet tall
>
> Set H_5: men taller than 6 feet

After the experiment is designed in detail, a random sample of the population is selected. The experimenter has no prior knowledge that the prospective sample will have subjects in every one of the sets, H_1, H_2, H_3, H_4, and H_5. It may turn out that the sample includes no man whose height is below $4\frac{1}{2}$ feet. The set H_1 will then be the empty set. This information would go into the report of the experiment unless the experimenter wished to redesign the experiment and deliberately seek subjects in the set H_1. The selection would then no longer be random, and such a change could make a difference in the conclusions.

EXERCISE 1.2

1. If $A = \{3,4,5,6,7\}$ and $B = \{3,6,9,12,15\}$
   a. Is $9 \in A$?
   b. Is $12 \notin A$?
   c. Is $3 \in A \cap B$?
   d. Is $5 \in A \cup B$?
   e. Is $10 \notin A \cup B$?
   f. Is $7 \in A \cap B$?

2. If $X = \{a,b,c,d,e,f\}$ and $Y = \{a,e,i,o,u\}$
   a. Is $d \in Y$?
   b. Is $f \notin X$?
   c. Is $i \in X \cap Y$?
   d. Is $c \in X \cup Y$?
   e. Is $g \notin X \cup Y$?
   f. Is $e \in X \cap Y$?

3. Copy the Venn diagram on the right.
   a. Shade the intersection $S \cap R$.
   b. Shade the combination $T \cup (S \cap R)$.
   c. Shade the set $(-S) \cap T$.

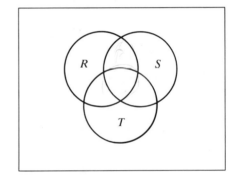

4. Copy the Venn diagram on the right.
   a. Shade the intersection $D \cap E \cap F$.
   b. Shade the combination $(D \cap F) \cup (D \cap E) \cup (F \cap E)$.
   c. Shade the set $-(D \cup E \cup F)$.

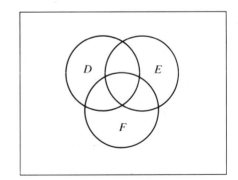

1.2 Set Combinations

▲ 5. To test a certain theory, a psychologist constructed two lists of words, one associated with colors and the other with emotional moods. Let

$$W_C = \text{the set of color words.}$$
$$W_M = \text{the set of mood words.}$$

A verbal description of the set $W_C \cap W_M$ is: "$W_C \cap W_M$ is a set of words associated with both colors and moods."

a. State a verbal description of the set $W_C \cup W_M$.

b. State a verbal description of the set $W_C - W_M$.

c. State a verbal description of the set $W_M - W_C$.

6. A simplified classification of prehistoric animals can be given by the sets

T = the set of prehistoric animals present in the temperate zone.

S = the set of prehistoric animals with short limbs.

a. Describe $T \cup S$ in words.

b. Describe $T \cap S$ in words.

c. Describe $T - S$ in words.

d. Let u, x, y, and z be animals in the set $T \cup S$ as designated below:

u = a finback, w = caseid, x = a therapsid, y = a dinosaur, and z = a cenozoic mammal. According to the article "Dinosaur Renaissance" by Robert T. Bakker in *Scientific American*, April 1975, we can write: $u \in S - T$, $w \in S - T$, $x \in T$, and $x \in S$, $y \in T - S$, and $z \in T - S$. Use this information to put the letters u, w, x, y, and z in their correct places in the Venn diagram on the right.

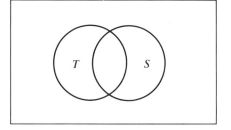

*7. All the sets in this problem refer to the following table on poverty households.

Table of the Annual Income (1967) of U.S. Households at the Poverty Level

Number of Members in the Household	Nonfarm Households Annual Income	Farm Households Annual Income
One member	$1635	$1145
Two members	$2115	$1475
Three members	$2600	$1815
Four members	$3335	$2345
Five members	$3930	$2755
Six members	$4410	$3090
Seven or more members	$5430	$3790

Source: *Economic Report of the President*, January 1969. Data from the Department of Health, Education, and Welfare.

a. Let A = the set of households that have exactly one or two members. Describe $-A$. If x is a family with more than seven members, is $x \in A$?

b. If B = the set of nonfarm households having two or more members, what is $A \cap B$? Do all farm families belong to $-B$?

c. If y is a nonfarm household with only one member, is $y \in B$?

d. If $C = \{h: h$ is a household with annual income greater than $2300\}$ and $D = \{h: h$ is a farm household with less than four members$\}$, describe $C \cap D$ in words.

e. Describe $A - D$ in words. Find the average income of the households in $A - D$.

f. Describe the set $A \cup D$ in words.

g. Describe the set $(A \cap C) \cup (C \cap B)$ in words.

*8. The data given here are from the U.S. Census Bureau (1970). They will be used in this and other problems.

1.2 Set Combinations

Alaska: Total Households: 88,300. Population: 300,000

Plumbing Facilities	Total	Urban	Rural
Flush toilets	74,500	42,700	31,800
No flush toilets	13,000	500	12,500
Kitchen			
Complete facilities	74,500	42,500	32,000
No complete facilities	13,500	800	12,700
Telephone			
Available	57,600	34,300	23,300
None	20,500	6,400	14,100

Arkansas: Total Households: 672,800. Population: 1,900,000

Plumbing Facilities	Total	Urban	Rural
Flush toilets	576,000	325,000	251,000
No flush toilet	92,000	9,500	82,700
Kitchen			
Complete facilities	585,000	320,000	264,500
No complete facilities	87,000	17,300	69,500
Telephone			
Available	463,500	260,000	203,500
None	152,500	56,500	96,000

Delaware: Total Households: 175,000. Population: 548,000

Plumbing Facilities	Total	Urban	Rural
Flush toilets	169,000	122,700	46,400
No flush toilets	5,200	300	4,800
Kitchen			
Complete facilities	169,700	122,300	47,400
No complete facilities	4,900	1,100	3,800
Telephone			
Available	149,000	108,500	40,300
None	16,000	9,400	6,500

North Dakota: Total Households: 200,300. Population: 618,000

Plumbing Facilities	Total	Urban	Rural
Flush toilets	176,800	80,800	96,000
No flush toilets	19,770	250	19,520
Kitchen			
Complete facilities	178,000	80,900	97,000
No complete facilities	22,000	2,800	19,200
Telephone			
Available	164,300	75,000	89,300
None	17,300	4,900	12,400

All sets refer to the households in these four states.

a. Let A = the set of all households that have flush toilets, and let B = the set of all households in Arkansas. How many elements are in $A \cup B$? in $A \cap B$?

b. Let $C = \{x: x$ is a rural household in either Delaware or North Dakota$\}$. Describe in words the sets $A \cup C$, $B \cup C$, $A \cap C$, and $B \cap C$.

c. Let D be all households with complete kitchen facilities. Describe $-(A \cap D)$, the complement of $A \cap D$, in words.
Describe $-A \cup -D$, the union of the complements of A and D, in words.

d. Let $S =$ the set of *all* households in *all four* states. What percentage of the elements in S are households that do not have access to a telephone? What percentage of S has no flush toilets? ▲

1.3

SUBSETS

If A is any set, then a **subset** C of A is a set within A. In other words, the statement that *the set C is a subset of A means that every element of C is an element of A* (or, there is no element of C that is not an element of A). See Figure 1.19.

Every set is a subset of itself, and the empty set is a subset of every set. This latter fact may not be obvious at first, but think about it. Let A be any set. If the empty set has no elements, how can it have elements outside of A? It can't. That is, all of its elements belong to A because it has no elements outside of A. Therefore, the empty set is a subset of A.

The notation for subsets is $\subseteq$. $A \subseteq B$ reads "A is a subset of B." (The symbol $\subseteq$ looks somewhat like the symbol $\in$ for is an element of; this might help remind

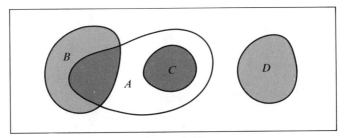

Figure 1.19
The set C is a subset of A. The set B is not a subset of A because B contains some elements that are not in A. The set D is not a subset of A because none of the elements of D is in A.

you that $A \subseteq B$ means that each element of A is an element of B.) $B \supseteq A$ is the notation for B contains A (as a subset).

Example 1.14 If $X = \{-1,8,7,3,5,0,17,-10\}$ and $Y = \{8,0,-1,3\}$,

    a. Is $X \subseteq Y$? Explain.

    b. Is $Y \subseteq X$? Explain.

Solution   **a.** No, X is not a subset of Y ($X \nsubseteq Y$) because there is some element of X that is not an element of Y. For example, 17 is in X but not in Y.

    b. Yes, $Y \subseteq X$ because all four elements of Y are in X.

Example 1.15 If $A = \{o,p,q,r,s,t,u\}$ and $B = \{a,e,i,o,u\}$,

    a. Is $A \cap B \subseteq A$?   **c.** Is $A \subseteq A \cup B$?

    b. Is $A \cap B \subseteq B$?   **d.** Is $B \subseteq A \cup B$?

Solution The answer to each question is yes.

    In general, the intersection, $A \cap B$, of two sets is a subset of either set A or B. See Figure 1.20a. Every element in the shaded area ($A \cap B$) is an element of set A and of set B. Also, any set is a subset of the union of that set with any other set. See Figure 1.20b.

    Rather than the notation $\subseteq$, some texts use the symbol $\subset$ for subsets. The difference between these two notations is not important to the material presented here. We remark only in passing that the extra bar in the former notation ($\subseteq$) is meant to emphasize the possible equality of the two sets:

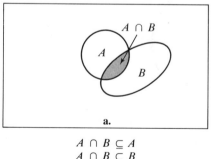
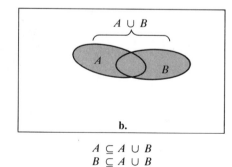

a.   b.

$A \cap B \subseteq A$ $A \subseteq A \cup B$
$A \cap B \subseteq B$ $B \subseteq A \cup B$

Figure 1.20

1.3 Subsets

$X \subseteq Y$ means X is a subset of Y (and possibly $X = Y$).

$X \subset Y$ means X is a subset of Y (but here X cannot equal Y).

Similarly, $B \supset A$ means B contains A as a subset, but B cannot be equal to A.*

More important than notation, however, is the application of subsets to logic. In a practical problem, when B is a subset of A, then there is some characteristic property of each element in B that allows the element to be classified as belonging to A. For example, if B is the set of men who are 6 feet tall, and A is the set of men who are over $5\frac{1}{2}$ feet tall, then every man in B must also be in A because the property of being 6 feet tall allows a man to be classified as belonging to the set of men who are over $5\frac{1}{2}$ feet tall.

We elaborate this theme in the following example.

Example 1.16 The set B of bicycles is a subset of the set C of vehicles that carry people, since all bicycles are vehicles that carry people. A mathematical statement that means the same thing is: *All bicycles are people carriers.* Three *logically equivalent* forms are stated below.

1. *Subset* form, $B \subseteq C$, or every element of B is an element of C:

   Every bicycle is a people carrier.

2. *If p then q* form:

   If x is a bicycle, then x is a people carrier.

3. The *p implies q* form:

   x is a bicycle implies x is a people carrier.

In general, the three forms listed above are not reversible; that is, it is not true that every people carrier is a bicycle. There are some people carriers that are not bicycles. This simply means that, given $B \subseteq C$, the subset B need not contain all of the elements of C. Note, however, that if an element is not in C, it cannot be in the subset B. This gives rise to the logical statement (in the above example): *If x is not a people carrier, then x is not a bicycle.* Such a statement is in the form *not q implies not p.*

---

*As often happens in mathematics, many kinds of notation exist for the same concept. Sometimes $A \subset B$ is denoted by $A \subsetneq B$ ($A \subseteq B$, but $A \neq B$).

Example 1.17 Let C be the set of concert pianists and let Y be the set of people who learned to play the piano when they were very young. Suppose the following statement is true:

S: *All concert pianists learned to play the piano when they were very young.*

a. Draw a Venn diagram describing the statement S.

b. Put statement S into the three forms: *subset* form, *if p then q* form, and *p implies q* form.

c. What conclusion can be drawn from the statement: Sam did not learn to play the piano when he was young?

Solution

a. Figure 1.21.

b. Every concert pianist learned to play the piano at an early age (subset form). If x is a concert pianist, then x learned to play piano at an early age (if p then q form). That x is a concert pianist implies that x learned to play piano at an early age (p implies q form).

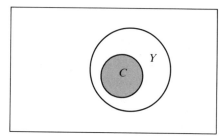

Figure 1.21

c. Conclusion: Sam is not a concert pianist.

UNIVERSAL SETS

Finally, a word about the universe. Any set that contains all the elements in any of the sets in a given problem is called a **universe of discourse,** or simply a **universe,** for that problem. In Example 1.15 involving the sets $A = \{o,p,q,r,s,t,u\}$ and $B = \{a,e,i,o,u\}$, one universe is the set $K = \{a,e,i,o,p,q,r,s,t,u\}$; another is the alphabet. There are also other possibilities.

EXERCISE 1.3

1. Let M, N, and P be sets such that $M \subseteq N$ and $N \subseteq P$.

    a. Sketch a single Venn diagram for these three sets.

1.3 Subsets

b. Does the diagram show that $M \subseteq P$?

2. In the Venn diagram shown here:

a. Is $x \in -R$?

b. Is $x \in -T$?

c. Is $y \in -T$?

d. Is $y \in -R$?

e. Is $z \in -R$?

f. Is $z \in -T$?

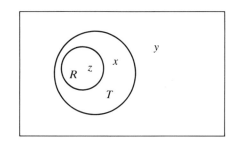

In the next two problems (3 and 4), let S = the set of all households in all four states in Problem 8 of Exercise 1.2 (page 24). Let A = the set of all households in Arkansas and F = the set of all Arkansas households with flush toilets.

3. True or false:

a. $S \subseteq A$.

b. $F - A$ is empty.

c. $A \subseteq F$.

d. $F \supseteq S$.

e. $A = A \cup F$.

f. $A \cap (F - A) = F$.

4. True or false:

a. $A \subseteq S$.

b. $A - S$ is empty.

c. $F = F \cap A$.

d. $S \subseteq S$.

e. $A \cup F \supseteq F$.

f. $(A - S) \cup S = A$.

5. Put the following statement into an if p then q form and draw a Venn diagram:
All languages require abstract symbols.

6. Put the following statement into an if p then q form and draw a Venn diagram:
All scientists know mathematics.

7. Let $X \subseteq Y$.

a. Draw a Venn diagram. Shade the area $-X$ and the area $-Y$.

b. Is $-Y \subseteq -X$?

c. If t is an element not in Y, can t be in X?

8. Suppose we know "If p then q." Which of the following statements may we conclude from this knowledge?

a. p implies q

b. *q* implies *p*

   c. not *q* implies not *p*

9. A set with two elements, {*a*,*b*}, has *four* distinct subsets. They are {*a*}, {*b*}, {*a*,*b*}, and the empty set (denoted by ∅).

   a. How many subsets does the set $S = \{a,b,c\}$ have? (Hint: The answer is *not* nine.)

   b. If two of the subsets are ∅ and {*a*,*b*,*c*}, list the rest of them.

10. How many subsets does a set with four elements have? If a set has five elements, how many subsets does it have?

11. Write a general formula for the number of subsets for a set with *n* elements.

*12. Use diagrams or definitions to show that for any two sets *A* and *B*:

   a. $(A \cap B) \cup (A \cap (-B)) = A$.

   b. $(A - B) \cap (B - A)$ is empty.

   c. $(A - B) \cup (A \cap B) \cup (B - A) = A \cup B$.

   d. $-(A \cup B) = (-A) \cap (-B)$.

   e. $-(A \cap B) = (-A) \cup (-B)$.

▲ *13. Let *s* be the statement "Every person who has gout has hyperuricemia." (Hyperuricemia is a high level of uric acid in the blood.)

   a. In the Venn diagram shown in the figure on the right, which of the two sets, *X* or *Y*, represents the people with hyperuricemia and which one represents people with gout? (Assume statement *s* is true.)

   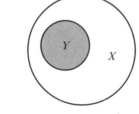

   b. Is it true that every person with a high uric acid level has gout?

   c. What statement can be made about elements that are in *X* but not in *Y*?

   d. In a study of 230 people with a high uric acid level, gout was the cause of the hyperuricemia in 28 cases. In 90 of the 230 people, the high uric acid level was due to the ingestion of certain foods and drugs: diuretics, high purines, and aspirin (set *D* in the figure on the right). Of these 90 people, only 14 of them have gout, the other 76 do not. What percentage of the 230 people neither have gout nor are among the ingestors of the diuretics?

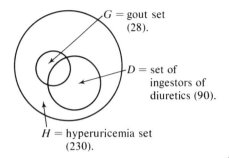

G = gout set (28).

D = set of ingestors of diuretics (90).

H = hyperuricemia set (230).

▲

2

INEQUALITIES AND SUMS

2.1

INEQUALITIES— BASIC CONCEPTS

Science needs to measure quantities. The physical sciences assign numerical values to time, speed, hardness, temperature, volume, electrical charge, and the extent of the universe. The life sciences gather data on neural impulses, cell respiration, age of trees, pain tolerance, biochemical reactions, dosage of medicine, reproduction, oxygen consumption, and the limit of human endurance. The social sciences need to measure capital, labor, package preference, number of votes, manpower, pollution levels, business sales and commissions, population growth, and the human capacity to learn. Of course, all these data and measurements must be put into some sort of order to be useful to the scientist so that he or she will be able to *compare* the various numerical values; look for patterns; and generate theories, formulas, and explanations.

A fundamental step in the process of ordering data and generating formulas is to write the quantities being measured as **variables** that can be represented by symbols (such as letters of the alphabet) in a formula or equation. For example, suppose a stockbroker determines his or her commission for buying and selling varying amounts of securities as follows: Let the money involved in the buy or sell order be represented by X. If X is between \$100 and \$2500, then the commission, denoted by C, is computed by using the formula

$$C = \$14.00 + \$(0.015) \cdot X. \qquad (2.1)$$

(The raised dot $\cdot$ is the symbol for multiplication.) If X is between \$2500 and \$20,000, the broker uses another formula to determine the commission:

$$C = \$25.50 + \$(0.0104) \cdot X. \qquad (2.2)$$

Example 2.1 Use the commission rates in Equations (2.1) and (2.2) to find the commission on a sale that involves stock worth **a.** $4200 and **b.** $2100.

Solution **a.** Since the amount of the sale $X = \$4200$, then X is between $2500 and $20,000; so we use Equation (2.2) above. The commission is

$$C = \$25.50 + \$(0.0104) \cdot (4200)$$
$$= \$25.50 + \$43.68 = \$69.18.$$

That is, the commission on a $4200 stock transaction is $69.18.

b. Here $X = \$2100$, which is less than $2500; so we use Equation (2.1):

$$C = \$14.00 + \$(0.015) \cdot (2100)$$
$$= \$14.00 + \$31.50 = \$45.50.$$

The commission on a $2100 stock transaction is $45.50.

Quantities such as volume, dosage, pollution, etc. are often written as variables (with letters such as x, y, and z) in formulas that *restrict* the set of numerical values over which the variables may range. In the brokerage commission given above, the variable X could range from $100 to $20,000. A different commission rate could be used for stocks worth more than $20,000 or less than $100. We shall discuss some of these rates later, but first we introduce the mathematical notation for the concepts of *more than* and *less than*.

The symbol $>$ means "is greater than" and the symbol $<$ means "is less than."
In Example 2.1a, the value of stock X was $4200; so we can write

$$X > \$2500 \ (X \text{ is greater than } \$2500), \text{ and}$$
$$X < \$20,000 \ (X \text{ is less than } \$20,000).$$

Both of these statements could be put together into one compact sentence

$$\$2500 < X < \$20,000,$$

which is read $2500 is less than X and X is less than $20,000, or X is between $2500 and $20,000.
In part **b** of the stockbroker example, $X = \$2100$, so we could write

$$\$100 < X < \$2500.$$

The symbols $>$ and $<$ are called **inequality symbols,** and an expression such as $y < 3$ is called an **inequality.** An expression such as $1 < z < 7$ is called a *double inequality* or a *betweenness statement*.

Sometimes students have difficulty remembering which symbol means *greater than* and which means *less than;* an easy way to get it right is as follows: The

2.1 Inequalities—Basic Concepts

bigger end of the symbol goes with the bigger number and the smaller end goes with the smaller number. This rule holds regardless of which symbol, $>$ or $<$, is used. For example, $\frac{1}{2} < 1$ and $1 > \frac{1}{2}$.

Example 2.2 Suppose that, in addition to Equations (2.1) and (2.2) above, you also know that for a stock sale involving less than $100, the commission rate is $15.50. That is,

$$C = \$15.50, \text{ for all } X < \$100. \quad (2.3)$$

Also, suppose that for a sale of stocks worth between $20,000 and $30,000, the commission rate is $95.50 plus 0.0069 times the stock value. For stocks worth between $30,000 and $300,000, the commission rate is $164.50 plus 0.0046 times the stock value. From these data and Equations (2.1), (2.2), and (2.3), write a five-equation formula for the broker's sales commission, using the inequality symbols where appropriate.

Solution Let X = the dollar value of the stock being sold and C = the commission in dollars, then

$$\left.\begin{array}{ll} C = 15.50 & \text{if } X < 100. \\ C = 14.00 + (0.015) \cdot X & \text{if } 100 < X < 2500. \\ C = 25.50 + (0.0104) \cdot X & \text{if } 2500 < X < 20{,}000. \\ C = 95.50 + (0.0069) \cdot X & \text{if } 20{,}000 < X < 30{,}000. \\ C = 164.50 + (0.0046) \cdot X & \text{if } 30{,}000 < X < 300{,}000. \end{array}\right\} \quad (2.4)$$

Note If the stock value falls exactly on one of the end values, such as 100, 2500, 20,000, or 30,000, then either of the two equations having that end value as a restriction on X may be used. For example, if $X = 2500$, we could use either Equation (2.1) or (2.2). (Check this for yourself.) ▲

Example 2.3
a. If W is the variable representing wages that fall within the income tax bracket from $15,000 to $20,000, write two inequalities for W. Also write the betweenness statement.

b. Suppose Y is a variable measuring the pollution from automobile exhausts. If the lower level of tolerance for a polluting car is 90 units, write an inequality for a polluting car that emits Y units.

c. Suppose that all real numbers are placed on a number line with the positive numbers in ascending order to the right of a zero point, and the negative numbers in descending order to the left of zero, as shown in Figure 2.1.

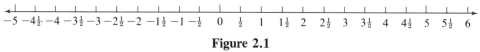

Figure 2.1
Number line.

On a number line such as this, *less than means to the left of* and *greater than means to the right of*. Insert the proper inequality symbol in place of the question mark in the following expressions:

$$-3 \ ? \ 0 \qquad -4 \ ? \ -2 \qquad -1\tfrac{1}{2} \ ? \ -2$$

$$2\tfrac{1}{2} \ ? \ 2 \qquad 2 \ ? \ 4 \qquad 2 \ ? \ -3?$$

Solution   a. $W > 15{,}000$ and $W < 20{,}000$. $15{,}000 < W < 20{,}000$.

b. $Y > 90$.

c. $-3 \ < \ 0 \qquad -4 \ < \ -2 \qquad -1\tfrac{1}{2} \ > \ -2$

$\ \ 2\tfrac{1}{2} \ > \ 2 \qquad\ \ 2 \ < \ 4 \qquad\ 2 \ > \ -3$

It is not always easy to tell when one number is greater than another, especially when the numbers are fractions with different denominators. For example, which is the larger of these two numbers: $31/17$ or $53/29$?

How are we to determine which is the larger of the two fractions? Subtract one from the other! If the result is positive, then the smaller was subtracted from the larger; but if the result is negative, then the larger was subtracted from the smaller. For the above example, we try $31/17 - 53/29 = ?$. Here, we need a common denominator of $17 \cdot 29$ in order to carry out the subtraction, so

$$\frac{31}{17}\frac{29}{29} - \frac{17}{17}\frac{53}{29} = \frac{899 - 901}{493} = \frac{-2}{493}.$$

Since division of -2 by 493 yields a negative number, the larger number was subtracted from the smaller. Therefore, $31/17 < 53/29$.

What if one fraction in the inequality has a variable in it? For example, which is larger $x/2$ or $2/3$? This problem cannot be solved because the answer depends on the size of the variable x. We *can*, however, solve a reverse type of problem; that is, determine a set of values for x that would make a given inequality true.

Example 2.4   Find all the values of the variable x that make the inequality $x/2 > 2/3$ true.

2.1 Inequalities—Basic Concepts

Solution Suppose x is any number for which the inequality $x/2 > 2/3$ is true. Then the subtraction of $2/3$ from $x/2$ must yield a positive number. Therefore, $x/2 - 2/3 > 0$. Combining fractions with a common denominator, we get $(3x - 4)/6 > 0$. Since 6 is positive, the fraction $(3x - 4)/6$ can be positive only if the numerator is also positive; that is, $3x - 4 > 0$, or $3x > 4$. Thus, we can use any number x as long as $3x$ is greater than 4. For example, x can be 5, or 2, or $1\frac{1}{2}$, or 17, etc. But it can't be 1, or $\frac{1}{2}$, etc. To get the set of *all* the solutions to the original inequality, we can divide both sides of $3x > 4$ by 3, thereby getting $x > 4/3$; that is, $x > 1\frac{1}{3}$. Hence, the inequality $x/2 > 2/3$ is solved by the set $S = \{x : x > 1\frac{1}{3}\}$. S is called the **solution set** for the given inequality.

To obtain the solution set for the above problem, we divided both sides of the inequality $3x > 4$ by 3. How do we know that we can do that? Are there any rules for dividing, multiplying, adding to, and subtracting from both sides of an inequality? Yes, and they are presented here as the **axioms of order**.

AXIOMS OF ORDER (Basic Rules for Inequalities)

Axiom 1 **(Trichotomy law for inequalities)**
If x is a number and y is a number, then one and only one of the following three statements is true: $x = y$, $x > y$, or $x < y$.

Axiom 2 **(Addition law for inequalities)**
If $x < y$, and z is any number, then $x + z < y + z$.

Axiom 3 **(Positive multiplication law for inequalities)**
If $x < y$ and z is any *positive* number ($z > 0$), then $x \cdot z < y \cdot z$.

Axiom 4 **(Transitive law for inequalities)**
If $x < y$ and $y < z$, then $x < z$.

Notes on Axiom 1 If some number a is not less than another number b, then we write $a \not< b$, and by Axiom 1, either $a = b$ or $a > b$. These two statements may be combined into a single one: $a \geqq b$, which reads "a is greater than or equal to b." This can also be written as $a \geq b$. Similarly, if $a \not> b$, then $a \leqq b$. One rarely writes $a \geqq b$ for $a \neq b$.

The *axioms of order* may remind the reader of the *properties of equals*. For review and comparison, we state those properties here:

PROPERTIES OF EQUALS (Basic Rules for Equalities)

Property 1 (Reflexive law)
If x is any number, then $x = x$.

Property 2 (Symmetric law)
If $x = y$, then $y = x$.

Property 3 (Transitive law for equalities)
If $x = y$ and $y = z$, then $x = z$.

Property 4 (Addition law for equalities)
If $x = y$ and z is any number, then $x + z = y + z$.

Property 5 (Multiplication law for equalities)
If $x = y$ and z is any number, then $x \cdot z = y \cdot z$.

Example 2.5 Solve for the variable in each question below:

  a. If $x \not> 7$, then what values can x have?

  b. If $x - 5 < 13$, find x.

  c. If $y/23 < 10$, find y.

  d. If $x < y$ and $y < 100$, write an inequality for the set of values for x.

  e. If $3x - 2 \geq 7$, find x.

Solution

  a. $x \not> 7$, then by Axiom 1, $x < 7$ or $x = 7$; that is, $x \leq 7$.

  b. $x - 5 < 13$; now, by Axiom 2 add 5 to both sides, getting $x - 5 + 5 < 13 + 5$, or $x < 18$.

  c. $y/23 < 10$. Since $23 > 0$, we can multiply both sides by 23 according to Axiom 3; therefore, $(y/23) \cdot 23 < (10) \cdot 23$; $y < 230$.

  d. If $x < y$ and $y < 100$, then by Axiom 4, $x < 100$.

  e. Combining the axioms of order and the properties of equals, we deduce from $3x - 2 \geq 7$ that $3x \geq 9$, so $x \geq 3$.

Example 2.6

  a. The U.S. House of Representatives has 435 members. A certain issue got x votes and won by a simple majority. Write an inequality for x.

  b. How many members are in the solution set in part **a**?

2.1 Inequalities—Basic Concepts

c. If an issue in the House passed with y votes and it won by more than a $2/3$ majority, write an inequality for y and tell how many members are in the solution set.

Solution
a. $x > 50$ percent of 435 ($217\frac{1}{2}$); therefore, in whole numbers, $x > 217$.

b. Here x can be any number in the set $\{218, 219, ..., 435\}$, which has 218 elements. So the inequality $x > 217$ has 218 different whole number solutions.

c. The inequality is obtained by $y > (2/3)(435) = 290$. Thus, $y > 290$. Its solution set is $\{y: y > 290$ and y is an integer$\} = \{291, 292, ..., 435\}$, which has 145 elements.

Sometimes, in comparing the sizes of two positive numbers, a and b, with $a < b$, we may wish to know how many times smaller a is than b. In such a case, we compute the ratio a/b.

Example 2.7 The number $1/350$ is how many times smaller than $1/200$?

Solution The ratio of the smaller to the larger is $(1/350)/(1/200)$, which equals $200/350 = 4/7$. Therefore, $1/350$ is $4/7$ of $1/200$.

SOME INEQUALITY THEOREMS

The axioms of order can be used to prove the following theorems on inequalities. Proof of Theorem 1 is obtained by an *indirect argument,* wherein the theorem is assumed to be false and contradictions of the axioms arise, thereby disproving the initial assumption that the theorem is false. The other theorems may similarly be proved.

Theorem 2.1 If $a < b$ and $a = c$, then $c < b$.

Proof For indirect argument, suppose the theorem is false; that is, suppose $a < b$ and $a = c$ as given, but $c \not< b$. Then by Axiom 1, either $c = b$ or $b < c$. First case: $c = b$. From the given, $a = c$; so if $c = b$, then $a = b$. However, by hypothesis $a < b$; so $a = b$ would be contrary to Axiom 1. Second case: $b < c$. From the given $a < b$; so if $b < c$, then by Axiom 4, $a < c$ which contradicts Axiom 1 with the given $a = c$. These contradictions arise from the supposition that $c \not< b$. Hence, $c < b$ as the theorem states.

Theorem 2.2 If $a \leq b$, then $a - c \leq b - c$ for any number c.

Theorem 2.3 If $a \leq 0$, then $-a \geq 0$.

Theorem 2.4 If $a < 0$, then $1/a < 0$; and if $a > 0$, then $1/a > 0$.

Theorem 2.5 If $a \leq b$, and $c < 0$, then $a \cdot c \geq b \cdot c$. (Multiplication by a negative number *reverses* the inequality.)

Theorem 2.6 If $a < b$ and $c < d$, then $a + c < b + d$.

Theorem 2.7 If $a > 0$ and $b > 0$ or if $a < 0$ and $b < 0$, then $a \cdot b > 0$.

Theorem 2.8 If all four of a, b, c, and d are positive, or all four are negative, then $a/b < c/d$ if and only if $a \cdot d < b \cdot c$.

Corollary If $0 < a < b$, or $a < b < 0$, then $1/a > 1/b$.

Theorem 2.9 If $a < b$, then $a < (a + b)/2 < b$.

EXERCISE 2.1

1. Use the stockbroker's commission Equations (2.4) to determine the commission to be charged on a sale of stock worth

   a. $25,000
   b. $50,000

2. Repeat Problem 1 for stocks worth

   a. $15,000
   b. $45,000

3. Which, if either, is the larger of the two numbers:

   a. $11/13$ or $113/134$?
   b. $3/7$ or $4/9$?
   c. $6/39$ or $51/331$?
   d. $91/210$ or $39/90$?

4. Which, if either, is the larger of the two numbers:

   a. $27/14$ or $25/13$?
   b. $90/312$ or $105/364$?
   c. $14/27$ or $13/25$?
   d. $7/11$ or $2/3$?

5. True or false (give reasons):

   a. Either $x = y$ or $x \neq y$.
   b. If $x > y$, then $x \neq y$.
   c. If $x > 0$ and $0 > y$, then $x > y$.
   d. If $x < y$, then $x + 3 < y + 2$.

2.1 Inequalities—Basic Concepts

6. True or false (give reasons):
   a. If $x \leq y$, then $x < y$.
   b. If $x > y$, then $x \leq y$.
   c. If $x > y$, then $x \geq y$.
   d. If $x < y$, then $y \not< x$.

7. Find the solution set for the inequality $8x + 4 > 20$.

8. Find the solution set for the inequality $3y - 17 < 19$.

▲ 9. A certain resolution in the U.S. House of Representatives received a majority greater than $2/3$. Let $x =$ the vote this resolution received, and write an inequality involving x. How many solutions does this inequality have? That is, how many elements are in the solution set? (There are 435 members in the House.)

10. A resolution in the U.S. House of Representatives has 25 sure votes. Let $x =$ the number of *additional* votes needed for the resolution to pass by a $2/3$ majority. Write an inequality involving x and find the solution set.

11. A nurse has an order to give x milligrams of a medicine, and the permissible dosage limits known to her are $45 \leq x \leq 50$. She has a solution that contains 75 milligrams per cubic centimeter. How much of a cubic centimeter can she give? (Select all the correct answers.)
    a. $2/3$ cc
    b. $1/2$ cc
    c. $3/4$ cc
    d. $1 1/2$ cc
    e. $65/100$ cc
    f. 6 cc

12. Another nurse has Zephrine solution in the strength $1/750$ and she wants to change it to a $1/1000$ solution.
    a. Which is the stronger solution?
    b. How much water should she add to 750 cc of the $1/750$ solution to convert it to $1/1000$?
    c. How much water should she add to 375 cc of the $1/750$ solution to convert it to $1/1000$?
    d. Show that the $1/1000$ solution is only $3/4$ as strong as the $1/750$. ▲

13. Use an inequality axiom to prove that no number can be greater than itself. (Assume the properties of equals in your proof.)

14. Assume $1 < 0$ and let x be any positive number ($x > 0$). By what axiom is it true that $x \cdot 1 < x \cdot 0$? Simplify this inequality to obtain $x < 0$. Explain what is wrong.

▲ 15. If Y_L = law enforcement budget and Y_P = parks and recreation budget, and the city manager has to contend with the budgetary constraint

$$\frac{1}{7} \cdot Y_L + Y_P \leq \$51{,}300,$$

what is the largest amount the parks and recreation department can get if $\$182{,}000 < Y_L$? ▲

16. Let I = an index that combines several factors x, y, and z of human personality. Suppose $I = 2x + 5y + z$. If measurements of the factors are $x > 30$, $y > 6$, and $z > 1$, find an inequality involving the index I.

17. a. Given $10 < 50$, show that $10 - 8 < 50 - 8$.

   b. Given $10 < 50$ and $3 < 45$, show that you cannot always subtract inequalities and keep the same order.

18. a. Given $x < 18$, show that $x - 10 < 8$.

   b. Given $14 < 55$ and $1 < 55$, show that you cannot always subtract inequalities and keep the order.

19. Prove the theorem: If $a < b$ and c is any number, then $a - c < b - c$.

   (Hint: $x - y$ means $x + -y$.)

20. Prove the theorem: If $a < b$ and $c < d$, then $a + c < b + d$.

   (Hint: Add c to both sides of $a < b$ and add b to both sides of $c < d$.)

*21. Prove: If $a \neq b$, then $a - b \neq 0$. (Use *indirect argument*.)

*22. Prove: If $a \neq 0$, then $a^2 > 0$. (Hint: Use Theorem 2.7.)

*23. Prove: If $a \neq b$, then $a^2 + b^2 > 2ab$.

*24. Prove Theorem 2.9 of this section. That is, if $a < b$, then $a < (a + b)/2 < b$. (Hint: Start with $a < b$ and add b to each side; repeat with a.)

2.2

INEQUALITIES AS CONSTRAINTS IN THE MANAGEMENT OF RESOURCES

Often in applied problems a variable may be assumed to range over a set of numbers up to some *limited* value. This means that the variable is *constrained*

2.2 Inequalities as Constraints in the Management of Resources

or bounded in the sense that it cannot get infinitely large. An example of a constrained variable is the amount for which an individual can write a check. The check amount is bounded by the checkwriter's bank balance or, through an arrangement with the bank, by some acceptable extended bank balance.

Suppose your bank balance is B dollars (B is itself a variable limited by your total income), and the bank will allow you to write checks for any amount up to $100 above your balance. If x is the amount for which you want to write a check, then your **constraint inequality** is $x \leq B + 100$, which may be written with both variables on the same side as $x - B \leq 100$.

Similar constraint problems exist for private industry, college administrators, and even government agencies because they have to operate within fixed budgets, not only of dollars, but also of manpower, building facilities, and equipment.

In this section, we shall discuss how constraint inequalities are used to describe resources and consumption in management.

Example 2.8 Suppose an enterprise has three departments, Administration, Research, and Production, that compete for 80 personnel positions. Let

w_1 = the number of workers to be hired in Administration.

w_2 = the number of workers to be hired in Research.

w_3 = the number of workers to be hired in Production.

(Note: The subscripted letters, w_1, w_2, ..., are used—rather than different letters, x, y, z, etc.—so that we can extend a problem of this sort to cases requiring several variables without running out of letters of the alphabet. Read w_1 as "w-one" or "w-sub-one," w_2 as "w-two," etc. Also, note that w_2 is *not* the same as w^2, which is $w \cdot w$.)

Now, since there are only 80 positions available, the hiring constraint is

$$w_1 + w_2 + w_3 \leq 80 \text{ (personnel constraint).} \tag{2.5}$$

Suppose, furthermore, that this enterprise uses computer time but is restricted to only 18 hours per day. An analysis shows that, on the average, each administrator requires $1/3$ hour of computer time per day. Each researcher needs 1 hour of computer time per day, and each production worker has a daily requirement of $1/8$ hour on the computer. The total computer time for each department is found by multiplying the number of workers in the department by the daily requirement of each worker. For example, the total Administration computer time is $(1/3) \cdot w_1$. The total departmental computer times for Research and Production are $1 \cdot w_2$ and $(1/8) \cdot w_3$, respectively. Therefore, the computer constraint is

$$\frac{1}{3} \cdot w_1 + 1 \cdot w_2 + \frac{1}{8} \cdot w_3 \leq 18 \text{ (computer time constraint).} \tag{2.6}$$

A third resource consumed by the three departments is working space. Suppose that the entire plant (all buildings) consists of 30,000 square feet. Each administrator uses 1000 square feet, each researcher uses 600 square feet, and each producer uses 300 square feet. The total square foot requirement for any department is found by multiplying the number of workers in the department by the square foot need of each worker. From this procedure, we find that the Administration requires $1000 \cdot w_1$ square feet, the research department needs $600 \cdot w_2$ square feet, and the production department needs $300 \cdot w_3$ square feet. The inequality describing the plant's constraint on assigned square footage is

$$1000w_1 + 600w_2 + 300w_3 \leq 30{,}000 \text{ (space constraint)}. \qquad (2.7)$$

In summary, there is a system of three resources (personnel, computer time, and working space) consumed by three departments (Administration, Research, and Production). This is called a *3-by-3 system* of inequalities; 3 resources, 3 consumers—3 rows, 3 columns, where a row is a resource and a column is a consumer.

Consumers / Resources	Administration		Research		Production		Capacity
Personnel Constraints	$1 \cdot w_1$	+	$1 \cdot w_2$	+	$1 \cdot w_3$	$\leq$	80
Computer Time Constraints	$\tfrac{1}{3} \cdot w_1$	+	$1 \cdot w_2$	+	$\tfrac{1}{8} \cdot w_3$	$\leq$	18
Working Space Constraints	$1000 \cdot w_1$	+	$600 \cdot w_2$	+	$300 \cdot w_3$	$\leq$	30,000

(2.8)

Each department consumes a certain amount of each resource; the consumption is "weighted" according to the **coefficient** of that department's work force. (Recall, from elementary algebra, that the coefficient of a variable x in an expression such as $7 \cdot x$ is the multiplier, 7.) For example, each of the w_1 people in Administration "consumes" one position of the 80, $\tfrac{1}{3}$ hour of the 18, and 1000 square feet of the 30,000. This means that for each administrator, 1 is the per person weight in the consumption of personnel positions, $\tfrac{1}{3}$ is the per person weight in the consumption of computer time, and 1000 is the per person weight in the consumption of floor space.

We determine the other departments' weighted use of resources in a similar

2.2 Inequalities as Constraints in the Management of Resources

fashion. The Research worker's consumption weights are 1 in personnel positions, 1 in computer time, and 600 in floor space. The Production worker's consumption weights are 1, $\frac{1}{8}$, and 300, respectively, in the three resources.

We have already mentioned that consumptions of resources are constrained by the total capacity of the resource, but as you have probably guessed, it may be true that not all resources are completely consumed. For example, it may happen that a high use of one resource could result in a low use of another. In this case, high use of floor space might mean low use of computer time, since a high consumption of working space means that there are a lot of administrators, thereby reducing the number of positions available to researchers, who are the heavy users of computer time. Or, if the company is just starting, there may be a low use of *all* resources.

Suppose this enterprise is conducting a study to see how it is using its resources. It wants to know: How many positions are unfilled? How much computer time is unused? How much floor space is unassigned?

The unused resources can be determined by introducing **slack variables**. A slack variable is the difference between the resource actually consumed and the total resource available. For example, if it turns out that on the first day of operation, only 3 administrators, 5 researchers, and 2 production workers have been hired (totaling 10 persons), then the company is 70 people short of full employment. Here the slack variable value of 70 in the personnel resource means that 70 of the 80 positions are unfilled. Suppose we let P be the slack variable in personnel positions. When $w_1 = 3$, $w_2 = 5$, and $w_3 = 2$, then $P = 70$. Thus,

$$w_1 + w_2 + w_3 + P = 80, \tag{2.9}$$

and we have been able to replace an *inequality* by an *equation*.

Equation (2.9) says the same thing as Inequality (2.5) because the variable P takes up the slack between the filled positions and the total capacity of 80 positions.

Before looking at the slack variables in the other constraints, let us state and prove an inequality theorem that justifies the process of changing the inequalities into equations by introducing an extra variable.

Theorem 2.10 (Slack variable theorem)
If $a \leq b$, then there exists a nonnegative number $s\,(s \geq 0)$ such that $a + s = b$.

Proof Since $a \leq b$ is given, then $b - a \geq 0$, by Theorem 2.2 (with $c = a$). Let $s = b - a$, then $b - a$ is nonnegative and $a + s = a + (b - a) = b$. That is, s is a nonnegative number such that $a + s = b$.

Corollary If $a \geq b$, then there exists a nonnegative number s such that $a - s = b$.

The corollary follows immediately from the theorem by adding s to b in the inequality $a \geq b$, then subtracting s from both sides, where s in this case is $a - b \geq 0$.

Now we return to the other resource constraint inequalities, the computer time and the work space. The 18 hours of computer time capacity may not be completely consumed by the three departments; that is, there would be some unused computer time. This can be translated into a slack variable equation by introducing a variable, say T, to stand for the unused computer time, from which we get the equation

$$\frac{1}{3} \cdot w_1 + 1 \cdot w_2 + \frac{1}{8} \cdot w_3 + T = 18. \tag{2.10}$$

Equation (2.10) replaces Inequality (2.6).

Also, the floor space constraint Inequality (2.7) can be replaced by the slack variable equation

$$1000 w_1 + 600 w_2 + 300 w_3 + F = 30{,}000, \tag{2.11}$$

where F is the slack variable for the unassigned floor space. Now we are ready to write the three constraint inequality systems (2.8) as the following system of *equations*.

Resources \ Consumers	Administration		Research		Production		Unfilled Positions		Unused Computer Time		Unassigned Square Feet	Capacity
Personnel	$1 \cdot w_1$	+	$1 \cdot w_2$	+	$1 \cdot w_3$	+	P					= 80
Computer Time	$\frac{1}{3} \cdot w_1$	+	$1 \cdot w_2$	+	$\frac{1}{8} \cdot w_3$			+	T			= 18
Floor Space	$1000 \cdot w_1$	+	$600 \cdot w_2$	+	$300 \cdot w_3$					+	F	= 30,000

(2.12)

Later, in Chapter 5, we shall seek values of variables (such as w_1, w_2, and w_3 in the above system) that will both satisfy the system of slack variable equations and minimize costs (or maximize profits) based upon the system. This will have to wait until we discuss methods for solving such systems. There is one important problem, however, that we can solve now, and that is to find the slack variables, given some data about each of the consumers (the departments).

Example 2.9 In the system (2.12) find the

  a. unfilled positions

  b. unused computer time

2.2 Inequalities as Constraints in the Management of Resources

c. unassigned floor space

given the following data for the three departments: There are 15 administrators, 7 researchers, and 24 production workers. That is, $w_1 = 15$, $w_2 = 7$, and $w_3 = 24$.

d. Discuss the possible hiring options based upon the information found in parts **a, b,** and **c.**

Solution

a. Substitute the given values for w_1, w_2, and w_3 into the personnel equation

$$1 \cdot (15) + 1 \cdot (7) + 1 \cdot (24) + P = 80$$

$$46 + P = 80; \text{ so, } P = 34.$$

There are 34 unfilled positions.

b. Substitute the given values for w_1, w_2, and w_3 into the computer equation, getting

$$\frac{1}{3} \cdot (15) + 1 \cdot (7) + \frac{1}{8} \cdot (24) + T = 18$$

$$15 + T = 18; \text{ so, } T = 3.$$

There are 3 hours of unused computer time.

c. Substitute the given values for w_1, w_2, and w_3 into the floor space equation, getting

$$1000 \cdot (15) + 600 \cdot (7) + 300 \cdot (24) + F = 30,000$$

$$26,400 + F = 30,000; \text{ so, } F = 3600.$$

The unassigned floor space is 3600 square feet.

d. Hiring options: The 34 positions open could never all be filled without exceeding the unused computer time or the unassigned floor space; therefore, there are more than enough unfilled positions to take care of any hiring decision. This allows us to concentrate on using up the other slack variables. No more than 3 researchers could be hired since there are only 3 hours of computer time left. And no more than 3 administrators can be hired because there are only 3600 square feet of space left. For the same reason, no more than 12 production workers could be hired. Making use of this analysis, we conclude that some of the options are:

  i. Increase administration by 3 and research staff by 1. This uses up all 3600 square feet of space, and 2 of the 3 hours of computer time (but only 4 of the 34 positions).

  ii. Hire 12 production workers. This uses up 12 of the 34 positions, only $1\frac{1}{2}$ of the 3 hours of computer time, and all 3600 square feet of space.

iii. Hire 8 production workers and 2 researchers, using up 10 positions, all 3 hours of computer time, and all 3600 square feet of space.

Several other combinations are possible: the choice would depend upon which of the slack variables are the most costly to leave unfilled. One obvious revelation of this slack variable analysis is the large number of unfilled personnel positions. This could call for a reduced consumption of one of the other resources to increase personnel. For example, reducing administration by 9 here would release 9000 square feet of space and 3 hours of computer time. This would allow the hiring of 24 more production workers. The decision depends on the comparative importance of administrators versus production workers.

Examples 2.8 and 2.9 define a method of setting up a management problem. This same method may be applied to a number of other management situations. Some of these include problems in ecology, medicine, and even farming.

▲ **Example 2.10** A farmer can grow two crops: corn and wheat. He has 400 acres of land. During the growing season, each acre of corn requires 10 hours of work and each acre of wheat requires 3 hours; his labor force consists of 5 workers who are available for 2000 hours during the season. The costs of seeds, fertilizers, irrigation, etc., for each acre of corn is \$32 and for wheat it is \$16. He can borrow up to \$8000 to cover these supply costs. Let x = the number of acres of corn and y = the number of acres of wheat. Answer the following questions:

a. Write three constraint inequalities, one for each of the resources; land, labor, and supplies.

b. Write the system of constraint inequalities in part **a** as a system of slack variable equations, and interpret the meaning of each slack variable.

c. Compute the slack variables for the data $x = 110$ and $y = 250$. Discuss several planting options.

Solution a.

	Corn	Wheat		Capacity	
Land	x +	y	$\leq$	400 (acres)	
Labor	$10x$ +	$3y$	$\leq$	2000 (hours)	(2.13)
Supplies	$32x$ +	$16y$	$\leq$	8000 (dollars)	

2.2 Inequalities as Constraints in the Management of Resources

b. Let s_1, s_2, and s_3 be the slack variables; then the equations are

$$\left. \begin{array}{l} x + y + s_1 = 400 \text{ acres.} \\ 10x + 3y + s_2 = 2000 \text{ hours.} \\ 32x + 16y + s_3 = 8000 \text{ dollars.} \end{array} \right\} \quad (2.14)$$

The meanings of the slack variables are s_1 = the unplanted acreage, s_2 = unused hours of labor, and s_3 = unused dollars in the line of credit.

c. If $x = 110$ and $y = 250$, then $s_1 = 40$ acres, $s_2 = 150$ hours, and $s_3 = \$480$. One planting option is to use up the 150 hours of labor and the \$480 by planting 15 more acres of corn. Alternatively, the farmer could plant 30 acres of wheat, thereby using up the rest of the \$480 loan, 90 of the 150 hours of labor, and 30 of the 40 unplanted acres. The final decision would depend on both the importance of using up the resources and the expected profit from each crop. ▲

All of the above practical problems were concerned with constraint inequalities in which the variables had to be less than a given capacity. But sometimes constraints may exist in which the variables *must* exceed some minimum quantity. For example, in the farmer's problem above, he may be required by contract or by the nature of his land to grow at least 50 acres of corn. Then his constraint would be $x \geq 50$ (minimum corn acreage requirement). This is not, strictly speaking, a resource-limiting consumption, but rather a *required minimum quantity*, limiting the system from consuming less than 50. We can refer to such an inequality as either a resource or a requirement. A slack variable equation can still replace a requirement inequality, but only if we use the corollary to Theorem 2.10.

That is, if we say s_4 is the slack variable representing the number of acres of corn *above* the minimum of 50 acres, then $x - s_4 = 50$. This could be a fourth equation in system (2.14). Then, using the data in Example 2.10 c, $x = 110$ and $y = 250$, we would get $110 - s_4 = 50$, or $s_4 = 60$. This indicates that the farmer is growing 60 acres of corn above the minimum requirement, and he may have reasons to choose an option that will reduce that excess amount.

Finally, in anticipation of the next section (on sums), we wish to state a slightly generalized form of the management problem.

Example 2.11 Suppose four variables, x_1, x_2, x_3, and x_4, satisfy two constraint inequalities: one a resource and one a requirement. Let the capacity of the resource be R_1 and the minimum for the requirement be R_2. For simplicity, we call both constraints "resources." Suppose the weights with which the variables consume the first resource are a_1, a_2, a_3, and a_4. The weights with which they consume the second resource are b_1, b_2, b_3, and b_4. (Actually here the weights with which the variables *consume the resource* means the weights with which they *meet the requirement*.) Then the 4-variable, 2-equation system of constraint inequalities is

$$\left. \begin{array}{ll} \text{Resource 1} & a_1 x_1 + a_2 x_2 + a_3 x_3 + a_4 x_4 \leq R_1. \\ \text{Resource 2} & b_1 x_1 + b_2 x_2 + b_3 x_3 + b_4 x_4 \geq R_2. \end{array} \right\} \quad (2.15)$$

If x_5 = the slack variable of unused units of Resource 1, and x_6 = the slack variable of excess units of resource (requirement) 2, then the 2 by 6 (2 resources, 6 consumers) system of equations is

$$\left. \begin{array}{ll} \text{Resource 1} & a_1 x_1 + a_2 x_2 + a_3 x_3 + a_4 x_4 + x_5 \quad\quad = R_1. \\ \text{Resource 2} & b_1 x_1 + b_2 x_2 + b_3 x_3 + b_4 x_4 \quad\quad - x_6 = R_2. \end{array} \right\} \quad (2.16)$$

System (2.16) is quite awkward to write; there must be an easier way. Luckily, there is an easier way—summation notation, and we will discuss it in the next section.

EXERCISE 2.2

1. Change the following system of inequalities to a system of slack variable equations:

$$\begin{aligned} x_1 + 3x_2 - x_3 + x_4 &\leq 7. \\ x_1 \quad\quad\quad + 2x_3 \quad\quad &\leq 10. \\ 5x_2 + x_3 + 2x_4 &\geq 3. \end{aligned}$$

2. Change the following system of inequalities to a system of slack variable equations:

$$\begin{aligned} x_1 + 5x_2 + x_3 &\geq 18. \\ x_1 + x_2 + 6x_3 &\leq 10. \end{aligned}$$

▲ 3. In an enterprise, departments A, B, and C have the following labor and payroll constraints:

Labor $\quad\quad x_1 + x_2 + x_3 \leq 100$ (personnel positions).

Payroll $\quad 300x_1 + 365x_2 + 400x_3 \leq 33{,}000$ (dollars).

2.2 Inequalities as Constraints in the Management of Resources

x_1, x_2, and x_3 are the number of workers in the three departments A, B, and C, respectively. The weights for the second resource (the coefficients 300, 365, and 400) are the per person salaries for the three departments.

a. Write these constraints in slack variable equations.

b. Explain what the slack variables mean.

4. Three industries, paper, lumber, and chemical, are consumers (in a certain area) of two resources, timber and water. The variables x_1, x_2, and x_3 are the three industries' sales in millions of dollars. The weights are the per sale consumption of the resources, which are constrained by federal regulations as follows:

Timber $\qquad 5x_1 + 20x_2 + 3x_3 \leq 500$ (square miles).

Water $\qquad 3x_1 + 6x_2 + 8x_3 \leq 400$ (acre-feet).

a. Write these constraints in slack variable equations.

b. Explain what the slack variables mean.

5. Refer to Problem 3. If the departments A, B, and C are staffed with 30 in A, 33 in B, and 28 in C, find the values of the slack variables. Find three possible hiring options that would consume some of the unused resources.

6. Refer to Problem 4. Suppose the sales (in millions of dollars) in the three industries are paper: $x_1 = 18$; lumber: $x_2 = 15$; and chemical: $x_3 = 25$. Find the slack variables and four possible options for increasing sales in the three industries.

7. A personnel manager hires workers for two shifts. The total work force cannot exceed 100 employees (work force constraint). The night shift must have at least 20 workers (minimum skeleton crew requirement). Day workers earn $30 and night workers earn $45; the payroll may not exceed $3200 (payroll constraint). The factory, community, and transportation interfacing requires that there are at least twice as many workers on days as night (plant operation constraint). Let x = the number of workers on the day shift and y = the night workers. Set up the constraint inequalities and convert them to slack variable equations.

8. A manufacturer can ship by rail, air, or truck lines. Let x = shipping mileage by rail, y = mileage by air, and z = mileage by truck. Suppose the following constraints hold: The trucks are owned by the company and must be used for at least 100,000 miles (minimum equipment requirement). Because of costs, rail shipping must not exceed one-fourth of the air mileage (cost constraint). Because of overseas commitments, air shipping requires at least 500,000 miles more than the total of both trucking and rail mileage (geographic constraint). Write the constraint inequalities and convert them to slack variable equations.

*9. A company has four departments. It hires w_1 workers in department I, w_2 workers in department II, w_3 in department III, and w_4 in department IV. There are three resources: personnel, floor space, and budget. The total personnel positions available are 100 (personnel resource). The total floor space available is 50,000 square feet (floor space resource). The total budget available is $610,000 (budget resource).

Each worker in department I uses 500 square feet of floor space and $5000 of the budget. Each worker in department II uses 600 square feet of floor space and $7500 of the budget. Each worker in department III uses 400 square feet of floor space and $8500 of the budget. Each worker in department IV uses 500 square feet of floor space and $9000 of the budget.

a. Write a 3-by-4 system of constraint inequalities (3 resources by 4 consumers; that is, 3 rows and 4 columns).

b. Write the system as a group of slack variable equations and explain what the slack variables mean.

c. If departments I, II, III, and IV hire 25, 21, 18, and 18 workers, respectively, discuss some of the possible hiring options left after these people are hired.

*10. A police station has four departments: Communication, Transportation, Patrol, and Special Unit. Let w_1 be the number of officers in Communication, w_2 the number in Transportation, w_3 the number in Patrol, and w_4 the number in Special Unit. This station can assign up to T_1 officers to these four departments. The departments must cover at least T_2 square miles of territory. (Thus, T_1 is the personnel resource and T_2 is the minimum area requirement.) The per person weights for meeting the area requirements are: A communication officer covers b_1 square miles; a transportation officer covers b_2 square miles; a patrol officer covers b_3 square miles, and a special unit officer covers b_4 square miles. Write a 2-by-4 system of constraint inequalities describing this situation. ▲

2.3

SUMMATION— AN INTRODUCTION

In system (2.16) of the previous section, the resource equations are

2.3 Summation—An Introduction

Resource 1 $\quad a_1 x_1 + a_2 x_2 + a_3 x_3 + a_4 x_4 + x_5 \quad\quad = R_1.$

Resource 2 $\quad b_1 x_1 + b_2 x_2 + b_3 x_3 + b_4 x_4 \quad\quad -1 \cdot x_6 = R_2.$

$\quad\quad\quad\quad\quad\quad\quad\quad\quad\quad\quad\quad\quad\quad\quad\quad\quad\quad\quad\quad\quad\quad\quad\quad\quad$ (2.16)

For uniformity, let us write both equations so that they will contain *all six* variables x_1, x_2, x_3, x_4, x_5, and x_6. We use the following device: If a variable is missing from the equation (as x_6 is missing from the first equation), then we include that variable with a zero coefficient. Doing this, we get

Resource 1 $\quad a_1 x_1 + a_2 x_2 + a_3 x_3 + a_4 x_4 + 1 \cdot x_5 + 0 \cdot x_6 = R_1.$

Resource 2 $\quad b_1 x_1 + b_2 x_2 + b_3 x_3 + b_4 x_4 + 0 \cdot x_5 + -1 \cdot x_6 = R_2.$

$\quad\quad\quad\quad\quad\quad\quad\quad\quad\quad\quad\quad\quad\quad\quad\quad\quad\quad\quad\quad\quad\quad\quad\quad\quad$ (2.17)

Furthermore, we can bring these equations into an even more uniform appearance as follows: In resource 1, let the coefficient 1 of x_5 be called a_5, and the coefficient 0 of x_6 be called a_6. In resource 2, let the coefficient 0 of x_5 be called b_5, and the coefficient -1 of x_6 be called b_6. In other words, if $a_5 = 1$, $a_6 = 0$, $b_5 = 0$, and $b_6 = -1$, then the system (2.17) may be written as

Resource 1 $\quad a_1 x_1 + a_2 x_2 + a_3 x_3 + a_4 x_4 + a_5 x_5 + a_6 x_6 = R_1.$

Resource 2 $\quad b_1 x_1 + b_2 x_2 + b_3 x_3 + b_4 x_4 + b_5 x_5 + b_6 x_6 = R_2.$

$\quad\quad\quad\quad\quad\quad\quad\quad\quad\quad\quad\quad\quad\quad\quad\quad\quad\quad\quad\quad\quad\quad\quad\quad\quad$ (2.18)

Notice that in system (2.18), each term in the first equation is of the form $a_i x_i$, with $i = 1, 2, 3, 4, 5$, and 6. Each term in the second equation is of the form $b_i x_i$ with the same six values for i. This suggests that it may be possible to simplify these equations by writing the left side of the first equation in (2.18) as a sum of $a_i x_i$ and the left side of the second equation as a sum of $b_i x_i$ for $i = 1, 2, 3, 4, 5$, and 6.

We now introduce such a *summation notation*, which uses the Greek letter for s (to stand for *sum*); it will make equations such as those in (2.18) easier to write.

The Greek letter Σ (sigma) is used as a signal telling you to find the sum. That is, the expression

$$\sum_{i=1}^{6} a_i x_i$$

means the sum of $a_i x_i$ for $i = 1, 2, 3, 4, 5$, and 6. In this notation, the system (2.18) can be written

Resource 1 $\quad\quad\quad \displaystyle\sum_{i=1}^{6} a_i x_i = R_1.$

Resource 2 $\quad\quad\quad \displaystyle\sum_{i=1}^{6} b_i x_i = R_2.$

$\quad\quad\quad\quad\quad\quad\quad\quad\quad\quad\quad\quad\quad\quad\quad\quad\quad\quad\quad\quad\quad\quad\quad\quad\quad$ (2.19)

We state the definition formally.

Definition **(Summation notation)**
Let n denote a positive integer. If u_i is a number for each integer i, where $i = 1, 2, 3, 4, ..., n$, then the sum

$$u_1 + u_2 + u_3 + u_4 + ... + u_n$$

may be written as

$$\sum_{i=1}^{n} u_i,$$

that is,

$$\sum_{i=1}^{n} u_i = u_1 + u_2 + u_3 + ... + u_n. \tag{2.20}$$

Example 2.12 **a.** If $n = 8$ and $u_i = 2 \cdot i$ for each integer i, where $i = 1, 2, 3, 4, 5, 6, 7,$ and 8, then

$$\sum_{i=1}^{n} u_i = \sum_{i=1}^{8} 2 \cdot i = 2 \cdot 1 + 2 \cdot 2 + 2 \cdot 3 + 2 \cdot 4 + 2 \cdot 5 + 2 \cdot 6 + 2 \cdot 7 + 2 \cdot 8$$

$$= 2 + 4 + 6 + 8 + 10 + 12 + 14 + 16 = 72.$$

b. If $n = 4$ and $u_i = 3i - 5$ for each integer i, where $i = 1, 2, 3,$ and 4, then

$$\sum_{i=1}^{n} u_i = \sum_{i=1}^{4} (3i - 5) = (3 - 5) + (6 - 5) + (9 - 5) + (12 - 5) = 10.$$

c. If $u_1 = 2$, $u_2 = 10$, $u_3 = -7$, $u_4 = -7$, $u_5 = 1/2$, and $u_6 = 13$, then

$$\sum_{i=1}^{6} u_i = 2 + 10 + -7 + -7 + 1/2 + 13 = 11 1/2.$$

The summation notation has four parts (shown in Figure 2.2): the summation sign, the lower index, the upper index, and the summand. The lower index does not always have to start at 1; it can start at any other integer.

The summand can be obtained from the index either by a formula such as in Example 2.12a and **b,** or from a list of values such as in **c** of that same example. Students frequently look for formulas to determine the values of the summand;

2.3 Summation—An Introduction

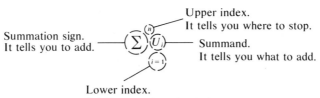

Figure 2.2
Anatomy of a summation.

they don't always exist and are not necessary. All one needs is an ordered set of numbers, a place to start, a place to stop, and the signal to add.

Example 2.13

a. Let c_1, c_2, c_3, c_4, and c_5 be the values of the five U.S. coins—penny, nickel, dime, quarter, and half-dollar. Find the sum (in cents)

$$\sum_{i=1}^{5} c_i.$$

b. A portfolio of five stocks is shown in the table. The stocks are numbered 1 through 5, and V_i is the value of the stock numbered i, where $i = 1, 2, 3, 4$, and 5. Find the total

$$\sum_{i=1}^{5} V_i.$$

Number	Stock Symbol	Value V_i
1	LK	$3500
2	CFC	1765
3	CUP	874
4	WSH	506
5	TA	1197

c. Let $u_i = 3 \cdot 1^i$, for $i = 1, 2, 3, 4, 5, 6, 7, 8, 9, 10$. Find

$$\sum_{i=1}^{10} u_i.$$

Solution

a. $\sum_{i=1}^{5} c_i = 1 + 5 + 10 + 25 + 50 = 91$ cents.

b. $\sum_{i=1}^{5} V_i = 3500 + 1765 + 874 + 506 + 1197 = \$7842$.

c. Here all of the terms u_i have the same value since $1^i = 1$ for all i.

$$u_1 = 3 \cdot 1^1 = 3 \cdot 1 = 3.$$
$$u_2 = 3 \cdot 1^2 = 3 \cdot 1 = 3.$$
$$u_3 = 3 \cdot 1^3 = 3 \cdot 1 = 3, \text{ etc.}$$

Therefore,

$$\sum_{i=1}^{10} 3 \cdot 1^i = 3 + 3 + \ldots + 3 = \text{sum of ten 3s} = 30.$$

An ordered set of values $u_1, u_2, u_3, \ldots, u_n$ is called a *sequence of n terms*, or simply a **sequence**. If all the values of a sequence are the same, as in Example 2.13c, where $u_i = 3$ for all i, then it is called a **constant sequence**.

EXERCISE 2.3

1. Write the following constraint inequalities in summation notation:

$$a_1 x_1 + a_2 x_2 + a_3 x_3 + a_4 x_4 \leqq R_1.$$
$$b_1 x_1 + b_2 x_2 + b_3 x_3 + b_4 x_4 \leqq R_2.$$
$$c_1 x_1 + c_2 x_2 + c_3 x_3 + c_4 x_4 \leqq R_3.$$

2. Write the following constraint equations in summation notation. First, introduce the necessary coefficients to make both equations contain all six variables.

$$p_1 y_1 + p_2 y_2 + p_3 y_3 + p_4 y_4 + y_5 \quad\quad = A_1.$$
$$q_1 y_1 + q_2 y_2 + q_3 y_3 + q_4 y_4 \quad\quad + y_6 = A_2.$$

3. Find the sums

   a. $\sum_{i=1}^{6} 1/i$
   b. $\sum_{i=1}^{3} 3i/(10 + i)$
   c. $\sum_{i=1}^{6} (2/i^2)$

4. Find the sums

   a. $\sum_{i=1}^{5} 5/(2 + i)$
   b. $\sum_{i=1}^{5} (4/2i)$
   c. $\sum_{i=1}^{5} (6 + i)/(2 + 2i)$

5. Find the sums

   a. $\sum_{k=1}^{10} [3 + (k - 1) \cdot 4]$
   b. $\sum_{k=1}^{10} (2 + 3k)$

2.3 Summation—An Introduction

6. Find the sums

    a. $\sum_{k=1}^{5} [1 + (k-1) \cdot 6]$

    b. $\sum_{k=1}^{7} [3 - (k+2) \cdot 5]$

*7. Find the sums

    a. $\sum_{k=1}^{5} 4 \cdot (2/3)^{k-1}$

    b. $\sum_{k=1}^{4} 6 \cdot (4/5)^{k}$

    c. $\sum_{k=1}^{7} 5 \cdot (2)^{k-1}$

*8. Find the sums

    a. $\sum_{k=1}^{3} 2 \cdot (5/6)^{k+1}$

    b. $\sum_{k=1}^{5} 5 \cdot (1/2)^{k}$

    c. $\sum_{k=1}^{5} 2 \cdot (3)^{k-1}$

▲ 9. In the book, *An Introduction to Mathematical Ecology* by E. C. Pielou (New York: John Wiley & Sons, 1969), a formula called Lloyd's index m of mean crowding is given as

$$m = \frac{\sum_{j=1}^{n} x_j \cdot (x_j - 1)}{P},$$

where P is the population, n is the number of living units, and x_j = the number of individuals living in the jth living unit. Then m is the measure of crowding of P individuals into n living units. Note that

$$P = \sum_{j=1}^{n} x_j.$$

   a. If six living units have the populations $x_1 = 4$, $x_2 = 7$, $x_3 = 10$, $x_4 = 1$, $x_5 = 6$, and $x_6 = 3$, find P and find m.

   b. If six living units have the populations $x_1 = 5$, $x_2 = 5$, $x_3 = 6$, $x_4 = 6$, $x_5 = 4$, $x_6 = 5$, find P and find m.

10. Refer to the formula in Problem 9. If there are ten living units and the number of individuals living in each is four, find the index m of mean crowding. ▲

11. Complete the $x_i y_i$ column in the table and find

    a. $\sum_{i=1}^{5} x_i$

    b. $\sum_{i=1}^{5} y_i$

    c. $\sum_{i=1}^{5} x_i \cdot y_i$

    d. $(\sum_{i=1}^{5} x_i) \cdot (\sum_{i=1}^{5} y_i)$

x_i	y_i	$x_i \cdot y_i$
2	1	2
3	4	12
5	−1	−5
−1	2	−2
0	6	0

12. Complete the $u_i \cdot v_i$ column in the table and find
    a. $\sum_{i=1}^{5} u_i$ b. $\sum_{i=1}^{5} v_i$ c. $\sum_{i=1}^{5} u_i v_i$ d. $(\sum_{i=1}^{5} u_i) \cdot (\sum_{i=1}^{5} v_i)$

u_i	v_i	$u_i \cdot v_i$
1	2	2
3	−1	−3
4	0	
−1	2	
6	3	

*13. The product of a sequence $u_1, u_2, u_3, \ldots, u_n$ is denoted by Π (Greek pi). That is,
$$\prod_{i=1}^{n} u_i = u_1 \cdot u_2 \cdot u_3 \cdot \ldots \cdot u_n.$$
a. Find $\Pi_{i=1}^{4} (2 + 3i)$.   b. Find $\Pi_{i=1}^{3} (1 + 2^i)$.

*14. Use the product notation in Problem 13 to write the product
$$\left(4 + \frac{2}{1}\right) \cdot \left(4 + \frac{2}{2}\right) \cdot \left(4 + \frac{2}{3}\right) \cdot \left(4 + \frac{2}{4}\right) \cdot \left(4 + \frac{2}{5}\right).$$

*15. Use the product notation in Problem 13 to write the product
$$(4 - 3^1) \cdot (4 - 3^2) \cdot (4 - 3^3) \cdot (4 - 3^4).$$

*16. Find the product $\Pi_{i=1}^{7} \dfrac{i}{i+1}$.

3

GRAPHS AND COORDINATE GEOMETRY

3.1

READING AND WRITING GRAPHS

A sheet of music is a graph; each note is located horizontally by the measure containing it and vertically by its position on the staff. A gasoline pump is a "graph machine." While it is pumping gasoline, it registers a series of number pairs—dollars versus gallons: $(x,y) = (x$ dollars, y gallons), which could be plotted on a sheet of graph paper.

In newspapers we find charts and graphs of stock market prices, sporting events, and weather data. Research journals in medicine, sociology, and economics are filled with graphs and tables communicating many facts to their readers. A graph can convey much useful information, but the reader must be careful so that he or she is not misled by it.

Example 3.1 The graph in Figure 3.1 below shows a comparison of the affects of marijuana and alcohol on (simulated) driving performances.

The graph appears to show that marijuana causes only about one-half as many errors as alcohol does. But look at the vertical scale! It runs from only 80 to 100; the 87 errors caused by marijuana are not very far below the 97 errors caused by alcohol. This is more apparent if we show this same graph with the vertical scale running from 0 to 100. (See Figure 3.2.)

The original chart was not meant to be misleading, and it may be true that marijuana is not as bad as alcohol in causing errors. The close relationship between the placebo and marijuana gives us a clue that this is probably a fact, but the overall impression in Figure 3.2 is so startlingly different from that of the original chart that we feel we should look twice before jumping to conclusions.

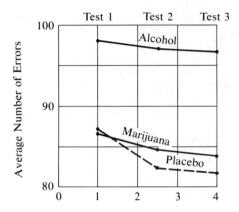

Figure 3.1
A comparison of alcohol and marijuana on simulated driving tests. From "Marihuana" by Lester Grinspoon. Copyright © 1969 by Scientific American, Inc. All rights reserved.

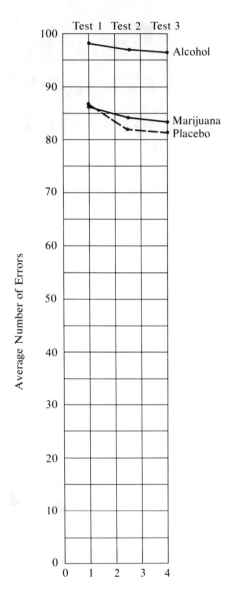

Figure 3.2

3.1 Reading and Writing Graphs

Most graphs, however, try to avoid ambiguities and to convey accurate information efficiently. After you learn how to analyze a graph, you can obtain more information than you may have previously thought was there and perhaps even more than the maker of the graph had intended.

As we have just seen from the above example, the first step in analyzing a graph is to examine the units on each scale to see if they are uniform in size or if they have been cut off, stretched, or shrunk.

Two other ideas for reading a graph are

1. Examine the area under the graph (that is, under the plotted line or curve). This area is expressed in units that come from multiplying together the units of the two scales. It represents the *total amount* of the quantities that the graph depicts.

2. Examine how fast the graph changes in the vertical direction in relation to a given change in the horizontal direction. This rate of change represents the growth or decline of one quantity with respect to the other.

The units for the rate of change equal the units of the vertical dimension divided by the units of the horizontal dimension.

▲ **Example 3.2** The British Forestry Commission's promotion of forestry and establishment of state forests are recorded in the graph in Figure 3.3. Each point on the graph

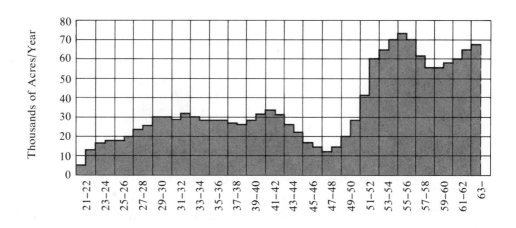

Figure 3.3
Graph of planting by the British Forestry Commission. Source: "Britain's Woodlands and Afforestation" by J. T. Coppock in *Geography*, July 1964. Reprinted with permission of the Forestry Commission.

represents two quantities: *years* and *acres of trees planted per year*. (The tree planting started in 1921.)

Answer the following questions concerning this graph.

a. How many thousands of acres were planted in 1945?

b. In what year were the most acres planted and how many acres were planted that year? In what year was the least planting done?

c. What does the area under the plotted line mean? For example, each square has a horizontal dimension of two years and a vertical dimension of 10,000 acres per year. Multiplying the dimensions together gives the area of each square in what units?

d. What is the approximate total acreage planted between the end of 1932 and the end of 1946? (Hint: Obtain the approximation by counting the whole squares and estimating the total of the fractions of squares under the curve.)

e. Roughly what percentage of the total acres (up through 1963) was planted before 1947?

Solution

a. About 15,000 acres

b. Maximum = approximately 72,000 acres in 1955. Minimum = 5 acres in 1920.

c. The area of each square is in units of 20,000 acres, since (2 years) times (10,000 acres/year) = 20,000 acres.

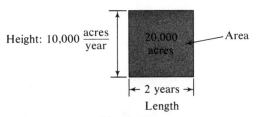

Figure 3.4

d. There are approximately $17\frac{1}{2}$ squares between these years and under the line; so, the total acreage planted is $(17\frac{1}{2}) \cdot (20,000 \text{ acres}) = 350,000$ acres (approximately).

e. From 1920 through 1946 there are approximately $30\frac{1}{2}$ squares, whereas from 1920 through 1963 there are approximately 74 squares; so, roughly $^{30.5}/_{74}$ of the total, or about 41 percent was planted before 1946. ▲

3.1 Reading and Writing Graphs

In the preceding examples, we discussed how one should read data from a graph. Now we will concentrate on the reverse problem: how to express data on a graph.

Any data that can be expressed as two columns of corresponding numbers (that is, two sets of variables) can be plotted on a graph. Very roughly, this is accomplished by letting one column of numbers correspond to points on the horizontal scale and the other to points on the vertical scale. In what follows, we will learn the details of this procedure, along with some of the ideas and language of a coordinate system.

CONVENTION ON THE *ORDER* OF THE VARIABLES

If we wish to locate the point (7,10) on a sheet of graph paper, the first number 7 corresponds to the number 7 on the horizontal scale, and the second number 10 corresponds to 10 on the vertical scale. (See Figure 3.5.) It is an unshakable convention that a number pair locating any point is *ordered* so that the first number corresponds to a number on the horizontal scale and the second number corresponds to a number on the vertical scale.

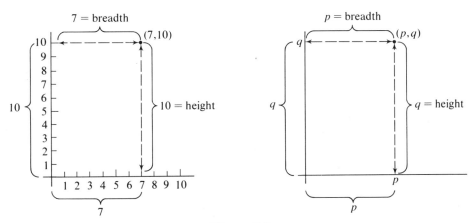

Figure 3.5

COORDINATES OF A POINT

If some point is located by an ordered pair of numbers (p,q), then this pair identifies the point, and the two numbers, together, are called the **coordinates** of the point.

The *first* number is called the **abscissa** of the point and the *second* number is called the **ordinate** of the point. The point (p,q) has abscissa p and ordinate q. The point (7,10) above has abscissa 7 and ordinate 10.

COORDINATE AXES

The horizontal scale is called the *axis of abscissas* and the vertical scale is called the *axis of ordinates*. The two scales, together, are called the *coordinate axes*. The set of all points in the plane determined by the coordinate axes is called the *coordinate plane*, or the *Cartesian plane* after René Descartes (1596–1650), the great philosopher and inventor of analytic geometry.

CONVENTION OF THE x-AXIS AND THE y-AXIS

Both axes have positive and negative values on them and they intersect at their zero point. On the vertical axis, the positive numbers run up the scale from zero and the negative numbers run down the scale from zero. On the horizontal axis the positive numbers run to the right from zero and the negative numbers run to the left. The intersection point (0,0) is called the **origin** of the coordinate plane. Usually, the axis of abscissas (the horizontal axis) is called the x-axis, and the axis of ordinates (the vertical axis) is called the y-axis. Frequently, points in the coordinate plane (the x-y plane) are located by x's with subscripts and y's with subscripts. For example, three points might be labeled as (x_1, y_1), (x_2, y_2), and (x_3, y_3).

The use of subscripts as just described is widespread in mathematics. Letters with subscripts, such as x_1, x_2, x_3, etc., representing numbers on the x-axis, and y_1, y_2, y_3, etc. on the y-axis give us an *infinite alphabet* for labeling numbers with letters.

Example 3.3 On a sheet of graph paper, plot and label the following points. (Mark the coordinate axes so that they range from -6 to $+6$.)

    a. $A = (3, 2\frac{1}{2})$; $B = (-2, -5)$; $C = (\pi, \sqrt{7})$.
        Note: π is approximately 3.14 and $\sqrt{7}$ is approximately 2.65.

    b. The point D, where D has abscissa -3 and ordinate 4.31.

    c. The points E_1 and E_2, where $E_1 = (x_1, y_1)$ and $E_2 = (x_2, y_2)$, and $x_1 = -2$, $x_2 = -4$, $y_1 = 3$, and $y_2 = 6$.

3.1 Reading and Writing Graphs

d. The point F, where $F = (x_3, y_3)$ and $x_3 = 4$, and

$$y_3 = \frac{1}{10}(x_3)^2 + \frac{1}{x_3} + 3.5.$$

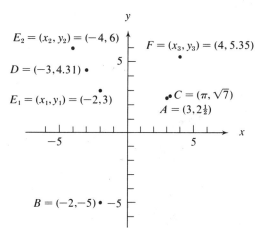

Figure 3.6

Solution **a.** See Figure 3.6

b. $D = (-3, 4.31)$.

c. $E_1 = (-2, 3)$ and $E_2 = (-4, 6)$.

d. $F = (4, 5.35)$ because if $x_3 = 4$, then

$$y_3 = \frac{1}{10}(4)^2 + \frac{1}{(4)} + 3.5 = \frac{16}{10} + 0.25 + 3.5 = 5.35.$$

A table of values can also be plotted as a graph.

Example 3.4 **a.** Plot the points whose coordinates are *tabulated* in the table on the right.

b. Draw a smooth curve through the points.

x	y
0.00	0.00
0.02	1.60
0.03	1.70
0.04	1.96
0.05	2.37
0.08	3.75

Solution Mark the *x*-axis from 0 to 0.10, and mark the *y*-axis from 0 to 4. From the table, the points to be plotted are (0,0); (0.02,1.60); (0.03,1.70); etc. Plotting these and drawing a smooth curve through them, we get the curve in Figure 3.7.

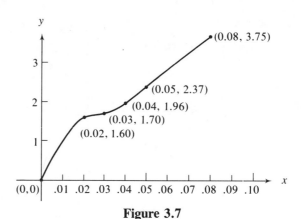

Figure 3.7

EXERCISE 3.1

▲ Problems 1–8 refer to the graph of the Gross National Product (GNP) and Consumer Price Index (CPI) on page 69. GNP is in terms of constant dollars and CPI is in terms of the 1949 dollar (= 100 cents). The graph is taken from the *1970 Historical Chart Book*, Board of Governors, Federal Reserve System.

1. An item costing $1.00 in 1949 cost 78¢ in 1945. How much did it cost in 1940? 1930? 1918?

2. How much would a $1.00 (1949) item cost in 1970? 1936? 1960?

3. a. What was the average price over the five-year period 1951–55? (Add together the price for each of the five years and divide by five.)

   b. What was the price change over the five-year period 1951–55? What was the average yearly price change?

4. a. What was the average price over the five-year period 1961–65?

   b. What was the change in price over 1961–65? What was the average yearly price change?

3.1 Reading and Writing Graphs

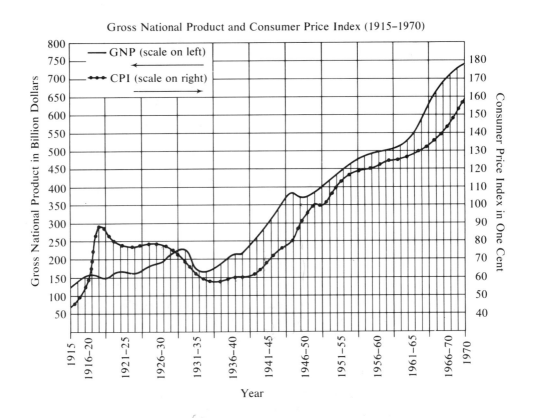

Gross National Product and Consumer Price Index (1915–1970)

 5. Find the GNP in 1949.

 6. Find the GNP in 1940.

 7. a. By how many billion dollars did the GNP increase over the 55-year period between 1915 and 1970?

   b. What was the yearly average growth in this period? (Find the difference between 1970 GNP and 1915 GNP and divide by 55.)

 8. a. By how many billion dollars did the GNP increase over the 20-year period between 1951 and 1970?

   b. What was the yearly average growth during this period?

For Problems 9–12 use the graph on page 70 about water flowing through a reservoir pipe.

 9. a. What was the velocity of water flow exactly 1 minute after opening the valve?

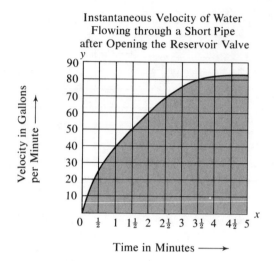

Instantaneous Velocity of Water Flowing through a Short Pipe after Opening the Reservoir Valve

Time in Minutes ⟶

b. What was the change in velocity of water flow between the end of the first minute and the end of the third minute?

10. a. What was the velocity of the water flow at exactly 2 minutes after the valve was opened?

b. What was the change in velocity of water between the end of the second minute and the end of $3\frac{1}{2}$ minutes?

11. a. What is the area of each square? (Also tell what the units for the square areas are.)

b. By counting the squares under the curve, find the total number of gallons that went through the pipe between $x = 1\frac{1}{2}$ and $x = 2\frac{1}{2}$ minutes.

c. How much water flowed during the first minute?

12. a. Assuming each square has an area representing 5 gallons, find the total number of gallons that flowed through the pipe between 1 minute and $3\frac{1}{2}$ minutes.

b. How much water flowed during the second minute (from the end of the first minute to the end of the second minute). ▲

13. Mark a pair of coordinate axes on a sheet of graph paper and locate the points $A_1 = (-1\frac{1}{2}, -4)$; $A_2 = (0, -1)$; $A_3 = (1\frac{1}{2}, 2)$; $A_4 = (3, 5)$; and $A_5 = (4\frac{1}{2}, 8)$. Can you find *one* rule that relates each point's ordinate to its abscissa?

3.1 Reading and Writing Graphs

14. On graph paper, plot the points $B_1 = (0,0)$; $B_2 = (0.8, 0.7)$; $B_3 = (1.6, 1.0)$; $B_4 = (2.4, 0.7)$; $B_5 = (3.1, 0)$; $B_6 = (3.9, -0.7)$; $B_7 = (4.7, -1.0)$; $B_8 = (5.5, -0.7)$; and $B_9 = (6.3, 0)$. Draw a smooth curve through them.

15. Plot the points $C_1 = (-5, 3)$; $C_2 = (-3, 3)$; $C_3 = (0, 3)$; $C_4 = (1\frac{1}{2}, 3)$; and $C_5 = (17, 3)$. What one statement can you make about all of these points?

16. Plot the points $D_1 = (x_1, 7)$; $D_2 = (x_2, 5)$; $D_3 = (x_3, 1\frac{1}{4})$; and $D_4 = (x_4, 0)$, where $x_i = -4$ for all $i = 1, 2, 3, 4$. (That is, $x_1 = x_2 = x_3 = x_4 = -4$.)

17. In the x-y plane, describe the following sets of points in terms of x and y:

    a. those whose abscissas are zero

    b. those whose ordinates are zero

18. a. On a sheet of graph paper, carefully plot the following twelve points: (0,0); (0.1, 0.31); (0.2, 0.45); (0.3, 0.55); (0.4, 0.63); (0.6, 0.76); (0.8, 0.89); (1.0, 1.0); (1.3, 1.14); (1.5, 1.22); (1.8, 1.34); and (2.0, 1.41).

    b. Draw a smooth curve through the plotted points.

    c. On your graph in part **b,** select the point whose abscissa is $x = 0.25$. What is the ordinate of that point? Do the same for $x = 0.64$.

    d. How are the ordinates and abscissas related?

19. Plot the points whose coordinates are tabulated in the table on the right. Draw a smooth curve through the points.

x	y
1.00	0.00
0.96	0.25
0.86	0.50
0.75	0.66
0.66	0.75
0.50	0.86
0.25	0.96
0.00	1.00

20. If x = sales and y = profits, plot the points in the table on the right and draw a smooth curve through them. From the curve, estimate the profit on sales of $3250.

x	y
500	115
1000	235
1500	335
2000	415
2500	475
3000	515
3500	535

▲ **21.** In a study of the incidence of a given disease, the data shown in the table on the right were collected where x = the age at diagnosis and y = the incidence of the disease per 1,000,000 population (of that age). Plot the points and draw a smooth curve through them. Estimate the incidence of the disease in 35 year olds.

x	y
10	4
20	32
30	80
40	56
50	64
60	75
70	87
80	142

22. In the table on the right, y = percent of automobiles exceeding the speed x on a given section of a main rural highway at slack periods. Plot the points and draw a smooth curve through them. Estimate the percentage of cars that exceed the 58 miles-per-hour speed.

x	y
40	98
45	95
50	86
55	73
60	55
65	28
70	9

*23. In the graph shown below, the area under the curve is shaded and the x-axis is marked off into six intervals. The area above each, starting at 0, is indicated. For example, above the interval from $x = 0$ to $x = 70$, the area is 0.02; from $x = 0$ to $x = 85$, the area is 0.16, etc.

  a. Find the area from 70 to 85.

  b. Find the area from 85 to 130.

  c. Find the area from 130 to 200.

  d. Complete the table on page 73. (The values in the x column correspond to the x values in the horizontal scale of the graph and the values in the z column correspond to areas from 0 to x.

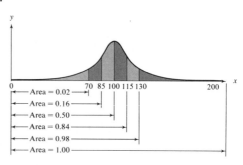

e. Plot the points in the table from part **d** and draw a smooth curve through them.

f. From the smooth curve plotted in part **e**, find z if $x = 90$. Is this value the area from 0 to 90 under the original curve? Explain.

g. From the graph in part **e**, find z if $x = 105$.

h. Using the graph in part **e**, find the area under the original graph from 0 to 120.

x	$z = $ area from 0 to x
0	0.00
70	0.02
85	0.16
100	0.50
.	.
.	.
.	.

3.2 RELATIONS AND FUNCTIONS

A graph in x-y plane represents a "picture" of a relationship between two quantities, x and y. For example, Figure 3.3 is a graph illustrating the relationship between y (the number of acres of forest planted per year) and x (the number of years of existence of the Forestry Commission). In this particular case, there is no equation to express the relationship. The Commission's records, rather than mathematical formulas, provide the information for determining how many trees were planted at a given time. In other words, the Commission's records express an association between two sets of variables, years and acres of planted trees. If x is an element from the set of years and y is an element from the set of acres, then the ordered pair (x,y) is a point in the British Forestry Commission's *relation* between years and acres.

In general, if X and Y are any two sets of variables (such as years and acres) and if some *description R* (a set of records, a table, a graph, or mathematical statement) expresses an association between X and Y by pairing elements x in X with elements y in Y, then the set of all such ordered pairs (x,y) is called the **relation** between X and Y satisfying description R.

For example, let a description R be the equation

$$x^2 + y^2 = 25. \tag{3.1}$$

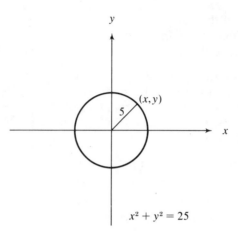

Figure 3.8

This equation expresses an association between two sets of variables (subsets of the x-axis and the y-axis). It happens that all points (x,y) that satisfy* this equation form a circle, which has its center at the origin and a radius of 5. This circle is the graph of the relation expressed by Equation (3.1). See Figure 3.8.

Sometimes, by abuse of notation, the description that states the association between the variables is itself called the relation. In the above example, we can say that the equation $x^2 + y^2 = 25$ is the relation between x and y, and that this relation is satisfied by the points on the circle shown in Figure 3.8.

This dual use of the term *relation* almost never produces any serious misunderstandings. In the following examples, we use the word in both senses. That is, we say the expression involving the two variables *and* the set of points (ordered pairs) satisfying such an expression are both the *relation* between the variables.

Example 3.5 a. The relation $y < x$ is satisfied by the set of all points (x,y) in the plane such that the ordinate y is less than the abscissa x. Its graph is the shaded area shown in Figure 3.9. The three sample points A, B, and C labeled in the figure are $A = (1,0)$; $B = (3,1)$; and $C = (-2,-3)$.

b. The set of all points in the x-y plane whose distance from the origin is 5 or less is a relation between a subset of the x-axis and a subset of the y-axis described by the inequality $x^2 + y^2 \leq 25$. This relation is the union of the circle in Figure 3.8 and the set of points on the interior of that circle. The graph is shown in Figure 3.10. The three sample points are

---

*The statement that the point (a,b) *satisfies* a given equation means that the equation is true when the coordinates a and b are substituted for x and y, respectively, in the equation.

3.2 Relations and Functions

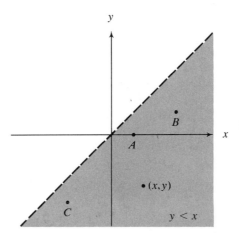

$$A = \left(\frac{5}{2}, \frac{5}{2}\right); B = \left(\frac{5}{4}, \frac{5 \cdot \sqrt{15}}{4}\right); \text{ and } C = \left(-\frac{5}{2}, -\frac{5}{3}\right).$$

Here the symbol "$\sqrt{}$" is the *square root*. $\sqrt{15}$ is a number whose square is 15.

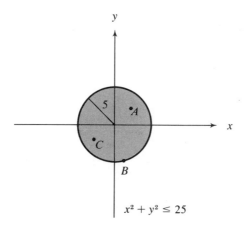

Figure 3.10

c. The relation $x^2 + y^2 > 25$ is satisfied by the set of all points (x,y) exterior to the circle with center at the origin and radius 5. Its graph is the shaded area in Figure 3.11. The three sample points are

$$A = (\sqrt{3}, \sqrt{23}); B = (4, -4); \text{ and } C = (-10, 5).$$

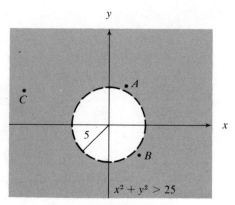

Figure 3.11

d. The relation described by the following table is the set of all points (x,y) for which x and y are numbers in *a certain row* of the table. Its graph is shown in Figure 3.12, and the three sample points are $A = (0,1)$; $B = (2,2)$; and $C = (2,6)$.

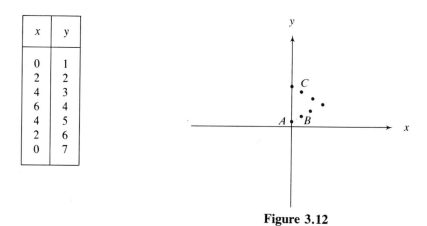

x	y
0	1
2	2
4	3
6	4
4	5
2	6
0	7

Figure 3.12

e. The relation $y = 1/x$ is satisfied by the set of all points (x,y) for which the ordinate of each point is the reciprocal of the abscissa. The variables are said to be **inversely** related in this case and the graph is shown in Figure 3.13. The three sample points are $A = (2, 1/2)$; $B = (-3, -1/3)$; and $C = (1/4, 4)$.

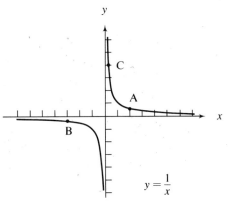

Figure 3.13

f. The equation $y = (x - 1)(x - 2)(x - 3)$ is satisfied by the set of points (x,y) in which the ordinate y is $(x - 1)(x - 2)(x - 3)$ for a given abscissa x. The graph is shown in Figure 3.14. The three points are $A = (1,0)$; $B = (2,0)$; and $C = (3,0)$.

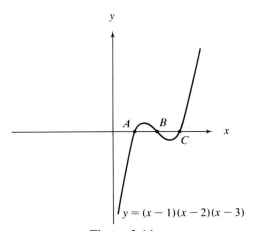

Figure 3.14

In applications to the sciences, relations between variables tell something about the way one measured quantity behaves with respect to another. For example, in economics, if we let R = revenue, w = demand, and p = price, then the equation

$$w = \frac{R}{p} \tag{3.2}$$

reflects the assumption that, for a fixed revenue, the demand w will decrease when the price p increases (w is inversely proportional to p). In this type of relation, it is important that a given value of one variable (say p) yields a *unique* value of the other variable w. That is, the same price should not yield two different demands.

In general, if a relation between two sets X and Y assigns a unique value y in set Y to a given value x in set X, then the relation is one of a special type called a *function;* the formal definition follows.

Definition **(Function)**
A function F from a set D to a set R is a collection of ordered pairs (d,r) with $d \in D$ and $r \in R$ such that no two distinct pairs in F have the same first term. The first set D is called the **domain** of the function and the second set R is called the **range.**

In the price-demand relation defined by Equation (3.2) above, p is paired with w; that is, (p,w) is an ordered pair in the relation. This relation is actually a function since it meets the requirement that no two distinct ordered pairs have the same first term. In other words, no price is associated with two different demands.

There is no prohibition, however, against two ordered pairs having the same *second* term. If we think of the first term of an ordered pair as a *cause* and the second term as an *effect,* then the restriction against having two of them with the same first term means that no *one* cause can produce *two* distinct effects. But it is possible for two different causes to produce the same effect. *Geometrically, the graph of a function must satisfy the condition that no two points are on the same vertical line.*

In Example 3.5, the relations described in parts **a, b, c,** and **d** are *not* functions because, in each case, some vertical line intersects the graph in two or more points. Look back at Figures 3.9 through 3.12. On the other hand, the relations in parts **e** and **f** of that same example *are* functions; each one passes the *vertical line test*—no vertical line contains more than one point of the graph.

Example 3.6 A library fine is a function of the number of days a book is overdue. The ordered pair (d,f) belongs to the function, if d = the number of days a book is overdue and f is the fine charged (in cents) for a book overdue by that many days. If

3.2 Relations and Functions

the library charges 5 cents per day for overdue books, then $f = 5 \cdot d$. Usually the total fine is limited to the cost of replacing the book; therefore, the range of the function is limited.

A shorthand symbolism for the statement that F is a function from set X to set Y is $F: X \to Y$. This may be read "F maps X into Y." The direction of the arrow indicates that the function is from X to Y; X is the domain and Y is the range. In the example from economics, Equation (3.2), the function F is from a set of numbers representing prices to a set of numbers representing demands (F maps prices into demands). Let P be the set of prices and W the set of demands; then we write $F: P \to W$.

Since functions are relations, they can be described by tables; by verbal statements; or by the most desirable method, equations. Equations have several advantages:

1. From an equation, we can usually compute the coordinates of any point by substitution into the equation. For example, if $y = x^3 - 3x + 1$, we can find y for any x we choose, not just for those that would be in a table.

2. An equation reveals what values of either variable are prohibited, which provides us with a method for finding the domain and range, if necessary. For example, suppose a function F has the equation

$$y = \sqrt{x} + \frac{1}{x}.$$

    Then for y to be real, x cannot be negative. (The square root of a negative number is not a real number.) Also, since division by zero is not allowed, x cannot be zero. This means that the above equation makes sense only if $x > 0$. This says that the domain of F must be a subset of the positive numbers.

3. We can learn to recognize certain forms of equations as representing certain types of graphs. For example, a *linear equation*, $y = a \cdot x + b$, (where x is variable and a and b are constants) is the equation of a line, as will be seen in the next chapter.

It is not always convenient or even possible to describe every function by an equation; therefore, verbal statements and numerical tables are also valuable ways to express the rules that associate the two variables in a function.

Frequently, when we are discussing functions, it is necessary to refer to a variable in the domain and the corresponding number in the range. Unfortunately, the terms *domain* and *range* are not very suggestive for the way we perceive the function; so, we return to two, somewhat older and more suggestive terms, *independent variable* and *dependent variable*.

Definition **(Independent variable and dependent variable)**
If $F:D \to R$, the numbers in the domain D are called the *independent variables*. The numbers in the range R are called the *dependent variables*.

Since the domain number is chosen first, it is *independently* selected; therefore, it is the independent variable. On the other hand, for a given pair (x,y) in the function, the value of the range number y *depends* on the value that was selected for the domain number x.

In the price-demand function $F:P \to W$, the equation is $w = R/p$. Pick any price p_0 at random from P, then w depends upon the choice from P; that is, w must be R/p_0. If this specific value of w is called w_0, then the graph of the function contains the point (p_0, w_0).

By tradition, one writes an equation for a function with the dependent variable on the left and all the independent variables and all constants on the right, if possible. If x is the independent variable and y is the dependent variable, we would write a function equation as follows:

$$y = \text{an expression involving } x. \tag{3.3}$$

Then to find a value for y, given a value for x, we simply substitute that value of x into the right-hand expression everywhere x appears. (Engineering students call this process "plugging in" the given value of x. They also call computing the value of y "turning the crank.")

Example 3.7 If F is a function whose equation is

$$w = \sqrt{t + 4} + \frac{(t - 30)^2}{100},$$

find w if

a. $t = -3$ b. $t = 5$ c. $t = 30$ d. $t = 38$ e. $t = 77$

f. What is the smallest value t can have?

g. If $F:T \to W$, describe the domain T as an inequality.

Solution   a. Plugging in $t = -3$ for t on the right (in both places where t appears), we get

$$w = \sqrt{(-3) + 4} + \frac{(-3 - 30)^2}{100}.$$

Now we turn the crank (carry out the calculations) and get

3.2 Relations and Functions

$$w = \sqrt{1} + \frac{(-33)^2}{100} = 1 + 10.89 = 11.89.$$

That is, for $t = -3$, we get $w = 11.89$.

b. For $t = 5$, $w = 9.25$.

c. If $t = 30$, then $w = \sqrt{34} \approx 5.83$ ($\sqrt{34}$ is approximately equal to 5.83).

d. For $t = 38$, $w = \sqrt{42} + 0.64 \approx 6.48 + 0.64 = 7.12$.

e. For $t = 77$, $w = 31.09$.

f. t cannot be less than -4. If $t < -4$, then $t + 4 < 0$ and the square root is not real.

g. The domain of F is the set of all numbers t, such that $t \geq -4$.

EXERCISE 3.2

1. Which of the following relations are *not* functions?

a.
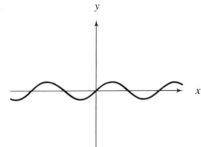

c.
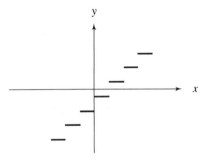

b.
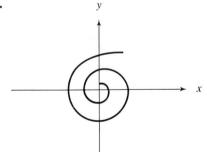

d.
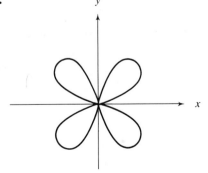

2. Which of the following relations are *not* functions?

a.
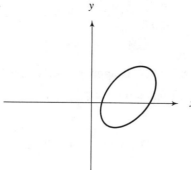

c.
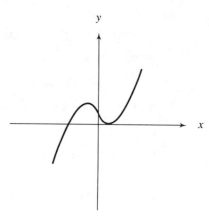

b.
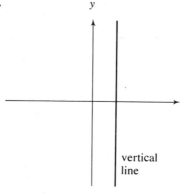

d.
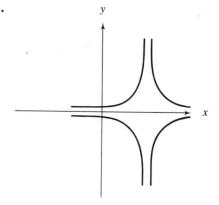

3. The graph of the relation $y^2 = x$ is shown to the right.

  a. Is this relation a function?

  b. Find two points whose abscissa = 1.

  c. From the equation, why can't $x < 0$?

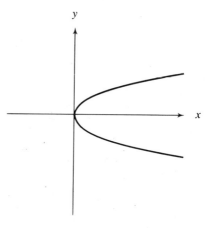

3.2 Relations and Functions

4. The graph of the equation $y = (1/2)^x$ is shown to the right.
   a. Is the relation a function?
   b. Find the value of y for $x = 3$.
   c. Find y for $x = -1$.
   d. Is there any restriction on what number x can be?

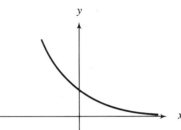

5. What restrictions are on x in the following equations?
   a. $y = 1/(x - 1)$
   b. $y = \sqrt{x + 10}$
   c. $y = [(x - 1)/(x - 2)] + \sqrt{x}$
   d. $y = \sqrt{1 - x^2}$

6. What restrictions are on x in the following equations?
   a. $y = \sqrt{1 + x^2}$
   b. $y = \sqrt{x + 1} + 1/(x - 4)$
   c. $y = (3 - x)/\sqrt{x^2 - 4}$
   d. $y = \sqrt{x^2 + 2x}$

▲ 7. Biologists can estimate the weight of an animal by its linear dimensions. The rule is that mass is proportional to cube of length. The equation describing this function is

$$M = k \cdot L^3,$$

where L is the overall length of the animal and M is the weight. k is a constant empirically derived and usually given in grams/cubic centimeter or pounds/cubic feet.

   a. If, for mice, k = 0.015 g/cm³, find the weight in grams of a mouse 12 cm long; of a mouse 10 cm long.
   b. If, for birds, k = 0.011 g/cm³, find the weight in grams of a bird 27 cm long. Find the weight of a 10-cm bird.
   c. If, for fat dogs, k = 0.04 g/cm³, find the weight in grams of a fat dog that is 2 feet (61 cm) long. If 454 grams = 1 pound, how much does this 2-foot fat dog weigh (in pounds)?

8. Refer to the formula in Problem 7.
   a. If, for fish, k = 0.0125 g/cm³, find the weight in grams of a 23-inch fish (1 inch = 2.54 cm).
   b. Find the weight in grams of a 4-foot fish.

c. Use the conversion 454 grams = 1 pound to find the weight of the fish in part **b** in pounds.

9. Refer to Problem 7 for the formula.

   For a whale that is 65 feet long and weighs 116 tons,

   a. Find k in pounds/cubic foot.

   b. Use the conversions: 1 pound = 454 grams and 1 foot = 30.48 cm to change the units of k in part **a** to grams/cubic centimeter.

10. Refer to Problem 7 for the formula.

    For people, $k = 0.0135$ g/cm³. Find the weight in pounds of a 6-foot person. ▲

11. The following system of equations is called a *recursive* or *hierarchical* sociological model. To evaluate the fourth variable, x_4, from a given value of the first variable, x_1, start with the *bottom* equation. Find the value of x_2, then use this value (of x_2) along with the given value of x_1 in the next higher equation to get x_3, and so forth, until you reach the top equation and determine the value of x_4. This system is useful in sociology because it can handle a number of interconnected variables.

    $$x_4 = 3x_1 + 4x_2 - x_3$$
    $$x_3 = 4x_1 + x_2$$
    $$x_2 = -8x_1 + 6$$

    a. Find x_4 if $x_1 = 0$.
    b. Find x_4 if $x_1 = \frac{1}{2}$.

12. Refer to Problem 11 for instructions. Given the following recursive sociological model:

    $$y_5 = -0.16y_1 - 0.02y_2 + 0.15y_3 - 0.40y_4$$
    $$y_4 = -0.04y_1 + 0.52y_3$$
    $$y_3 = 0.50y_1 + y_2 + 1$$
    $$y_2 = -0.40y_1$$

    a. Find y_5 if $y_1 = 0$.
    b. Find y_5 if $y_1 = 10$.

3.2 Relations and Functions

▲ *13. The maximum flow of traffic entering a highway is a function of four variables as described by the equation (derived by experiments*)

$$Q_{max} = \frac{108(w + e)\left(1 - \dfrac{p}{3}\right)}{\left(1 + \dfrac{w}{L}\right)},$$

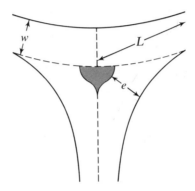

where:

Q_{max} = maximum flow of traffic (in cars/hour)

w = width of the weaving section

e = width of entry into the weaving section

L = length of weaving section

p = proportion of weaving traffic to the total traffic in the section

The *weaving section* is that part of the road where drivers in one lane seek gaps in the other lane so that they can cross over into that lane.

  a. If $w = 30$, $e = 30$, and $p = \tfrac{2}{3}$, what must the length of the weaving section be to allow a flow of 4480 cars per hour?

  b. Using the same width, entry, and proportion as in part **a**, why is it impossible, no matter what length L is, to have a flow of 5040 cars per hour?

  c. How does a decrease in the proportion of weaving traffic to the total traffic affect the flow? ▲

*See Wohl, M. and Martin, B. V. *Traffic Systems Analysis for Engineers and Planners* (New York: McGraw-Hill, 1967).

3.3

FUNCTION NOTATION

We now introduce the *f of x* notation for functions.

Definition (Function notation)
If F is a function and (x,y) is an ordered pair in F, then the dependent variable y paired with the given independent variable x is denoted by $f(x)$.

In other words, $f(x)$ denotes the *value of the function at x*. Read $f(x)$ as "f of x" or "f at x." Note that $f(x)$ does *not* mean f times x.

Example 3.8 Suppose F is the function defined by the following table.

a. What is y at $x = 5$?
b. What is the value of the function at 3?
c. What is $f(6)$?
d. What is $f(7)$?
e. Show that $f(9) + f(10) = f(8)$.

x	$y = f(x)$
0	7
1	2
2	5
3	10
4	−1
5	4
6	0
7	11
8	2
9	−3
10	5

Solution

a. y at $x = 5$ is the value in the y column that corresponds to 5 in the x column; that is, $y = 4$ at $x = 5$.

b. The value of the function at 3 means the value of y at $x = 3$. That is, $y = 10$ at $x = 3$.

c. $f(6)$ is the value of y at $x = 6$; that is, $f(6) = 0$ [because 6 in the x column is paired with 0 in the $f(x)$ column].

d. $f(7) = 11$.

3.3 Function Notation

e. To show that $f(9) + f(10) = f(8)$, notice that $f(9) = -3$, $f(10) = 5$, and $f(8) = 2$. Therefore, $f(9) + f(10) = -3 + 5 = 2 = f(8)$.

The above example illustrates the use of the $f(x)$ notation in a table, but by far its greatest use is for functions defined by equations. In the previous section, for example, Equation (3.3), $y =$ an expression involving x, was given as a general equation for functions. Now we can write this as $y = f(x)$.

We will sometimes loosely speak of $f(x)$ as a function. What we really mean is that the collection F of ordered pairs $[x, f(x)]$ is the function. When we say "find the value of the function at x_0," we mean find the value y_0 such that $(x_0, y_0) \in F$. That is, $[x_0, f(x_0)] \in F$.

Example 3.9 Let the equation

$$f(x) = 3 \cdot x^2 - \left(\frac{1}{x}\right) + 4$$

define a function $F = \{[x, f(x)]\}$.

a. Find $f(1)$.

b. Find $f(7)$.

c. Find $f(1/20)$.

d. Does $f(0)$ exist?

e. If $a \neq 0$, find $f(a)$.

f. What is the domain of the function F?

Solution

a. To get $f(1)$, substitute 1 for x everyplace x occurs in the expression $3x^2 - (1/x) + 4$. We get $f(1) = 3 \cdot 1^2 - (1/1) + 4 = 6$; $f(1) = 6$.

b. Similarly, $f(7) = 3 \cdot 7^2 - (1/7) + 4 = 3 \cdot 49 + 3\frac{6}{7} = 150\frac{6}{7}$. So, $f(7) = 150\frac{6}{7}$.

c. $f(1/20) = -15^{397}/_{400}$.

d. Does $f(0)$ exist? No, because $f(0)$ would require the substitution of $x = 0$ everyplace x appears in $3x^2 - (1/x) + 4$. But we *cannot* put $x = 0$ in the term $1/x$ since $1/0$ is not a number (division by zero is forbidden).

e. $f(a) = 3a^2 - (1/a) + 4$.

f. The domain of F is the set of numbers x for which $f(x)$ exists. In this case, the only restriction is that $x \neq 0$. Therefore, the domain of F is all of the x-axis, except for 0.

Sometimes we will use the dependent variable itself in function notation; for example, we can write $y = y(x)$, and we say y is a function of x. Function notation is convenient for expressing relations (such as rate of growth) between various points of the function.

Example 3.10 In the function $y(x) = 3x + 10$:

a. Find $y(10)$ and $y(5)$.
b. Find the change in y when x goes from 5 to 10.
c. Find the rate of change in y with respect to x as x goes from 5 to 10.
d. Find $y(a)$ and $y(b)$.
e. Find the change in y as x goes from a to b.
f. Find the rate of change in y with respect to x as x goes from a to b.

Solution

a. $y(10) = 3 \cdot 10 + 10 = 40$, and $y(5) = 3 \cdot 5 + 10 = 25$.

b. The change in y between $x = 5$ and $x = 10$ is the difference between $y(10)$ and $y(5)$. That is, the change in $y = y(10) - y(5) = 40 - 25 = 15$.

c. The rate of change in y with respect to x is the ratio of the change in y to the change in x, or

$$\frac{\text{change in } y}{\text{change in } x} = \frac{y(10) - y(5)}{10 - 5} = \frac{15}{5} = 3.$$

d. $y(a) = 3a + 10$, $y(b) = 3b + 10$.

e. Change in $y = y(b) - y(a) = 3b + 10 - (3a + 10) = 3b + 10 - 3a - 10$
$= 3b - 3a = 3(b - a)$.

f. Rate of change:

$$\frac{\text{change in } y}{\text{change in } x} = \frac{y(b) - y(a)}{b - a} = \frac{3(b - a)}{b - a} = 3.$$

The rate of change in one variable with respect to another has several applications; two prominent ones are *growth* and *marginal analysis*. Rate of growth is a concept that has obvious importance in biology, anthropology, and psychology. Marginal analysis, used in economics and management, is concerned with the rates of change in *output* caused by small (marginal) changes in *input*. For example, if we assume

3.3 Function Notation

that revenue R is a function of sales x, $R = R(x)$, then the marginal revenue MR is the rate of change in revenue per unit change in sales. If x_1 and x_2 are two different sales, then the marginal revenue in going from selling x_1 units to selling x_2 units is

$$MR = \frac{R(x_2) - R(x_1)}{x_2 - x_1} \tag{3.4}$$

The difference between two values of a variable is sometimes denoted by the Greek letter Δ. So, $R(x_2) - R(x_1)$ is written ΔR, and $x_2 - x_1$ is written Δx. ΔR is read "delta R" or "change in R." Similarly, Δx is read "delta x" or "change in x." Now the marginal revenue Equation (3.4) may be written as

$$MR = \frac{\Delta R}{\Delta x}.$$

Similarly, if we assume that production cost C is a function of x items produced, then $C = C(x)$, and the marginal cost MC is

$$MC = \frac{\Delta C}{\Delta x},$$

where $\Delta C = C(x_2) - C(x_1)$, and $\Delta x = x_2 - x_1$. The marginal cost represents the rate of change in cost per unit of change in production.

The essential question answered by marginal analysis is: Will a change in the independent variable produce a proportionally favorable or unfavorable change in the dependent variable?

Example 3.11 Let revenue R be a function of the sale of x items, $R = R(x)$, and suppose the function is defined by the following equation:

$$R(x) = 200 \cdot x - \left(\frac{1}{10}\right) \cdot x^2,$$

where R is in dollars.

a. Find the revenue R from the sale of 100 items, from the sale of 110 items.
b. Find ΔR for $x_1 = 100$ and $x_2 = 110$.
c. Find the marginal revenue MR between $x_1 = 100$ and $x_2 = 110$.
d. Find the revenue from the sale of 250 items and 260 items.

e. Find the marginal revenue MR between 250 and 260 items.

f. Show that if $x_1 = 1000$ and $x_2 = 1010$, the marginal revenue (between these two values) is *negative*. What does this mean?

Solution

a. $R(100) = \$19{,}000$, $R(110) = \$20{,}790$.

b. $\Delta R = R(110) - R(100) = \$1790$.

c. $MR = \Delta R/\Delta x = 1790/10 = \$179$.

This means that between 100 and 110, each additional sale of an item brings in, on the average, $179 more.

d. $R(250) = \$43{,}750$, $R(260) = \$45{,}240$.

e. $MR = \Delta R/\Delta x = (45{,}240 - 43{,}750)/(260 - 250) = 1490/10 = \$149$.

This means that between 250 and 260, each additional sale of an item brings in, on the average, $149 more.

f. $[R(1010) - R(1000)]/(1010 - 1000) = (99{,}990 - 100{,}000)/10 = -10/10 = -\$1$.

This means that between 1000 and 1010, each additional sale of an item brings in *minus* one dollar. That is, money is *lost* by increased sales. This is because the revenue function here is one of *diminishing returns*, and 1000 is the point at which the returns start declining for increased sales. Now let's see what happens if some quantity, say revenue, depends upon *two variables* (such as the sale of x items of one product and y sales of another). ▲

A function from a *two-dimensional set D* to a set R is called a *function of two variables*. A simple example of this occurs in computing scores in a sporting event, such as basketball, where the two variables might be $g =$ the number of field goals and $t =$ the number of free throws. Then the basketball score S is a function of both g and t. The following equation describes the relationship

$$S = 2g + t. \qquad (3.5)$$

For instance, if a team makes 29 field goals and 13 free throws, its score is 71. [This is found by letting $g = 29$ and $t = 13$ and substituting into into Equation (3.5): $S = 2 \cdot 29 + 13 = 58 + 13 = 71$.] Here the function is the collection of ordered pairs $[(g,t), S]$, where the first term (g,t) is the number of field goals and the number of free throws, and the second term S is the score. Each point (g,t) determines a unique score S. If $(g,t) = (35,15)$, then $S = 85$.

3.3 Function Notation

Example 3.12 Find S if $(g,t) = (22,8)$. Find S if $(g,t) = (19,14)$.

Solution For $(22,8)$, $S = 52$. For $(19,14)$, $S = 52$.

The use of functional notation for two-variable functions is similar to that for functions of one variable. Thus, Equation (3.5) may be written $S(g,t) = 2g + t$. Notation for functions of more than two variables can also be written in a similar manner.

Example 3.13 In Chapter 2, Example 2.8 dealt with an enterprise with three departments, Administration, Research, and Production, and three resources, personnel, computer time, and floor space. Each of the three resources can be expressed as functions of the number of workers in the three departments, as follows. Let w_1, w_2, and w_3 be the numbers of workers in the three departments, Administration, Research, and Production, respectively. Then each of the three functions P, C, and F given below is a function of the three variables w_1, w_2, and w_3.

$P =$ personnel resources function: $P(w_1, w_2, w_3) = w_1 + w_2 + w_3$.

$C =$ computer time function: $C(w_1, w_2, w_3) = \dfrac{1}{3} \cdot w_1 + w_2 + \dfrac{1}{8} \cdot w_3$.

$F =$ floor space function: $F(w_1, w_2, w_3) = 1000w_1 + 600w_2 + 300w_3$.

If we are given a set of values for the independent variables w_1, w_2, and w_3, then we could find the value of each of the resource functions P, C, and F.

a. Let $w_1 = 3$, $w_2 = 5$, and $w_3 = 2$; find the values of P, C, and F.

b. Find $P(15,7,24)$; $C(15,7,24)$; and $F(15,17,24)$.

Solution a. $P(3,5,2) = 3 + 5 + 2 = 10$.
$C(3,5,2) = \frac{1}{3} \cdot 3 + 5 + \frac{1}{8} \cdot 2 = 6\frac{1}{4}$.
$F(3,5,2) = 1000 \cdot 3 + 600 \cdot 5 + 300 \cdot 2 = 6600$.

b. $P(15,7,24) = 46$.
$C(15,7,24) = 15$.
$F(15,7,24) = 26{,}400$.

EXERCISE 3.3

1. Let G be the function defined by the table below.
   a. Find y at $x = 0.8$.
   b. Find $g(0.4)$.
   c. Find $g(1.6)$.
   d. Find $g(2.4) - g(1.6)$.
   e. Find $g(2.4)/g(1.6)$ to two decimal places.

x	$y = g(x)$
0.0	1.00
0.4	1.49
0.8	2.23
1.2	3.32
1.6	4.95
2.0	7.39
2.4	11.02
2.8	16.44

2. Let H be the function defined by the table below.
   a. Find y at $x = 2.5$.
   b. Find $h(3.5)$.
   c. Find $h(4.0)$.
   d. Find $h(1.0) + h(3.0)$.

x	$y = h(x)$
1.0	0.79
1.5	0.98
2.0	1.11
2.5	1.19
3.0	1.25
3.5	1.29
4.0	1.33

3. If $f(x) = 3x^2 - 5x + 4$, find
   a. $f(0)$
   b. $f(1)$
   c. $f(2)$
   d. $f(2/3)$
   e. $f(1/2)$
   f. $f(-1/2)$

4. If $g(x) = \sqrt{x} + x^2 + 1$, find
   a. $g(0)$
   b. $g(1)$
   c. $g(9)$
   d. $g(4)$
   e. $g(2)$, approximately (to three decimal places)
   f. $g(3)$, approximately (to three decimal places)

3.3 Function Notation

5. Let the equation

$$f(x) = \frac{(x-1)}{(x-2)(x-3)}$$

define the function $F = \{[x, f(x)]\}$.
   a. Find $f(1)$.
   b. Find $f(5)$.
   c. Find $f(3/2)$.
   d. Does $f(2)$ exist?
   e. What happens to $f(x)$ as x gets close to 2?
   f. What is the domain of F?

6. Let the equation

$$p(x) = 1 - \sqrt{1-x^2}$$

define the function $P = \{[x, p(x)]\}$.
   a. Find $p(0)$.
   b. Find $p(1)$.
   c. Find $p(-1)$.
   d. Find $p(1/2)$
   e. Does $p(2)$ exist?
   f. What is the domain of P?

7. For $f(x) = 4 - 5x$:
   a. Find $f(0)$ and $f(6)$.
   b. Find the change in f between $x = 0$ and $x = 6$.
   c. Find the rate of change in f with respect to x as x goes from 0 to 6.
   d. What value of x makes $f(x) = 0$?

8. For $g(x) = (7 - 3x)/4$:
   a. Find $g(-3)$ and $g(9)$.
   b. Find the change in g between $x = -3$ and $x = 9$.
   c. Find the rate of change in g with respect to x as x goes from -3 to 9.

9. For $g(x) = 5x^2$:
   a. Find $g(2)$ and $g(4)$.
   b. For $x_1 = 2$ and $x_2 = 4$, find Δg [$\Delta g = g(x_2) - g(x_1)$].
   c. Find the rate of change in g with respect to x between $x_1 = 2$ and $x_2 = 4$.
   d. If $x_1 = a$ and $x_2 = b$, find Δg and Δx.
   e. Find the rate $\Delta g / \Delta x$ for $x_1 = a$ and $x_2 = b$. Simplify as much as possible.

10. For $f(t) = t^2 + 4t$:

   a. Find $f(3)$ and $f(8)$.

   b. For $t_1 = 3$ and $t_2 = 8$, find Δf.

   c. Find the rate of change in f with respect to t between $t_1 = 3$ and $t_2 = 8$.

   d. If $t_1 = a$ and $t_2 = b$, find Δf and Δt.

   e. Find the rate $\Delta f / \Delta t$ for $t_1 = a$ and $t_2 = b$. Simplify as much as possible.

▲ 11. Let revenue R be a function of the sale of x items, $R = R(x)$, and $R(x) = 500x - x^2$.

   a. Find the revenue from the sale of 100 items and 150 items.

   b. Find ΔR for $x_1 = 100$ and $x_2 = 150$.

   c. Find the marginal revenue MR between 150 and 100.

   d. Find $R(200)$ and $R(250)$.

   e. Find ΔR for $x_1 = 200$ and for $x_2 = 250$.

   f. Find the marginal revenue MR between 200 and 250.

   g. Find $R(300)$, ΔR for $x_1 = 250$ and $x_2 = 300$. Is the MR negative? What does this mean?

12. If cost $C = C(x)$ in producing x items is defined by $C(x) = 1000 + 75x$:

   a. Find the cost of producing 10 items, 40 items, and 100 items.

   b. Find the average cost of producing 10 items, 40 items, and 100 items. (Hint: The average cost of producing 10 items can be found from $C(10)/10$, the cost of 10 items divided by 10.)

   c. In general, what is the average cost of producing x items? What value does the average cost approach as the number of items gets larger and larger. (Hint: Try to find the average cost for 1000 items, for 500,000 items, for 1,000,000 items. Doesn't it appear that the average cost is approaching some limit?)

   d. For $x_1 = a$ and $x_2 = b$, find ΔC, and the marginal cost MC.

13. Let Q = total energy supply to lungs, heart, and intestines of an organism. The volume of the intestines is a function $V(Q)$ with the equation

$$V(Q) = \frac{\sqrt{k \cdot (Q + 1)}}{\sqrt{\pi}},$$

where k is a constant to be determined for a given organism.

3.3 Function Notation

a. Find $V(1)$.

b. Find $V(8)$.

c. Show that to *triple* the amount of energy from 5 to 15 requires about a 63 percent increase in the intestine's volume. (Hint: Show that $V(15)/V(5)$ is approximately 1.63).

14. The function with equation

$$z(x) = \frac{x - 35}{7.4}$$

is a standardized score for data with a mean of 35 and standard deviation of 7.4. Find $z(22)$, $z(61)$, $z(6)$, $z(40)$, and $z(35)$.

15. A milkman can choose from various combinations of houses and apartments for his route. Each type of delivery has its advantage over the other. For example, he can deliver to several apartments on one stop, but on the other hand, each house stop gives him a larger order. Let x = the number of houses and y = the number of apartments on any given route. He has six routes to choose from:

$R_1 = (15,80)$ $R_3 = (50,50)$ $R_5 = (35,64)$

$R_2 = (71,12)$ $R_4 = (40,60)$ $R_6 = (22,75)$

That is, the first route $R_1 = (15,80)$ means there are 15 houses and 80 apartments on that route, etc. Say that he makes, on the average, $4 per month per house and $3 per month per apartment. The function $z = z(x,y)$ expresses his profit: $z(x,y) = 4x + 3y$. Compute $z(x,y)$ for each of the routes and determine which one gives him the largest profit.

16. A farmer's profit for growing x acres of corn, y acres of wheat, and z acres of hay is the function $P(x,y,z)$ defined in the equation $P(x,y,z) = 65x + 54y + 38z$ in dollars.

    a. Find his profit in 1975 when he grew 120 acres of corn, 150 acres of wheat, and 200 acres of hay.

    b. In 1976 the crop distribution was $x = 125$, $y = 140$, and $z = 150$. What was his profit?

    c. In 1977 it was $(x,y,z) = (100,200,108)$. Find P. ▲

17. If $f(u,v,w,x) = 3u + (x/v) - w^2 - \sqrt{x}$:

    a. Find $f(7, 1/2, 8, 81)$.

    b. Find $f(9,7,6,9)$.

18. If $g(x_1, x_2, x_3, x_4) = (x_1 - 1)^2 + [x_1/(x_2 + 8)] - x_3 \cdot x_4 + 6$:
   a. Find $g(3,2,9,4)$.
   b. Find $g(3,0,5,2)$.

4
LINES

4.1

LINEAR EQUATIONS

A straight line is the shortest distance between two points; it is also the simplest "curve" through them on a graph. Indeed, the straight line is so simple that we need to know only *two things* about it in order to define it. If we are given the coordinates of *any two points* of a straight line, then we can write an equation for the line. Or if we are given *just one point and the slope* (rate of change in one variable with respect to the other), we can also write its equation.

Perhaps this is one reason that straight lines (or, more properly, **linear functions**) are widely used in the sciences. Admittedly, the stock market or any other natural phenomenon may be more likely to oscillate up and down as in Figure 4.1b than to follow a straight line, but the ease with which linear equations can be derived, as well as their geometric simplicity still make linear functions useful. Straight lines also serve as short-run approximations to many curves.

A straight line in the *x-y* plane is the graph of the equation

$$A \cdot x + B \cdot y = C,$$

where A, B, and C are constants, and A and B are not both zero. Some examples are

(I) $\quad 3x + 7y = -10; \quad$ A = 3, B = 7, and C = -10.

(II) $\quad -2x + \frac{1}{2} \cdot y = 1; \quad$ A = -2, B = $\frac{1}{2}$, and C = 1.

(III) $\quad y = -15; \quad$ A = 0, B = 1, and C = -15.

(IV) $\quad x = 8; \quad$ A = 1, B = 0, and C = 8.

The graphs of these equations are shown in Figure 4.2.

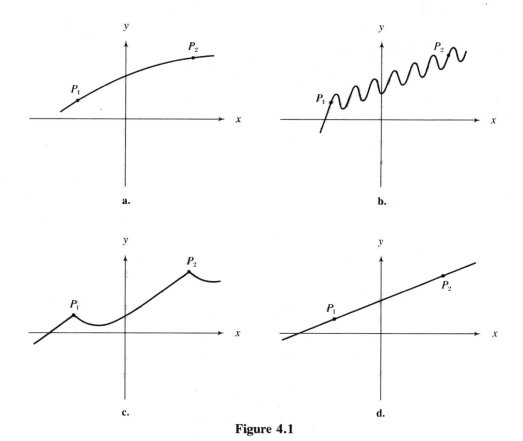

Figure 4.1

Notice that in all but one case, the line is the graph of a function. The exceptional case is (IV), where the graph of $x = 8$ is a vertical line. When a line equation defines a function (that is, it is *not* a vertical line), then that function is called **linear,** and any linear function in the x-y plane may be put into the form

$$y(x) = m \cdot x + b,$$

where m and b are constants and $y(x)$ is function notation (see Section 3.3).

How did we get these graphs for the line equations? For any given linear function in x and y, evaluate the dependent variable y at two different values of the independent variable x. Plot the two resulting points on graph paper, and use a ruler to draw a line through those two points. For example, let $y = -3x + 10$. To draw the graph, first let $x = 6$. Then $y(6) = -8$, so $(6,-8)$ is one point. Second, evaluate y at some other value of x, say $x = 10$. Then since $y(10) = -20$, $(10,-20)$ is another point. (Any other two values of x could have been chosen.) Now, plot the two points $(6,-8)$ and $(10,-20)$ as in Figure 4.3a. Draw a straight line through the two plotted points, as in Figure 4.3b.

4.1 Linear Equations

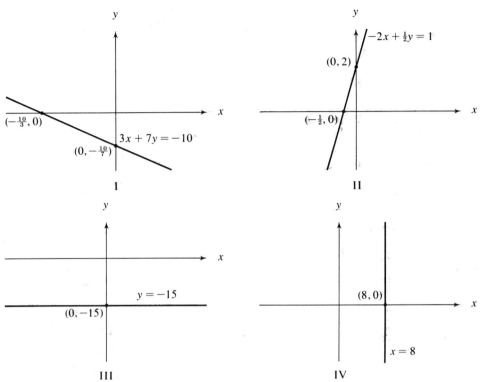

Figure 4.2
Some line equations and their graphs.

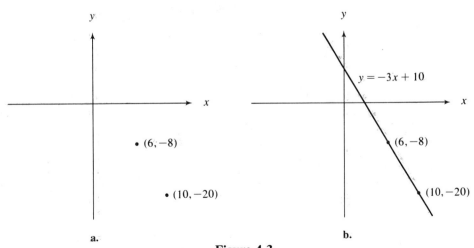

Figure 4.3
Drawing the graph of the linear function $y = -3x + 10$.

Example 4.1   For each of the following linear functions $y = y(x)$, evaluate y at some two values of x, plot the resulting points, and draw a straight line through them.

  a. $y = -5x + 15$.

  b. $y = 7x - 2$.

  c. $y = 4$.

  d. Put the following equation into the form $y = mx + b$ and proceed as above: $5x + 6y - 10 = 0$.

Solution   a. Choose two values x_1 and x_2 of x, say $x_1 = 0$ and $x_2 = 3$. Then $y(0) = 15$ and $y(3) = 0$; therefore, two points on the line are $(0,15)$ and $(3,0)$. See Figure 4.4a.

b. Let $x = 0$; then $y = -2$; so $(0,-2)$ is on the graph. If $x = -1$, then $y = -9$; so $(-1,-9)$ is on the graph. See Figure 4.4b.

c. $y(1) = 4$ and $y(5) = 4$, so $(1,4)$ and $(5,4)$ are on the graph. See Figure 4.4c.

d. From $5x + 6y - 10 = 0$, we get the equation $y = -5/6 \cdot x + 5/3$. Two points on this line are $(8,-5)$ and $(2,0)$. See Figure 4.4d.

The above examples illustrate how to get two points on a line if we already know the equation, but now we want to discuss the reverse problem. How do we write a *line equation* from knowing just two points? We need to consider two cases.

First, if the two points are on the same vertical line, such as $(3,5)$ and $(3,10)$, then the entire line is described by an equation in which x is set equal to the

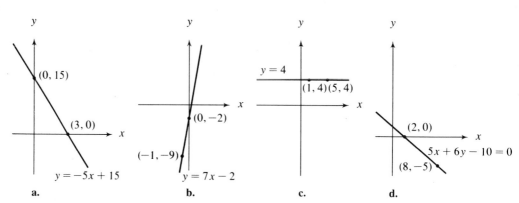

Figure 4.4

4.1 Linear Equations

abscissa of either point. In this example, $x = 3$ is the *vertical line equation* for the line passing through the two points (3,5) and (3,10).

Second, if the two points are not in the same vertical line, then the rate of change in y with respect to the change in x is constant between any two points of the graph. Thus, if $P_1 = (x_1, y_1)$ and $P_2 = (x_2, y_2)$ are any two points on a given line, with $x_1 \neq x_2$, then there is a constant m such that $\Delta y / \Delta x = m$; that is,

$$\frac{y_2 - y_1}{x_2 - x_1} = m. \tag{4.1}$$

Therefore, if (x,y) is any other point on the line through P_1 and P_2, then the ratio

$$\frac{y - y_1}{x - x_1}$$

is also this same number m. In other words,

$$\frac{y - y_1}{x - x_1} = \frac{y_2 - y_1}{x_2 - x_1}. \tag{4.2}$$

Similarly,

$$\frac{y - y_2}{x - x_2} = \frac{y_2 - y_1}{x_2 - x_1}. \tag{4.3}$$

See Figure 4.5.

Equations (4.2) and (4.3) are linear equations relating the two variables x and y to the given points P_1 and P_2, but as the equations now stand, the point (x,y) cannot be P_1 [in Equation (4.2)], and cannot be P_2 [in Equation (4.3)]. If we rewrite these equations as follows, we will eliminate this restriction and see that they both represent an equation of the line passing through the points P_1 and P_2:

$$y - y_1 = \frac{y_2 - y_1}{x_2 - x_1} \cdot (x - x_1) \tag{4.4}$$

and

$$y - y_2 = \frac{y_2 - y_1}{x_2 - x_1} \cdot (x - x_2). \tag{4.5}$$

If we substitute $x = x_1$ into either equation, we get $y = y_1$, and if we put $x = x_2$ into either equation, then we get $y = y_2$. Therefore, both points (x_1, y_1) and (x_2, y_2) have coordinates satisfying Equations (4.4) and (4.5). Each of these equations is called the *two-point form of the line equation*. See Figure 4.6.

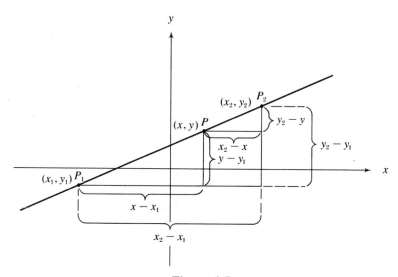

Figure 4.5
Constant growth rate: $(y_2 - y_1)/(x_2 - x_1) = (y_2 - y)/(x_2 - x)$
$= (y - y_1)/(x - x_1)$.

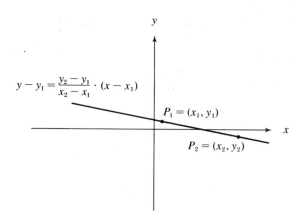

Figure 4.6
An equation for a line through the points P_1 and P_2.

4.1 Linear Equations

Example 4.2 Write an equation for the line passing through the points $(-3,7)$ and $(4,6)$. Use this *line equation* to find y when $x = 10$.

Solution Here let $(x_1, y_1) = (-3,7)$ and $(x_2, y_2) = (4,6)$. Then by Equation (4.4),

$$y - 7 = \frac{6 - 7}{4 - (-3)} \cdot [x - (-3)]$$

or

$$y - 7 = -\frac{1}{7} \cdot (x + 3). \tag{4.6}$$

If we had used Equation (4.5) instead of Equation (4.4), then we would have gotten

$$y - 6 = -\frac{1}{7} \cdot (x - 4). \tag{4.7}$$

Equations (4.6) and (4.7) are *both* equations for the *same* line. Each one can be rewritten as the equation

$$y = -\frac{1}{7} \cdot x + \frac{46}{7}. \tag{4.8}$$

Now, by any one of the three equations (4.6), (4.7), or (4.8), $y = 36/7$ if $x = 10$.

▲ **Example 4.3** Suppose that eight sales produce a profit of $5.00, and eleven sales produce a $13.50 profit. Write a *line equation* for profit as a linear function of sales. Predict the profit on fifteen sales.

Solution Here, the two points are $(8,5)$ and $(11, 13.5)$. Let the two variables be s and p for sales and profit, respectively. One line equation is

$$p - 5 = \frac{13.5 - 5}{11 - 8} \cdot (s - 8) = \frac{8.5}{3} \cdot (s - 8).$$

The profit on fifteen sales is then $p(15)$, or $24.83.

Example 4.4 Two well-known points of conversion from temperature measured in degrees Celsius (centigrade) to Fahrenheit are the freezing point of water: 0°C, 32°F, and the boiling point of water: 100°C, 212°F. Interpret this relation as a straight line in the C-F plane, with variables c and f. The two points are $(0,32)$ and $(100,212)$. Write an equation relating c and f. Here, c is the independent variable and f

is the dependent variable. Use the equation to find f when $c = 37$. Also find the point where f and c are the same number.

Solution In the C-F plane

$$f - 32 = \frac{212 - 32}{100 - 0} \cdot (c - 0) = \frac{180}{100} \cdot c.$$

That is,

$$f - 32 = \frac{9}{5} \cdot c, \text{ or } f = \frac{9}{5} \cdot c + 32. \qquad (4.9)$$

Now, if $c = 37$, then by Equation (4.9), $f = 32 + (9/5) \cdot 37 = 98.6°F$. (In other words, 37°C is normal body temperature.) To find where c and f are the same, replace f by c in Equation (4.9). Then

$$c = \frac{9}{5} \cdot c + 32, \text{ or } -\frac{4}{5} \cdot c = 32, \text{ or } c = -40.$$

That is, $-40°C = -40°F$. ▲

EXERCISE 4.1

In Problems 1-4 draw the lines (graphs) of the given equations.

1. a. $y = 3x - 7$.
    b. $y = x + 2$.
2. a. $y = 2x + 2$.
    b. $y = \frac{1}{2} \cdot x + \frac{3}{4}$.
3. a. $3y = 6x + 4$.
    b. $-3x + y - 4 = 0$.
4. a. $2y - 7x + 1 = 0$.
    b. $1/(x - y) = \frac{1}{3}$.
5. Let $x =$ family income and $c = c(x)$ be consumption in dollars (including tax) by a family earning x dollars. Suppose the function $c(x)$ is linear and described by the equation $c = 3000 + 0.7x$, with $x \geq 10,000$.

    a. What is the consumption by a family earning $11,000? $25,000? $18,000?

4.1 Linear Equations

b. Savings, $s = s(x)$, is proportional to the difference between income and consumption. Suppose the savings equation is $s(x) = 0.4[x - c(x)]$. Find the savings by a family earning $11,000, $25,000, $18,000.

▲ 6. In human beings between the ages of 20 and 45 years, the rate of wound healing w is a linear function of the age t of the individual. Then $w = w(t)$ is measured in square centimeters per day and t is measured in years. The linear equation is $w = 1.82 - 0.0357\,t$ square centimeters/day.

a. Find the rate at which a wound heals in a 23-year-old person. 1 SQ CM /DAY

b. What is the daily area of wound healing in a 35-year-old person? .57 SQ CM/DAY

c. Find $w(45)$. = .2135

d. Your friend has a three-square-centimeter wound that takes four days to heal. How old is your friend? 30 YRS OLD ▲

7. a. Write an equation for the line L passing through the points $(0,0)$ and $(7,-1)$.

b. Find the coordinates of the point on L with an abscissa of -8000.

8. a. Write an equation for the line K through the points $(-3,4)$ and $(-2,-3)$.

b. If P is a point on K and the abscissa of P is 0, find the ordinate of P.

9. a. Write an equation for the line H through the points $(0, 7/2)$ and $(12/5, 0)$.

b. Find a point on H for which the abscissa and ordinate are equal to each other.

10. a. Write an equation for the line G through the points $(0,1)$ and $(2,-3)$.

b. Find a point on G for which the ordinate is three times the abscissa.

▲ 11. **Oil Slick** On February 3, 1969, after six days of seepage from offshore wells, an oil slick covered 80 square miles of ocean near Santa Barbara, California. On February 9, the wells were finally capped and on that day the oil slick covered 104 square miles. Assume that from February 3 to 9 the area of the oil slick is a linear function of time. (Before February 3, the seepage was not a straight-line curve, but from that time until the capping date, the linear model fits very well. See the figure below.)

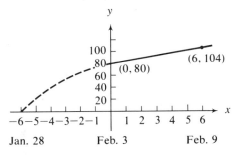

a. Write a linear equation in the x-y plane, with x = days of seepage and y = square miles of ocean surface covered. Let x = 0 on February 3.

b. Find y on February 5 (x = 2).

c. If 1160 gallons cover one square mile, how many gallons were in the slick on February 7?

12. You can buy one gram of hashish for $5.25 and twenty-eight grams for $80.31.

    a. Write a linear equation with price as a function of weight.

    b. How much does ten grams cost?

    c. How much does half a gram cost?

13. A certain mutual fund has enough capital to buy 1000 shares of Greyhound and 200 shares of Xerox, or it can buy 400 shares of Greyhound and 350 shares of Xerox.

    a. Write a linear equation representing all the possible combinations.

    b. If it buys 1600 shares of Greyhound, how many Xerox can it buy?

    c. If both companies had a stock rise of 4 dollars per share last year, what would have been the best combination to have had? Isn't hindsight wonderful?

14. Assume that farm income I in dollars is a linear function of farm size x in acres. Suppose that a 310-acre farm has an income of $7460 and a 700-acre farm has an income of $29,300. Also suppose that a linear equation is valid only for the domain $300 \leq x \leq 1500$.

    a. Find the income of a 1000-acre farm.

    b. Find the average income per acre in a 300-acre farm compared to the average income per acre in a 1500-acre farm. ▲

4.2

POINT-SLOPE FORM OF THE LINE EQUATION

In Chapter 3, the rate of change between two points (x_1, y_1) and (x_2, y_2) of a curve was defined as $\Delta y / \Delta x$, or $(y_2 - y_1)/(x_2 - x_1)$. As mentioned in the last section, if the curve is a straight line, then the rate of change between any two of its points is constant. This constant rate is called the **slope** of the line. For

4.2 Point-Slope Form of the Line Equation

example, the line through the points $(-2,4)$ and $(3,5)$ has slope $(5 - 4)/[3 - (-2)] = 1/5$. Slope is usually denoted by the letter m; so the slope of this line is $m = 1/5$.

Although slope is defined in terms of two points, it *is* the ratio of one given quantity *per* another quantity, and in some practical problems, it is given as such a ratio. For example, the price of eggs—given as a certain amount of *money per dozen*—is the slope of a line relating money to dozens of eggs.

In general, just knowing the slope of a line is not enough information for writing the equation of that line, but if the slope and just one point are known, then the equation can be determined.

Let (a,b) be any point of a given line with slope m [see Equation (4.1)]. Then the equation of that line is

$$y - b = m \cdot (x - a). \tag{4.10}$$

Note that this equation does not work for vertical lines, because m must be a finite number, and a vertical line has *infinite* slope.

Example 4.5 If a bar in San Francisco charges $1.75 per drink and a $4.00 cover charge, write a line equation with x = number of drinks and y = the cost of each excursion into this bar.

Solution Here the slope is revealed by the given 1.75 *per* drink, so $m = 1.75$. Since the cover charge is the price you would pay even if you bought no drinks, when $x = 0$, $y = 4$. Therefore, the known point (a,b) is $(0,4)$. This point, $(a,b) = (0,4)$, with $m = 1.75$ substituted into Equation (4.10) yields $y - 4 = 1.75(x - 0)$, or $y = 4 + 1.75x$.

The reverse problem, finding the slope given the equation, can be solved by using the equation to find the coordinates of two points on the line.

Example 4.6 From the line equation $3x + 4y + 7 = 0$:

a. Find two points on the line.

b. Find the slope of the line.

Solution a. We arbitrarily select any value of x and use the equation to determine y. This produces one point on the line. Then we select a different value of x and determine y; this gives us a second point. Let $x = 0$, then $3 \cdot 0 + 4y + 7 = 0$, or $y = -7/4$. Let $x = 5$, then $3 \cdot 5 + 4y + 7 = 0$, or $y = -22/4$. Therefore two points on the line are $(0, -7/4)$ and $(5, -22/4)$.

lope and from the two points found in part **a**,

$$\frac{\frac{-22}{4} - \left(\frac{-7}{4}\right)}{5 - 0} = \frac{\frac{-15}{4}}{5} = \frac{-15}{20} = -\frac{3}{4}.$$

EXERCISE 4.2

1. The slope of a given line L is $m = -1/2$ and L contains the point $(3,7)$. Write an equation for L.

2. The slope of a given line K is $m = 12/7$ and K contains the point $(-1/2, 10)$. Write an equation for K.

3. Use the *slope definition* $m = (y_2 - y_1)/(x_2 - x_1)$ to find the slope of a line through the two points $(x_1, y_1) = (3,4)$ and $(x_2, y_2) = (5,7)$.

4. Use the *slope definition* in Problem 3 to find the slope of the line through the two points $(-1,6)$ and $(8,-2)$.

5. Find two points on the line whose equation is $3x - 7y + 8 = 0$. Find the slope of the line.

6. Find two points on the line whose equation is $14x + 6y - 15 = 0$. Find the slope of the line.

7. What lines have zero slope?

8. What lines have *infinite* slope?

9. Two *different* lines with the same slope are parallel to each other. Show that the line K with equation $6x + 2y = 3$ and the line L with equation $y = 1 - 3x$ have the same slope and yet are not the same line. (For the latter, find some point that is on one line but not on the other.)

10. Show that $3x + 7y - 6 = 0$ and $5x + 11\,2/3\,y + 10 = 0$ are parallel lines (they have the same slope but are not the same line).

▲ 11. **Pollution-Free Electric Car** The Tokyo Shibaura Electric Co. and the Yuasa Battery Co. have announced the joint development of a sodium-sulfur battery. Their experimental model has an output of 165 watt-hours and can drive a one-ton pollution-free car (at normal speeds) 180 miles on a single charge. Experiments show that such a car can be driven $11/6$ mile per watt-hour on each charge.

a. Write a linear equation relating distance run to the battery output capacity, using x = output and y = distance on one charge. Hint: The slope is $11/6$ and one point is $(165, 180)$.

b. These companies plan to mass-produce by 1976 a battery that has an output of 300 watt-hours. How far will this battery drive a car on a single charge?

c. If the 300-watt-hour battery is rechargeable 1000 times, how far can a car be driven before the battery must be replaced?

12. Decaying Bark Versus Salmon Embryo If bark from logs is deposited in a stream, it begins to decay; and the decaying process consumes oxygen in the water. These same streams are usually a source of salmon. When fertilized salmon eggs have to compete for the oxygen with the decaying bark, the salmon eggs lose. The linear relation between the amount of decaying bark present in the water and the number of salmon embryos surviving is as follows: When the bark concentration is one percent of the volume, then 52 percent of the eggs survive. The percent of surviving eggs drops 5.5 percent for each percent increase in the bark. (Thus, the slope is -5.5.)

a. Write a linear equation with y = percent of eggs surviving and x = percent of bark concentration.

b. Find the percent of salmon embryos that would have survived if there was no bark at all in the stream. (Note: We are using the terms "eggs," "embryos," and "fertilized eggs" interchangeably.)

c. Find the percent of eggs surviving if the bark concentration reaches ten percent. ▲

4.3
SIMULTANEOUS LINEAR EQUATIONS

The *line equation* is a very useful tool in applied problems involving only one linear function of the independent variable, but many real problems are concerned with two or more functions of the same variable. For example, both spending and saving are functions of income. Most mathematical models in the social and life sciences require the use of several functions simultaneously. When the different functions are all linear, then the model is called a **system of linear equations.**

The simplest system is one with two equations, and the following example presents various methods for solving a *two-equation system.* Systems with three linear equations are slightly more tedious to solve than systems with two equations, but the techniques are not much different. Solving general systems with any number of equations in as many variables is better done by determinants, vectors, and matrices. These more complex topics of linear analysis will be found in Chapters 6 and 7.

Example 4.7 In a certain office there are personnel consisting of only secretaries and executives. Each secretary has one desk and two filing cabinets, and each executive has two desks and one filing cabinet. There is a total of 28 desks and 32 filing cabinets in the office. How many secretaries are there? How many executives?

Solution As a rule, the best approach to word problems is to start by answering the numerical question posed in the problem with a letter of the alphabet: Let x be the unknown.

Let $x =$ the number of secretaries and $y =$ the number of executives. Since each secretary has two filing cabinets and each executive has one, then the number of filing cabinets generated by x secretaries is $2x$ and the number generated by y executives is y. The total number of filing cabinets, 32, is $2x + y$. That is, $2x + y = 32$. Similarly, y executives generate $2y$ desks and x secretaries generate x desks, so the 28 desks given is the total $x + 2y$. Thus $x + 2y = 28$. The pair of simultaneous linear equations

$$2x + y = 32$$
$$x + 2y = 28$$

represents a pair of lines, each with infinitely many points on it, but we seek the point (x_1, y_1) that is on both lines; that is, we want the coordinates x_1 and y_1 to satisfy both equations simultaneously. When we have found these coordinates, then we have *solved* the system of equations.

There are at least four methods for solving systems of linear equations:

1. Elimination of the variables, one at a time, by combined operations on both equations.
2. Substitution—that is, solving an equation for one variable in terms of the others and substituting into the other equations.
3. Determinants.
4. Vectors and matrices—some pretty flowers growing on the determinant weeds.

We will now demonstrate the use of methods 1 and 2 for this problem, but we will save methods 3 and 4 for later.

4.3 Simultaneous Linear Equations

Method 1 *Elimination of a variable between equations* Start by deciding which variable to eliminate. Then multiply the equations by whatever number it takes to make the coefficients of that variable add up to zero (or *subtract down to zero*) when the equations are combined.

The two equations in this problem are

$$2x + y = 32 \tag{4.11}$$

and

$$x + 2y = 28. \tag{4.12}$$

Let us decide to eliminate y first. To make the coefficients of y such that subtracting one equation from the other will result in a zero, we need to make these coefficients the same in both equations. Multiplying both sides of Equation (4.11) by the number 2, we get $2 \cdot (2x + y) = 2(32)$, or

$$4x + 2y = 64. \tag{4.13}$$

But now Equation (4.13) has the same y coefficient as Equation (4.12), so we can eliminate y by just subtracting Equation (4.12) from Equation (4.13):

$$\begin{array}{r} 4x + 2y = 64 \\ -(\ x + 2y = 28) \\ \hline 3x + 0 = 36 \end{array}.$$

We have succeeded in eliminating* y, and we now have $3x = 36$, or $x = 12$.

Now to eliminate x, multiply by whatever it takes to force both of the equations to have the same x coefficients, and then subtract.

Multiplying Equation (4.12) on both sides by 2, we get

$$2(x + 2y) = 2(28), \quad \text{or} \quad 2x + 4y = 56. \tag{4.14}$$

Subtracting Equation (4.11) from Equation (4.14),

$$\begin{array}{r} 2x + 4y = 56 \\ -(2x + \ y = 32) \\ \hline 3y = 24 \end{array},$$

we get $y = 8$. The answer is 12 secretaries and 8 executives.

---

*What has actually happened, of course, is that one system of equations has been replaced by another equivalent one, but the replacement equation has a zero coefficient for the y term. This is what is usually meant by "eliminating" a variable.

Check

12 secretaries have	12 desks	and	24 filing cabinets
8 executives have	16 desks	and	8 filing cabinets
Total personnel has	28 desks	and	32 filing cabinets

Method 2 *Substitution* Solve one equation for one variable and substitute into the other equation. In the Equation (4.11), $2x + y = 32$, solving for y yields

$$y = 32 - 2x. \tag{4.15}$$

Substitution of $y = 32 - 2x$ into Equation (4.12), $x + 2y = 28$, gives

$$x + 2(32 - 2x) = 28$$
$$x + 64 - 4x = 28$$
$$-3x = 28 - 64$$
$$-3x = -36 \quad \text{or} \quad x = 12.$$

Now substitute $x = 12$ into Equation (4.15) to find $y = 32 - 2 \cdot (12) = 32 - 24 = 8$.

This system of Equations [(4.11) and (4.12)] is a mathematical model that describes the consumption by secretaries and executives of two resources, desks and filing cabinets. Systems of equations can provide models for economics and business because they describe such things as competition for resources, input-output phenomena, and break-even points.

These topics are really just special cases of problems in which two functions can represent conflicting quantities. For example, one straight line may represent a quantity that increases with respect to the independent variable, such as income increases with an increase in number of items produced. Yet, in the same system, another line may represent a quantity that counteracts the first, such as increasing costs or taxes with increased production. The points on these opposing lines have ordinates that measure opposing quantities. The vertical distances between two points on such lines is a positive or a negative quantity representing either a *gain* or a *loss*. See Figure 4.7.

The area between two such lines is the gain or loss throughout the domain of x values. See Figure 4.8.

When the cost line and the income line cross, the difference between the y values is zero. There is neither a profit nor a loss at this point of intersection. This is the point at which the company breaks even, and it divides the profit area from the loss area. See Figure 4.9.

4.3 Simultaneous Linear Equations

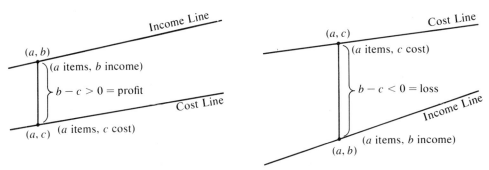

Figure 4.7

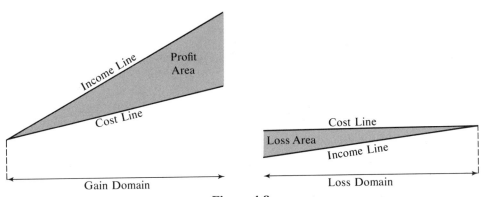

Figure 4.8

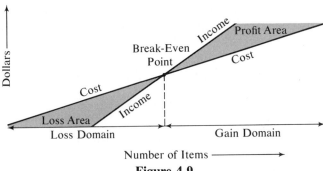

Figure 4.9

▲ **Example 4.8** A manufacturer can use one of two production methods in making certain transistors. Method I costs 20 dollars to set up and produces the transistors for 30 cents each. Method II costs 100 dollars to set up and produces transistors for 25 cents each. The transistors are sold at 80 cents each.

a. Write the equations of the three lines.

b. Find the break-even point for each method.

c. How many transistors must be produced for both methods to have the same average costs per transistor?

d. If he can never sell more than 1475 of these transistors, which is the best production method?

e. If he could sell as many as 2000 transistors, which would be the best method?

Solution a. The equations of the three lines are obtained by using the point-slope formula. For method I the point is (0,20) and the slope is 0.30 (since the initial cost of dollars $20 is at zero items and the items cost 30 cents each). For method II the point is (0,100) and the slope is 0.25, and for the revenue line the point is (0,0) and the slope is 0.80. Let x = the number of items and y the number of dollars, then:

(I)   The equation of method I is $y_I - 20 = 0.30x$.

(II)   The equation of method II is $y_{II} - 100 = 0.25x$.

(R)   The equation of the revenue line is $y_R = 0.80x$.

b. Solving Equations (I) and (R) simultaneously yields $x = 40$ and $y_I = 32$. Equations (II) and (R) solved simultaneously yield $x = 181.81$ and $y_{II} = 145.45$. He must make and sell 40 items by method I to break even and 182 items by method II to break even.

c. Both methods have the same cost when $y_I = y_{II}$, which is at $x = 1600$.

d. For fewer than 1600, method I is cheaper.

e. For more than 1600, method II is cheaper. ▲

EXERCISE 4.3

1. Find the values of x and y satisfying the equations

$$x + y = 17$$
$$3x - y = 8$$

2. Solve the simultaneous equations

$$3x + 7y = 16$$
$$x + 11y = 4$$

3. Solve the system

$$2x + 3y = 5$$
$$7x + 11y = 13$$

4. Solve the system

$$5x + 5y = 4$$
$$19x - 3y = 2$$

▲ 5. Two types of insects, A and B, visit a wheat-farming area. Each type-A insect consumes one cent worth of wheat and each type-B insect consumes seven cents worth. Together they devour $300,000 worth of the crop.

   a. Write a line equation describing this statement. (Use x for the number of type-A insects and y for the number of type-B insects.)

   Insect spray is applied; this spray is known to kill 50 percent of the type-A insects and 75 percent of the type-B insects. The total number (both types) killed by the spray was estimated at 10 million.

   b. Write a line equation describing this statement (use the same x and y as in part a).

   c. Use the two line equations in parts a and b to find how many of each type originally invaded the farms.

6. The stock market report claims "gaining stocks lead declining stocks by 380 issues, with the gainers having a 5 to 3 ratio over the decliners." How many stocks were gaining and how many were declining? ▲

7. (I) $\quad 3x + y + z = 2$
   (II) $\quad x - y + z = 8$
   (III) $\quad 2x + 3y - z = 10$

   Solve these three simultaneous equations for x, y, and z as follows. Add Equation I to II, to get one equation in x and z. Multiply Equation II by 3 and add this Equation III to get a second equation in x and z, then solve these two equations in x and z for x and z. Use the two values for x and z in any one of the three original equations to find the value of y.

8. (I) $\quad x + y + z = 100$
   (II) $\quad x - y + 2z = 75$
   (III) $\quad 75x - 5y + z = 2$

   Solve for x, y, and z as follows. Add Equations I and II; this gives one equation in x and z. Then add Equation III to five times Equation I to get another equation in x and z. Solve this *two-equation system* for x and z, then use these values to find y from any of the three equations.

9. Draw the following three lines in the x-y plane and solve for the three points of intersection: $K: x = 5$, $L: 3x + y = 6$, and $M: y = x - 1$. Hint: Solve the three two-equation systems K and L, K and M, M and L.

10. Draw the following three lines in the x-y plane and solve for the three points of intersection: $K: x - 15y = -74$, $L: x + y = 20$, and $M: 7x - 11y = 140$. See the hint in Problem 9.

▲ 11. A business hires men and women for the same job. Each man works 40 hours per week and earns 200 dollars. Each woman works 26 hours per week and earns 90 dollars. The company shows a weekly payroll of $4030 for a total of 862 hours of work.

   a. How many men are employed? How many women?

   b. What percent of the men's hourly wages do the women get?

12. Suppose the authorities of the Golden Gate Bridge decide to encourage car pools and reduce the number of automobiles driving across the bridge. They adopt the following plan: Cars with four or more adults go across toll-free. Cars with only two or three adult riders pay 25 cents. Cars with only one adult, the driver, pay one dollar.

   In one morning 1200 cars pass through a given toll gate, and the attendant collected $505. There were 66 more cars paying one dollar than there were cars paying no toll.

4.3 Simultaneous Linear Equations

a. How many of each type of auto toll fees were collected?

b. Find the least possible number of adults who could have gone through the toll gate that morning.

13. A city can install either a bus system or a streetcar system. To run the bus system, the daily cost is $12,000 plus $130 per vehicle. To run the streetcar system, the daily fixed cost is $9000 and the per vehicle cost is $160. The daily revenue from either system is $330 per vehicle.

    a. Let x = the number of vehicles in daily operation, y_B the daily cost of a bus system operating x buses, y_S = the daily cost of operating x streetcars, and y_R = the daily revenue from x vehicles (either type). Write the three line equations using this notation.

    b. If the city runs 50 vehicles daily, what profit or loss does it experience from buses? From streetcars?

    c. What is the break-even point for buses? Streetcars?

    d. For what value of x will the daily cost of both mass transit systems be the same?

    e. If the city can never run more than 100 vehicles, which system provides the most profitable operations? ▲

5
INTRODUCTION TO LINEAR PROGRAMMING

5.1

INEQUALITIES IN THE X-Y PLANE

A city manager has a budget that must be shared among various departments (law enforcement, fire protection, parks and recreation, street maintenance, sanitation, etc.). The costs in these departments are variables with a range of values, but because the budget is finite, the range is restricted. Each cost influences and competes with the others. For this reason, they are said to constrain each other and the budget can be described as a *function of several independent variables subject to constraints*.

The constraints are given in the form of *inequalities* with the property that when one of the variables increases, at least one of the others is forced to decrease. For example, money spent on recreation reduces the budget for law enforcement, but perhaps the existence of recreational facilities also reduces the actual cost of law enforcement. Suppose that the most money available each year for both law enforcement and recreation is

$$\frac{513}{10} x \text{ dollars,}$$

where x is the population of the city. Then the constraint inequality might be

$$\frac{1}{7} y_L + y_P \leq \frac{513}{10} x,$$

where y_L is the law enforcement budget and y_P is the parks budget. For a fixed population (say, $x = 1100$), the point (y_P, y_L) can range in the y_P-y_L plane throughout the shaded region given in Figure 5.1.

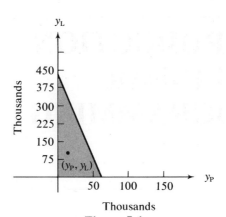

Figure 5.1
Region of constraint on the variables y_P and y_L.

The problem of the city manager is to select, for all the variables, the combination of values that will maximize the services and minimize the costs. This is the *management problem*; it can take many forms.

A farmer manages a farm whose variables are the amounts of each crop that he should grow. The constraints are his land size and suitability, irrigation and fertilization costs, market demand, and government price support for the crop. The farmer's problem is to maximize his profit subject to the given constraints.

The given constraints in these and other cases are inequalities. In this chapter, we will concentrate on *management*-type problems in which the constraint inequalities involve only two variables. The two-variable systems of inequalities are the easiest ones to picture geometrically, and the geometry of linear inequalities plays an important role in explaining *linear programming*, a topic to be introduced in the next section.

To see what is meant by "the geometry of inequalities," let us start with the line equation $y = mx + b$, and replace the $=$ sign by an inequality sign, $<$, $>$, $\leq$, or $\geq$. The resulting inequality defines a region of the plane, with the line from the original line equation as a boundary. (See Figure 5.2.)

If L is the line whose equation is $y = mx + b$, then the region defined by

$y = mx + b$ is the set of all points on L.

$y < mx + b$ is the set of all points below L.

$y > mx + b$ is the set of all points above L.

$y \leq mx + b$ is the set of all points either below or on L.

$y \geq mx + b$ is the set of all points either above or on L.

5.1 Inequalities in the X-Y Plane

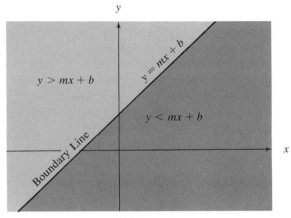

Figure 5.2

We now give several examples.

Example 5.1 Each point above the x-axis has a positive ordinate but could have either a positive or negative abscissa. The entire *upper half plane* may be described as the set of all points (x,y) with positive ordinates; that is, the set of all points (x,y) with $y > 0$. (See Figure 5.3.)

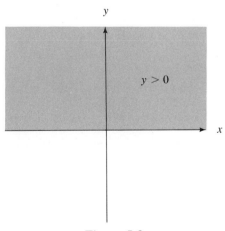

Figure 5.3

Example 5.2 The set of all points (x,y) where $y > -3$ is shown in Figure 5.4. It is all the points above the line $y = -3$; the line $y = -3$ is the boundary to the region $y > -3$.

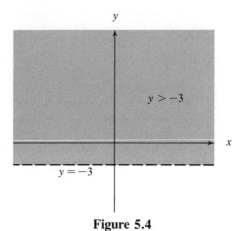

Figure 5.4

Example 5.3 All points of the plane such that $-3 < y \leq 15$ are located between the lines $y = -3$ and $y = 15$, including those points on the line $y = 15$ but not those on the line $y = -3$ (Figure 5.5).

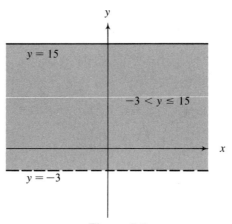

Figure 5.5

5.1 Inequalities in the X-Y Plane

Example 5.4 All the points of the x-y plane to the right of the y-axis have $x > 0$ (Figure 5.6).

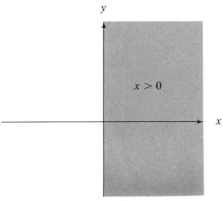

Figure 5.6

Example 5.5 All points in the x-y plane with $3 \leq x < 10$ lie between the lines $x = 3$ and $x = 10$, including $x = 3$ and excluding $x = 10$, as suggested in Figure 5.7.

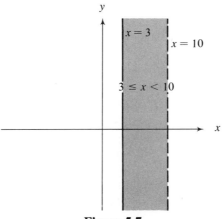

Figure 5.7

Example 5.6 The set of points (x,y) such that both $2 \leq y \leq 6$ and $3 \leq x < 10$ is the set of all points that are between and including the horizontal lines $y = 2$ and $y = 6$,

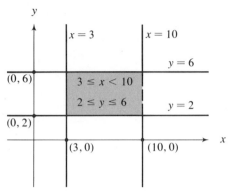

Figure 5.8

and between the vertical lines $x = 3$ and $x = 10$, excluding the line $x = 10$. (See Figure 5.8.)

Example 5.7 Where are the points (x,y) such that $3x - 2y < -5$? This is tricky. (Watch out for the negative coefficient of y.) Adding $2y$ to both sides we get $3x < -5 + 2y$ or $-5 + 2y > 3x$. Now adding 5 to both sides, we get $2y > 3x + 5$, and finally, after dividing both sides by 2,

$$y > \left(\frac{3}{2}\right)x + \frac{5}{2}.$$

The region is shown in Figure 5.9.

Example 5.8 Where are all the points of the x-y plane for which $y \leq x$, $y \geq -3x$, and $x \leq 7$? The three lines $y = x$, $y = -3x$, and $x = 7$ form the boundary for the region as sketched in Figure 5.10. Note that the corner points are found by solving the *boundary line equations two at a time.* Thus, the simultaneous solution to $y = x$ and $y = -3x$ is $(x,y) = (0,0)$. The simultaneous solution of $y = -3x$ and $x = 7$ is $(x,y) = (7,-21)$. The simultaneous solution of $x = 7$ and $y = x$ is $(x,y) = (7,7)$.

Why do we need sketches of the region defined by sets of linear inequalities? The reason is that, without the sketch, we will not know which points are the actual corner points of the region (the corner points, as we shall soon see, play a most important role in linear programming). Unfortunately, boundary lines do

5.1 Inequalities in the X-Y Plane

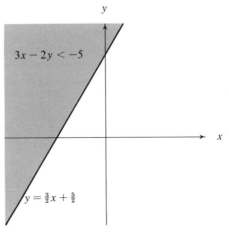

Figure 5.9

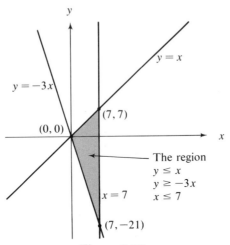

Figure 5.10

not always intersect in corner points. Two boundary lines could intersect in a point that is cut off from being a corner point by some other constraint line. For example, in Figure 5.11 below, L and K are boundary lines of the shaded region, but their intersection (point P) is *not* a corner point, since another boundary line M cuts off P from the rest of the region. (P is then called an **infeasible** solution.)

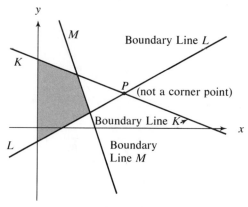

Figure 5.11

Example 5.9 For the following inequalities:

$$3x + 7y \leq 42$$
$$x - y \leq 5$$
$$x \geq 0$$
$$y \geq 1$$

  a. Draw the boundary lines.
  b. Shade the region defined by the inequalities.
  c. Show that at least one pair of boundary lines has an intersection at a point that is not a corner point of the region.
  d. Find the coordinates of the corner points.

Solution a. First we write the equations of the boundary lines:

$$3x + 7y = 42.$$
$$x - y = 5.$$
$$x = 0.$$
$$y = 1.$$

Now, using methods presented in Chapter 4, we draw the lines as shown in Figure 5.12.

5.1 Inequalities in the X-Y Plane

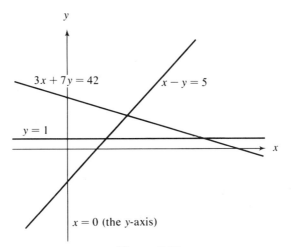

Figure 5.12
Four intersecting boundary lines.

b. Next, we shade the region according to the given inequalities. Thus, we shade the area that is, simultaneously, *below* the line $3x + 7y = 42$, *above* the line $x - y = 5$, to the *right* of the line $x = 0$, and *above* the line $y = 1$. See Figure 5.13.

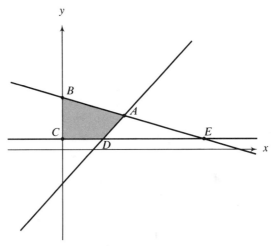

Figure 5.13
Shaded region and corner points.

c. The point labeled E in Figure 5.13 is the intersection of the two boundary lines $3x + 7y = 42$ and $y = 1$, but E is not a corner point. The coordinates of E are $(35/3, 1)$.

d. The four corner points of the region are labeled A, B, C, and D. A is the intersection of the lines $3x + 7y = 42$ and $x - y = 5$. The other corner points are obtained from the boundary lines in pairs as follows:

$$\text{Point } B: \quad 3x + 7y = 42$$
$$x = 0$$
$$\text{Point } C: \quad x = 0$$
$$y = 1$$
$$\text{Point } D: \quad y = 1$$
$$x - y = 5$$

When we solve each of the above pairs of lines, we get the coordinates of the four corner points: $A = (7.7, 2.7)$; $B - (0,6)$; $C = (0,1)$; and $D = (6,1)$.

We knew *exactly which* pairs of lines to solve in order to get the corner points because we had the graph of the region to guide us. In a later section, we will learn a method for finding the corner points of the region without actually drawing the graphs of the lines. Such a method involves an analysis of the *solutions for all possible pairs of boundary lines.*

EXERCISE 5.1

1. Sketch the lines and shade the region of the plane defined by the inequalities $x \leq 6$, $y \geq 2$, and $y \leq x$.

2. Sketch the lines and shade the region of the plane defined by the inequalities $y \leq 2x + 2$, $y \geq 1$, and $x \leq 5$.

3. Find the coordinates of the corner points of the region in Problem 1.

4. Find the coordinates of the corner points of the region in Problem 2.

5. Shade the side of the line described by $-3x - 2y \leq 4$.

6. Shade the side of the line described by $4x - 3y \geq 10$.

5.1 Inequalities in the X-Y Plane

7. Sketch the boundary lines, shade the region, and find the coordinates of the corner points for the region defined by the inequalities

$$2x + y \leq 10$$
$$3x - y \leq 6$$
$$x \leq 7$$
$$y \geq -4.$$

8. Sketch the boundary lines, shade the region, and find the coordinates of the corner points for the region defined by the inequalities

$$x \geq 0$$
$$y \geq 0$$
$$x + 2y \leq 20$$
$$x + y \leq 15$$
$$-2x + y \leq 5.$$

9. Write a set of four inequalities describing the shaded region in the figure below.

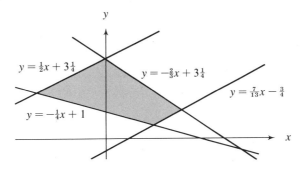

10. Write a set of five inequalities describing the shaded region in the figure below.

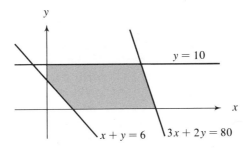

▲ 11. In a manufacturing process, two types of items, types I and II, are produced. Let x = the number of type I items produced and y = the number of type II items produced. Suppose a nonnegative number of each type is produced, the sum of the two types does not exceed 250, and the number of type II items is less than 75 added to twice the number of the type I items. Write inequalities describing this situation. Graph the region of the plane and find the production limits (the corner points).

12. A personnel manager hires workers for two shifts (day and night). The total work force cannot exceed 100 employees (work force constraint); the night shift must have at least 20 workers (skeleton crew constraint); day workers earn 30 dollars and night workers earn 45 dollars; and the payroll may not exceed 3200 dollars (payroll constraint). The factory, community, and transportation interfacing requires that there are at least twice as many workers on days as nights (plant operation constraint). Write inequalities describing this situation and graph the region of the plane. Find the extreme points in this management system (the corner points). ▲

5.2

GEOMETRIC LINEAR PROGRAMMING

We have stated that the *management problem* is one of selecting the right combination of variables for which some function (such as cost or profit) will be minimum or maximum.

In a problem with only two variables, the region of the plane gives the constraints on the variables, and the profit or cost that is dependent upon the constrained variables is computed at each point within the region. The points of the plane that make the profit maximum are then selected as the solution to the management problem.

An analogue to the management problem is the following: If a room with an irregularly shaped floor has a slanting ceiling, then the *plane region is the floor plan* and the variables are allowed to range all over the floor. The *ceiling is the profit or cost function* and there is exactly one point of the ceiling above each point on the floor. The height of the ceiling above a given point on the floor is the value of the profit at that point. See Figure 5.14.

Finding the maximum profit is the same as finding the highest point of the ceiling. If the floor is **convex** (a line drawn from any one point of the floor to any other point lies entirely with the floor plan), then the highest or lowest point of the ceiling occurs at some corner of the room. See Figure 5.15. This means that *we need to check the value of the profit function only at those points that*

5.2 Geometric Linear Programming

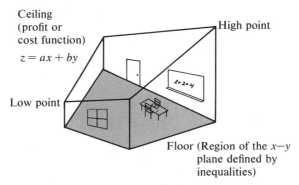

Figure 5.14
Room analogy for linear programming problems.

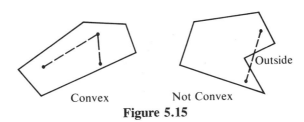

Figure 5.15

are corners of the convex region. The corner points producing this maximum profit are called the *solutions* to the management problem, and the process required for finding them is called **linear programming.**

Algebraically, linear programming is the process of solving a set of linear equations and inequalities simultaneously. Geometrically, it is the process of finding the

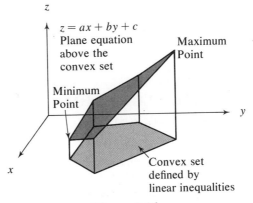

Figure 5.16

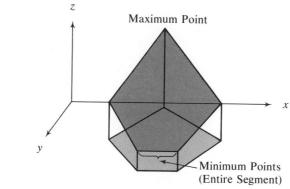

Figure 5.17

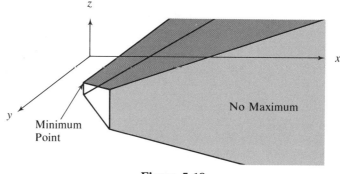

Figure 5.18

maximum or minimum points on a plane function directly above a convex set. See Figures 5.16, 5.17, and 5.18.

In practice, we combine these processes by using geometry to tell us what calculations to make and then using algebra to make those calculations. Here are the steps of an **algorithm** for geometric linear programming. (An algorithm is a set of instructions for accomplishing some computational goal.)

GEOMETRIC LINEAR PROGRAMMING ALGORITHM (FOR TWO VARIABLES)

To find the maximum or minimum value of a plane function $z = z(x,y)$ in which the variables x and y are subject to a set of linear constraint inequalities (given that the inequalities define a convex set), do the following steps:

5.2 Geometric Linear Programming

Step 1 Write the boundary line equations.

Step 2 Draw these lines on graph paper.

Step 3 Shade the region of the plane according to the given inequalities, and check that it is convex.

Step 4 Find the coordinates of the corner points of this region.

Step 5 Evaluate the function $z(x,y)$ at each of these corner points and identify the maximum value of z and the minimum value of z.

NOTES ON THE FIVE-STEP ALGORITHM

1. The function $z = z(x,y)$ to be maximized or minimized is called the **objective function**. It represents the ceiling in the room analogy of Figure 5.14.

2. The shaded region is a convex set and is the domain of the objective function. This domain represents the floor plan in the room analogy.

3. The corner points of the region are called **feasible solutions**.*

4. The corner point at which the maximum value of z occurs is called the **maximizing solution**. The corner point at which the minimum value of z occurs is called the **minimizing solution**. They are also called the **extreme** points of the region.

Example 5.10 Find the maximum and minimum points of the plane function defined by
$$z(x,y) = 2x + 3y$$
with (x,y) subject to the constraints
$$x \geq 0$$
$$y \geq 0$$
$$y \leq 3x + 1$$
$$y \leq 5 - x.$$

---

*Actually, all the points either on the boundary or interior to the domain of the objective function are "feasible," but we will usually apply this term only to the corner points.

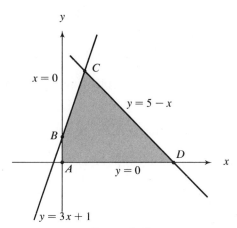

Figure 5.19

Solution The graphs of the *line equations* $x = 0$, $y = 0$, $y = 3x + 1$, and $y = 5 - x$ are drawn in Figure 5.19. The convex set defined by the inequalities is shaded and the corner points are labeled A, B, C, and D. Now all we need to do is find the coordinates of the corner points and evaluate the objective function $z(x,y)$. The corner points have coordinates $A = (0,0)$; $B = (0,1)$; $C = (1,4)$; and $D = (5,0)$. Evaluating the plane function $z = 2x + 3y$ at these points, we get $z(0,0) = 0$; $z(0,1) = 3$; $z(1,4) = 14$; and $z(5,0) = 10$. Hence, $z = 0$ is the minimum at A, and $z = 14$ is the maximum at C.

Example 5.11 A milkman has a route including apartments and suburban houses. He can make up to 100 deliveries. There are at most 60 apartments. He must deliver to at least five houses and at least eight apartments. The number of his house deliveries is not more than 75 minus one-half of his apartment deliveries. His monthly profit is $4 per house and $3 per apartment. To how many houses and how many apartments should he deliver in order to maximize his profit?

Solution As usual for word problems, we start by assigning variables to the unknowns. Let x = the number of houses and y = the number of apartments. The given constraints are $x + y \leq 100$, $8 \leq y \leq 60$, $5 \leq x$, and $x \leq 75 - \frac{1}{2}y$. The function for which we want to find the maximum is $z = 4x + 3y$. First we sketch the convex set defined by the inequalities and label the corner points. See Figure 5.20. The maximum value of $z = 4x + 3y$ will occur at one of the corner points A, B, C, D, or E. We compute the value of z at these five points to see which is the maximum.

5.2 Geometric Linear Programming

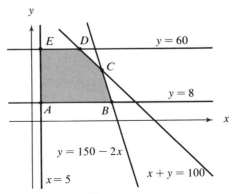

Figure 5.20

At $A = (5,8)$, $z = 44$.
At $B = (71,8)$, $z = 308$.
At $C = (50,50)$, $z = 350$.
At $D = (40,60)$, $z = 340$.
At $E = (5,60)$, $z = 200$.

The maximum value of z occurs at $C = (50,50)$, Therefore, the combination that maximizes the profit is 50 houses and 50 apartments. ▲

EXERCISE 5.2

In all linear programming problems, the variables x and y are assumed to be nonnegative. The practical nature of the problems requires this automatic constraint on the two variables.

1. Given the constraint inequalities

$$y \geq \left(\frac{3}{7}\right)x - \frac{8}{7}$$

$$y \leq -\left(\frac{3}{4}\right)x + 3$$

and the objective function

$$z = 10x + 40y,$$

a. Graph and shade the convex region defined by the inequalities.

b. Write the equations of the four boundary lines. (Remember that the inequalities $x \geq 0$ and $y \geq 0$ are assumed to be part of every linear programming problem.)

c. Find the coordinates of the corner points of the region.

d. Determine the maximum and minimum values of the objective function.

2. Given the constraints

$$x - 2y \leq 4$$
$$-8x + 6y \leq 9$$
$$x + y \geq 1$$
$$x + y \leq 5$$

and the objective function

$$z = 7x + 8y + 10,$$

repeat parts **a, b, c,** and **d** of Problem 1. In this case, part **b** requires six boundary lines instead of four.

3. Maximize

$$z = 6x - 2y + 5$$

subject to

$$x + y \geq 10$$
$$x + y \leq 20$$
$$-x + 2y \leq 10.$$

4. Minimize

$$z = 8x - 7y$$

subject to

$$x + y \geq 15$$
$$x - y \leq 10$$
$$-x + 2y \leq 20.$$

▲ 5. A farmer can grow x acres of crop A and y acres of crop B. He has only 175 acres ($x + y \leq 175$). Twenty-five of the acres are suitable only for crop

5.2 Geometric Linear Programming

A and thirty of the acres are suitable only for crop B ($25 \leq x \leq 145$ and $30 \leq y \leq 150$). The cost of irrigation, fertilizers, etc., to grow crop A is $16 per acre, and it costs $8 per acre to grow crop B. The farmer must borrow money to cover these costs and he can borrow no more than $2000 ($16x + 8y \leq 2000$). He can make a profit of $60 per acre from crop A and a profit of $50 per acre from crop B. So, the objective function (profit) is $z = 60x + 50y$.

a. How many acres of crop A and how many acres of crop B should he grow for maximum profit?

b. Suppose that next year the government adds $15 per acre price support to crop B so that its profit per acre is $65. If crop A remains as before, how many acres of each should he grow?

6. A hospital has 100 rooms that are large enough to be made into double or triple rooms. The management can spend no more than $76,800 in remodeling and furnishing the rooms. Each double room costs $700 and each triple room costs $1100 to remodel and furnish. Patients demand doubles at least twice as frequently as they want triples and the management decides to consider this demand in their planning. If the daily income on a triple is $40 per bed ($120 per room) and daily income on a double is $50 per bed, find the combination of doubles and triples that maximizes the daily income.

7. Refer to Problem 11 in Exercise 5.1. If the profit from the production of x type I items and y type II items is defined as

$$z = 1000 \cdot x + 980 \cdot y,$$

find the maximum profit subject to the constraints stated in that problem.

8. Refer to Problem 12 in Exercise 5.1. If each day worker produces two units and each night worker produces four units, maximize production subject to the constraints stated in that problem.

9. A radiologist uses two radioactive isotopes A and B (such as Hg^{203} and Tc^{99M}) to locate tumors in some body organs. He uses x millicuries of A and y millicuries of B. The constraints are

a. The total millicuries (mc) cannot exceed 11.

b. The mc for B cannot be greater than the mc for A.

c. The mc for B must be no less than 4 mc minus $2/9$ times the mc for A.

The objective function is described as follows: The dosage is computed in millirads of radiation. Each mc of B produces 12 millirads and each mc of A produces 613 millirads. Find the *minimum* dosage needed to meet the constraints.

10. A mathematical learning model could identify two types of activities A and B as essential to the learning of certain skills, such as playing chess, playing the piano, winning debates, or solving mathematics problems. One possible learning model can be stated in terms of linear programming. For such a model to be realistic, of course, experiments would be necessary, and the hypothetical problem presented here is meant only to demonstrate the *possibility* of a linear programming model.

Let activity A be the study of *theory* in a given subject and let B stand for *practice*. Let x = the number of hours per day devoted to A and y = the daily number of hours devoted to B. For the objective function, assume that a test has been developed in which the test score S is a function of x and y as follows:

$$S(x,y) = 26x + 14y.$$

The constraints are

a. Because you have another job, you cannot devote more than 5 hours per day to (the total of both) these activities.

b. To establish a firm theoretical foundation, you should study theory for at least one-fourth as much time as you practice.

c. To develop full practical understanding, the daily practice time minus daily theory-time should never be less than $1\frac{1}{2}$ hours.

d. Due to fatigue, you cannot practice more than $3\frac{1}{2}$ hours per day.

Complete the following problems:

a. Maximize the objective function subject to the above constraints.

b. Suppose you decide to ignore constraint d, and assume you can practice four hours per day and still meet the other constraints, how would your test score be affected? ▲

5.3

THE USE OF SLACK VARIABLES TO FIND CORNER POINTS

It is very time-consuming and awkward to have to draw the boundary lines and shade the convex region defined by the inequalities just to learn which points are at the corners of the region. We would like to omit the graphs, not only

5.3 The Use of Slack Variables to Find Corner Points

because they are time-consuming in the two-dimensional situation, but also because it is impossible to draw them in problems with more than two variables. A better procedure would be one that would allow us to find the feasible solutions (the corner points) *without graphing the region*. Yet the geometric algorithm, given in the previous section, requires a picture of the region before we can find the coordinates of the points used in evaluating the objective function. So, we really need the information that the graph gives us, but what else can we do to get it?

We will develop a different procedure; the first step is to consider every possible combination of boundary lines taken two at a time. Then we use a new method for telling which ones are the feasible solutions.

Example 5.12 Given the following set of constraint inequalities:

$$\left.\begin{array}{r} x \geq 0 \\ y \geq 0 \\ x + y \leq 10 \\ y \leq x - 5. \end{array}\right\} \quad (5.1)$$

Without drawing the graphs, find all possible points of intersection of the boundary lines, L_1, L_2, L_3, and L_4:

$$\left.\begin{array}{rll} L_1: & x & = 0 \\ L_2: & y & = 0 \\ L_3: & x + y & = 10 \\ L_4: & y & = x - 5 \end{array}\right\} \quad (5.2)$$

by considering them two at a time.

Solution If we consider these four lines two at a time, there will be six such combinations yielding six possible points, P_1, P_2, P_3, P_4, P_5, and P_6. These points come from the following pairs of equations:

$$\left.\begin{array}{l} P_1 \text{ from } L_1: x = 0, L_2: y = 0. \\ P_2 \text{ from } L_1: x = 0, L_3: x + y = 10. \\ P_3 \text{ from } L_1: x = 0, L_4: y = x - 5. \\ P_4 \text{ from } L_2: y = 0, L_3: x + y = 10. \\ P_5 \text{ from } L_2: y = 0, L_4: y = x - 5. \\ P_6 \text{ from } L_3: x + y = 10, L_4: y = x - 5. \end{array}\right\} \quad (5.3)$$

The coordinates of these points, obtained from solving the equations in system (5.3), are

$$P_1 = (0,0); \ P_2 = (0,10); \ P_3 = (0,-5); \ P_4 = (10,0);$$
$$P_5 = (5,0); \ P_6 = \left(7\frac{1}{2}, 2\frac{1}{2}\right). \quad \quad (5.4)$$

But which of these are corner points? We can't tell yet. This first step of using the boundary line equations does not give us enough information to identify the corner points. We still need to make use of the information provided by the inequalities, but how? The answer is by using slack variables!

In Section 2.2, we stated the following theorem: If $a \leq b$, then there is a nonnegative number s such that $a + s = b$. Its corollary was: If $a \geq b$, then there is a nonnegative number s such that $a - s = b$.

The interpretation given to slack variables in Chapter 2 was that when an inequality represented the constraint on a resource, then the slack variable measured the amount of the unused resource. We still use this interpretation, but now we will add another meaning to slack variables, a geometric one. Geometrically, the slack variable can tell us which side of the boundary line is defined by a linear inequality. Then by using the nonnegative values of a slack variable, we can find the feasible solutions without a graph.

When a linear inequality such as $x + y \leq 10$ is converted to the slack variable equation $x + y + s = 10$, then s is a number that measures the directed distance from a point (x,y) to the line $x + y = 10$. The sign of s is positive or negative depending upon which side of the line the point (x,y) lies. See Figure 5.21. If

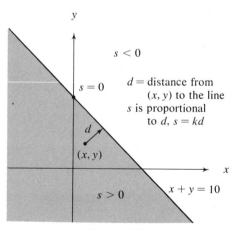

Figure 5.21

5.3 The Use of Slack Variables to Find Corner Points

$s = 0$, then (x,y) is on the line. If $s > 0$ then (x,y) is on the side of the line defined by the inequality. But, if $s < 0$, then *the point is on the wrong side of the line* (that is, not on the side defined by the inequality). Keep this in mind and recall the geometric algorithm for linear programming.

In the geometric algorithm, all the points in the shaded region (and, therefore, the corner points) *had to be on the correct side of every boundary line;* that is, on the side defined by the constraint inequalities. Now the connection between the corner points and the slack variables is almost obvious. When the boundary lines are taken in pairs, if any intersection point has coordinates that result in a negative slack variable, then that point cannot be a corner point.

Furthermore, since the variables x and y are themselves assumed to be nonnegative, then we may conclude the following statement as a method of defining corner points. *If either of the variables x, y, or if any of the slack variables has a negative value in a given solution to a system of slack variable equations, then that solution is not feasible; that is, the point (x,y) defined by the solution is not a corner point.*

Let us return to Example 5.12 and illustrate the use of the above principle to find which of the six points are corner points. From the inequalities in system (5.1), write the system of slack variable equations:

$$
\left.
\begin{aligned}
\text{(I)} \quad & x - s_1 = 0 \\
\text{(II)} \quad & y - s_2 = 0 \\
\text{(III)} \quad & x + y + s_3 = 10 \\
\text{(IV)} \quad & -x + y + s_4 = -5.
\end{aligned}
\right\} \quad (5.5)
$$

Now, we examine each of the six points P_1, P_2, P_3, P_4, P_5, and P_6, whose coordinates were displayed in Equation (5.4).

If we require that the point P_1 satisfy the system (5.5) above, then, since $P_1 = (0,0)$, we will get $x = 0$, $y = 0$, $s_1 = 0$, $s_2 = 0$, $s_3 = 10$, and $s_4 = -5$. The point $(0,0)$ makes one of the slack variables, s_4, negative; so the point $(0,0)$ is on the wrong side of one of the boundary lines, and it cannot be a corner point of the shaded region.

What about the point P_2? From $P_2 = (0,10)$ and substitution into system (5.5), we notice that from Equation (IV), $s_4 = -15$, again a negative value of a slack variable; so $(0,10)$ is not a corner point.

We also see that P_3, which is $(0,-5)$, is infeasible since the resulting value of y is negative.

If we substitute the coordinates of $P_4 = (10,0)$ into the system, we get $x = 10$, $y = 0$, $s_1 = 10$, $s_2 = 0$, $s_3 = 0$, and $s_4 = 5$. Since none of these values is negative, this is a feasible solution and the point $(10,0)$ is a corner point.

What about the point P_5? From $P_5 = (5,0)$, we find by substitution into system

(5.5) that no variable has a negative value. P_5 is therefore a corner point. Also, P_6 produces a feasible solution to system (5.5), so P_6 is a corner point.

Not only have we found the corner points without drawing a graph, but we also know from the value of the slack variable how much resource is unused at the corner points. This is a bonus of this method. One other important fringe benefit from using slack variables is that they provide us with a method for solving the system of inequalities in the first place. This method depends upon the fact that when a slack variable is zero, the points defined by the corresponding *slack variable equation* are all on the boundary line. As in Figure 5.21, when $s = 0$, the equation $x + y + s = 10$ is the line $x + y = 10$. This means that *everytime we set a slack variable equal to zero, we get one of the boundary lines*, and if we set *two* slack variables equal to zero, we get *two* of the boundary lines. So, if we did not already have the solutions to a system, such as in system (5.5), then we could get them by setting the variables (including x and y) equal to zero two at a time, thus creating pairs of boundary lines to be solved. This process (together with the requirement that all the variables have nonnegative values) is called the **slack variable method** for finding feasible solutions.

Example 5.13 Find the feasible solutions by the slack variable method to the following system of constraint inequalities:

$$\left.\begin{aligned} x &\geq 0 \\ y &\geq 0 \\ \tfrac{1}{2}x + y &\leq 10 \\ x + y &\leq 15. \end{aligned}\right\} \quad (5.6)$$

Solution Computation will be easier if we suppress (until further notice) the $x \geq 0$ and $y \geq 0$ conditions and concentrate on the inequalities

$$\left.\begin{aligned} \tfrac{1}{2}x + y &\leq 10 \\ x + y &\leq 15. \end{aligned}\right\} \quad (5.7)$$

Write the inequalities in (5.7) as the system of slack variable equations

$$\left.\begin{aligned} \tfrac{1}{2}x + y + s_1 &= 10 \\ x + y \phantom{{}+s_1} + s_2 &= 15. \end{aligned}\right\} \quad (5.8)$$

This is a system of two equations in four variables (x, y, s_1, and s_2). We consider all pairs of equations in the following way: Set two of the variables equal to

5.3 The Use of Slack Variables to Find Corner Points

zero to reduce this to a system of two equations in two variables. Since there are six ways to set two of the four variables equal to zero, we will get six such two-by-two systems. The six ways to set two of these four variables equal to zero are:

$$
\begin{aligned}
&\text{(I)} && x = 0, y = 0. \\
&\text{(II)} && x = 0, s_1 = 0. \\
&\text{(III)} && x = 0, s_2 = 0. \\
&\text{(IV)} && y = 0, s_1 = 0. \\
&\text{(V)} && y = 0, s_2 = 0. \\
&\text{(VI)} && s_1 = 0, s_2 = 0.
\end{aligned}
\quad (5.9)
$$

Each one of these ways changes the system (5.8) into a two-equation, two-variable system. These two-by-two systems are (numbered in the same order):

$$
\begin{aligned}
&\text{(I)} && s_1 = 10, s_2 = 15. \\
&\text{(II)} && y = 10, y + s_2 = 15. \\
&\text{(III)} && y + s_1 = 10, y = 15. \\
&\text{(IV)} && \tfrac{1}{2}x = 10, x + s_2 = 15. \\
&\text{(V)} && \tfrac{1}{2}x + s_1 = 10, x = 15. \\
&\text{(VI)} && \tfrac{1}{2}x + y = 10, x + y = 15.
\end{aligned}
\quad (5.10)
$$

The solutions to these six systems are (including the two variables that were set equal to zero):

$$
\begin{aligned}
&\text{(I)} && x = 0, y = 0, s_1 = 10, s_2 = 15. \\
&\text{(II)} && x = 0, y = 10, s_1 = 0, s_2 = 5. \\
&\text{(III)} && x = 0, y = 15, s_1 = -5, s_2 = 0. \\
&\text{(IV)} && x = 20, y = 0, s_1 = 0, s_2 = -5. \\
&\text{(V)} && x = 15, y = 0, s_1 = 2\tfrac{1}{2}, s_2 = 0. \\
&\text{(VI)} && x = 10, y = 5, s_1 = 0, s_2 = 0.
\end{aligned}
\quad (5.11)
$$

We can see that systems (I), (II), (V), and (VI) in (5.11) yield feasible solutions. Therefore, the coordinates of x and y in those systems provide the corner points (I): (0,0); (II): (0,10); (V): (15,0); and (VI): (10,5).

Systems (III) and (IV), on the other hand, yield solutions in which at least one variable is negative; hence, they are infeasible solutions. Thus, the points (III): (0,15) and (IV): (20,0) are *not* corner points.

Notice that at the corner points we can read off the amount of unused resources. For example, in (I) the variables x and y consume none of the resource and the entire system is "in reserve" with all of the resource remaining in the slack variables s_1 and s_2.

Before completely abandoning the security of the geometric approach, let us take a look at the graph of system (5.6) to verify that we have indeed found the corner points. See Figure 5.22 in which the points are labeled to correspond to the six systems.

Each system in (5.10) contains two equations and two of the original four variables in system (5.8). In other words, each two-by-two system in (5.10) is derived from the two-by-four system (5.8) by leaving out two variables. The "left out" variables in any system are called the **nonbasic variables,** and the variables "left in" the system are called the **basic variables.** The nonbasic variables are the ones that were deliberately set equal to zero. For example, from (5.10), system (I) is the system in (5.8) with x and y made nonbasic (they were set equal to zero at the outset). The variables s_1 and s_2 are basic; they are the ones for which the system is to be solved. Similarly, take system (V) from (5.10); it is the system in (5.8) with y and s_2 as the nonbasic variables and x and s_1 as the basic ones. This definition is also used for systems with more inequalities and more variables.

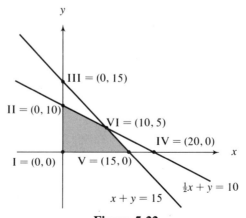

Figure 5.22

5.3 The Use of Slack Variables to Find Corner Points

Example 5.13, which has only two inequalities, is fairly easy to work with but, unfortunately, the more inequalities there are, the more complicated the problem becomes. Even in the more complicated problems, however, the procedure is similar to the simpler cases. For example, if the system has three inequalities in the x-y plane (other than the nonnegative requirements on x and y), we still get two lines by making two of the variables at a time nonbasic (set equal to zero), but the resulting system will be a *three-by-three system* of three equations with three (basic) variables.

Example 5.14 For the following constraint inequalities, find the feasible solutions by the slack variable method. Then draw a graph to verify that all the corner points have been identified. (Assume $x \geq 0$ and $y \geq 0$.)

$$\left. \begin{array}{r} x + y \geq 10 \\ 2x + 3y \leq 40 \\ -x + y \leq 0 \end{array} \right\} \quad (5.12)$$

Solution The slack variable equations are

$$\left. \begin{array}{r} x + y - s_1 = 10 \\ 2x + 3y + s_2 = 40 \\ -x + y + s_3 = 0. \end{array} \right\} \quad (5.13)$$

This is a system of three equations with five variables. We get a point in the two-dimensional plane from the intersection of two lines; hence, we propose to make two of the variables at a time nonbasic.

There are ten possible ways to set two of the five variables equal to zero:

$$\left. \begin{array}{ll} \text{(I)} & x = y = 0. \\ \text{(II)} & x = s_1 = 0. \\ \text{(III)} & x = s_2 = 0. \\ \text{(IV)} & x = s_3 = 0. \\ \text{(V)} & y = s_1 = 0. \\ \text{(VI)} & y = s_2 = 0. \\ \text{(VII)} & y = s_3 = 0. \\ \text{(VIII)} & s_1 = s_2 = 0. \\ \text{(IX)} & s_1 = s_3 = 0. \\ \text{(X)} & s_2 = s_3 = 0. \end{array} \right\} \quad (5.14)$$

Each choice in (5.14) changes the system (5.13) into a three-equation, three-variable system. These ten three-by-three systems are:

$$
\left.\begin{array}{ll}
\text{(I)} & -s_1 = 10,\ s_2 = 40,\ s_3 = 0. \\
\text{(II)} & y = 10,\ 3y + s_2 = 40,\ y + s_3 = 0. \\
\text{(III)} & y - s_1 = 10,\ 3y = 40,\ y + s_3 = 0. \\
\text{(IV)} & y - s_1 = 10,\ 3y + s_2 = 40,\ y = 0. \\
\text{(V)} & x = 10,\ 2x + s_2 = 40,\ -x + s_3 = 0. \\
\text{(VI)} & x - s_1 = 10,\ 2x = 40,\ -x + s_3 = 0. \\
\text{(VII)} & x - s_1 = 10,\ 2x + s_2 = 40,\ -x = 0. \\
\text{(VIII)} & x + y = 10,\ 2x + 3y = 40,\ -x + y + s_3 = 0. \\
\text{(IX)} & x + y = 10,\ 2x + 3y + s_2 = 40,\ -x + y = 0. \\
\text{(X)} & x + y - s_1 = 10,\ 2x + 3y = 40,\ -x + y = 0.
\end{array}\right\} \quad (5.15)
$$

The solutions to these ten systems (including the two nonbasic variables) are:

(I) $\quad x = 0,\ y = 0,\ s_1 = -10,\ s_2 = 40,\ s_3 = 0.$ INFEASIBLE

(II) $\quad x = 0,\ y = 10,\ s_1 = 0,\ s_2 = 10,\ s_3 = -10.$ INFEASIBLE

(III) $\quad x = 0,\ y = \dfrac{40}{3},\ s_1 = \dfrac{10}{3},\ s_2 = 0,\ s_3 = \dfrac{-40}{3}.$ INFEASIBLE

(IV) $\quad x = 0,\ y = 0,\ s_1 = -10,\ s_2 = 40,\ s_3 = 0.$ INFEASIBLE

(V) $\quad x = 10,\ y = 0,\ s_1 = 0,\ s_2 = 20,\ s_3 = 10.$ FEASIBLE; (10,0) is a corner point

(VI) $\quad x = 20,\ y = 0,\ s_1 = 10,\ s_2 = 0,\ s_3 = 20.$ FEASIBLE; (20,0) is a corner point

(VII) $\quad x = 0,\ y = 0,\ s_1 = -10,\ s_2 = 40,\ s_3 = 0.$ INFEASIBLE

(VIII) $\quad x = -10,\ y = 20,\ s_1 = 0,\ s_2 = 0,\ s_3 = -30.$ INFEASIBLE

(IX) $\quad x = 5,\ y = 5,\ s_1 = 0,\ s_2 = 15,\ s_3 = 0.$ FEASIBLE; (5,5) is a corner point

(X) $\quad x = 8,\ y = 8,\ s_1 = 6,\ s_2 = 0,\ s_3 = 0.$ FEASIBLE; (8,8) is a corner point

Notice that three of the solutions (I, IV, and VII) are exactly the same; this means that the x-y point they define, (0,0), must be on three pairs of boundary lines, as we shall see in the drawing.

The second part of the problem calls for the graph to verify the above results. See Figure 5.23 for this otherwise unnecessary graph.

5.3 The Use of Slack Variables to Find Corner Points

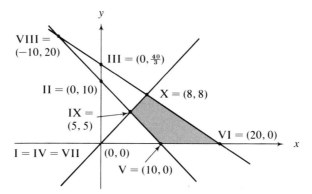

Figure 5.23

While it may not be apparent from the above examples, the slack variable method *does* have several advantages over the geometric method. First, it provides a systematic procedure that can be programmed for solution by computer. Second, it reveals more information; for example, it tells the amount of resource kept in reserve at each feasible point. Third, it can be applied to problems of more than two variables. (We will introduce some of these in the exercises of this section and the next section, but they will be more fully discussed in a later chapter.)

The disadvantage, of course, is that in the slack variable method, we have to compute the coordinates of so many "useless" points (the infeasible solutions). But, we can look forward to a way to correct this disadvantage in the chapter on the simplex method. (The simplex method is a slack variable method that allows us to proceed from one feasible solution to another through the systematic replacement of basic variables by the "best" nonbasic variables.)

EXERCISE 5.3

Use the slack variable method to find the feasible solutions without drawing a graph (assume x and y to be nonnegative).

1.
$$3x + 4y \leq 10$$
$$7x - y \leq 5$$

2. $$4x - 3y \leq 3$$
$$-x + 2y \leq 4$$

3. For the inequality
$$5w + 2x + y + 3z \leq 1,$$
introduce a slack variable, getting one equation in five unknowns. Make four of them at a time nonbasic and find all the feasible solutions.

4. For the inequality
$$-u + 6x + \frac{1}{2}y - 7w + 2z \leq 20,$$
introduce a slack variable, getting one equation in six variables. Make five of them at a time nonbasic and find all possible feasible solutions.

5. Consider the following system of inequalities:

(I) $\qquad x + y + z \leq 3$

(II) $\qquad 2x + y - z \leq 1$

(III) $\qquad x + y \geq 2$

(IV) $\qquad -3x + 4y \leq 5.$

   a. Transform it into a system of slack variable equations by introducing the slack variables s_1, s_2, s_3, and s_4 into Inequalities (I), (II), (III), and (IV), respectively. The result will be four equations in seven variables. Select the variables three at a time as nonbasic for each of the following cases.

   b. If s_1, s_2, and s_3 are nonbasic, solve the system that has (x,y,z,s_4) as its basis (basic variables). Is this solution feasible?

   c. If y, s_2, and s_3 are nonbasic, solve for (x, z, s_1, s_4). Is this solution feasible?

   d. If x, y, and s_3 are nonbasic, is there a feasible solution (z, s_1, s_2, s_4)?

6. Consider the following system of inequalities:

(I) $\qquad x + y + z \leq 5$

(II) $\qquad 3x + 2y - 2z \leq 4$

(III) $\qquad x \quad\ \ + 2z \geq 1$

(IV) $\qquad y - z \leq 2.$

   a. Transform it into a system of slack variable equations by introducing the slack variables s_1, s_2, s_3, and s_4 into Inequalities (I), (II), (III), and (IV),

5.3 The Use of Slack Variables to Find Corner Points

respectively. Select three variables at a time as nonbasic for the following cases.

b. If s_1, s_2, and s_3 are nonbasic and x, y, z, and s_4 are basic, is the solution obtained feasible?

c. Set x, s_2, and s_3 nonbasic, and solve for (y,z,s_1,s_4). Is this solution feasible?

d. If x, y, and s_3 are nonbasic, is the solution (z,s_1,s_2,s_4) feasible?

7. For the system of slack variable equations given below, make two of the variables at a time nonbasic by letting them be zero as indicated in the following table. Use the solutions you get as table entries, and determine from the table whether or not the solution is feasible. The solutions for five of the systems have been filled in.

$$x + 2y - s_1 = 4$$
$$x + y + s_2 = 6$$
$$x + s_3 = 3$$
$$-x + y + s_4 = 3$$

System	x	y	s_1	s_2	s_3	s_4	Feasible?
I	0	0					
II	0		0				
III	0	6	8	0	3	−3	No
IV	0	—	—	—	0	—	Impossible
V	0					0	
VI		0	0				
VII		0		0			
VIII		0			0		
IX		0				0	
X	8	−2	0	0	−5	13	No
XI			0	0			
XII	−2/3	7/3	0	13/3	11/3	0	No
XIII				0	0		
XIV	3/2	9/2	13/2	0	3/2	0	Yes
XV					0	0	

8. For the system of slack variable equations given below, make two of the variables at a time nonbasic and fill in the following table of ten systems. The solutions for four of them have been given.

$$y + s_1 = 15$$
$$2x + y + s_2 = 40$$
$$-3x + 10y - s_3 = 50$$

System	x	y	s_1	s_2	s_3	Feasible?
I	0	0	15	40	-50	No
II	0	15	0	25	100	Yes
III	0			0		
IV	0				0	
V	—	0	0	—	—	Impossible
VI		0		0		
VII		0			0	
VIII			0	0		
IX			0		0	
X	350/23	220/23	125/23	0	0	Yes

▲ 9. A manufacturer can ship by rail, air, or truck lines. Let x = mileage of rail shipping, y = air mileage, and z = truck mileage, and suppose the following constraints hold. The trucks are owned by the company and must be used for at least 100,000 miles. Rail shipping mileage must not exceed one-fourth of the air mileage. Air shipping mileage is at least 500,000 miles more than the combined total of the trucking and rail mileage.

   a. Write the constraint inequalities and convert to slack variable equations.

   b. By making three of the six variables nonbasic, find one feasible solution.

10. A petroleum refinery can produce gasoline, asphalt, or heating oil. Let x = barrels of gasoline, y = barrels of asphalt, and z = barrels of heating oil, and suppose the following conditions hold. Total production cannot exceed 100 million barrels (write as $x + y + z \leq 100$). The amount of gasoline produced cannot be more than 20 million barrels plus the combined total of the heating oil and asphalt.

   a. Write the constraint inequalities as slack variable equations.

   b. By making three of the five variables nonbasic, find one feasible solution.

▲

5.4

THE SLACK VARIABLE LINEAR PROGRAMMING METHOD

So far, we have applied the slack variable method only to finding the basic feasible solutions for a system of constraints. Now, we want to use the information obtained by that method for finding the maximum or minimum value of the objective function, subject to the given constraints. First, we need to see how the objective function is affected by the slack variables.

Curiously enough, there is no effect. Nothing happens to the objective function when slack variables are introduced into it. The reason is that the slack variables represent *unused resources* or *unfulfilled requirements;* therefore, they contribute nothing to the cost function* and nothing to the profit function.

Example 5.15 We return to an example of an earlier chapter. In a given enterprise consisting of three departments, Administration, Research, and Production, let w_1 = the number of workers in Administration, w_2 = the number of workers in Research, and w_3 = the number of workers in Production. Suppose that the total work force cannot exceed 80, then the personnel constraint inequality is

$$w_1 + w_2 + w_3 \leq 80. \tag{5.16}$$

Consider, as the objective function, the total salary S which we wish to minimize, subject to Equation (5.16) and some other, not yet named, constraints.

Say that administrators get a monthly salary of $3000, researchers get $1500, and production workers get $1000. Then the objective function may be written

$$S(w_1, w_2, w_3) = 3000w_1 + 1500w_2 + 1000w_3. \tag{5.17}$$

The coefficients of the objective function will be unaffected by any increase or decrease in the labor force. (A researcher gets $1500 if there is one, or two, or any number up to 80, researchers.) All these coefficients remain as they are, regardless of the number of unfilled positions.

Let us denote the number of such unfilled positions by the letter P. Then the constraint inequality (5.16) becomes the slack variable equation $w_1 + w_2 + w_3 + P = 80$. Any linear programming problem that includes this equation will have at least the four variables w_1, w_2, w_3, and P in its domain. So we must write the salary S as a function of them; that is, $S = S(w_1, w_2, w_3, P)$.

---

*It is conceivable, of course, that unused resources could cost something. Idle machinery can require maintenance, for example. But this type of cost could easily be put into a fixed-cost category.

But what weight shall we assign to P? To put it another way, what salaries should the unfilled positions get? Zero seems appropriate! So, let us say that the coefficient of P in the salary function is 0, then

$$S(w_1, w_2, w_3, P) = 3000w_1 + 1500w_2 + 1000w_3 + 0 \cdot P.$$

This equation incorporates the slack variable for the personnel resource in (5.16) into the objective function, but now suppose that this enterprise has a second resource—computer time.

If there are 18 hours of computer time available daily and each administrator uses $1/3$ of an hour, each researcher uses 1 hour, and each production worker uses $1/8$ of an hour, then the constraint inequality is

$$\frac{1}{3} \cdot w_1 + 1 \cdot w_2 + \frac{1}{8} \cdot w_3 \leq 18.$$

Let the slack variable T stand for the daily unused computer time. Then the above constraint, written as a slack variable equation, is

$$\frac{1}{3} \cdot w_1 + 1 \cdot w_2 + \frac{1}{8} \cdot w_3 + T = 18.$$

Unused computer time deserves a zero coefficient when it comes to costs, but can the computer time constraint be put into the salary function? Obviously not, but we *can* define a more general cost function C, which would include both salaries and computer costs:

$$C(w_1, w_2, w_3) = S(w_1, w_2, w_3) + Y(w_1, w_2, w_3), \tag{5.18}$$

where S is the salary function defined in Equation (5.17) and Y is a cost function for computer time to be described below. Before giving this cost function, however, we need to put the computer costs on a monthly basis to correspond to the monthly salary.

Suppose there are 24 work days per month. Each administrator uses $1/3 \cdot 24 = 8$ hours of computer time per month; each researcher, who consumes one hour per day, uses 24 hours per month; and the production worker has a monthly usage of $1/8 \cdot 24$ or 3 hours of computer time.

If computer time costs \$100 per hour, then the monthly use by each administrator is \$800 worth; for each researcher, it is \$2400; and for each production worker, it is \$300. The monthly cost function Y for computer time is

$$Y(w_1, w_2, w_3) = 800w_1 + 2400w_2 + 300w_3. \tag{5.19}$$

Now the total cost function in Equation (5.18) can be obtained by adding the salaries in Equation (5.17) to the computer time costs in Equation (5.19):

$$C(w_1, w_2, w_3) = 3000w_1 + 1500w_2 + 1000w_3 + 800w_1 + 2400w_2 + 300w_3.$$

5.4 The Slack Variable Linear Programming Method

Combining terms, we get

$$C(w_1, w_2, w_3) = 3800w_1 + 3900w_2 + 1300w_3.$$

The slack variables from both constraints can be brought into this cost function. Simply make the coefficients of the unused computer time zero, just as was done for the unfilled positions. The cost for both salaries and computer time, as a function of all five variables (w_1, w_2, w_3, P, and T), is

$$C(w_1, w_2, w_3, P, T) = 3800w_1 + 3900w_2 + 1300w_3 + 0 \cdot P + 0 \cdot T.$$

Similarly, if there are other constraints, we can modify the objective function to accommodate, with zero coefficients, all the slack variables introduced into the constraints. But, why is it necessary to put the slack variables into the objective function, if they only have zero coefficients? The answer is that we want to get a *general method* for working with *all* the variables uniformly, without making special allowances for the slack variables; so they must also have coefficients.

Now we are ready to state the slack variable method for linear programming problems. The algorithm given below works for any number of inequalities in any number of variables, but we will use it for only a few (two or three) inequalities in a few (two, three, or four) variables because, for any more than that, the number of systems to be solved becomes enormous. (For example, a slack variable system with four equations and eight variables yields 70 basic systems to be solved, and it is very likely that many of them will be infeasible.) The large number of possibly infeasible solutions justifies staying away from large systems at this time; later we will discuss a method that avoids the infeasible solutions. (See Chapter 9 on the simplex method.)

SLACK VARIABLE ALGORITHM FOR LINEAR PROGRAMMING

Let n and k be positive integers. Given a set of n constraint inequalities in k variables that defines an appropriate domain for a linear objective function, we can find the maximum or minimum value of the objective function by the following steps.

Step 1 Introduce n slack variables, one into each of the n inequalities. This results in an n-by-$n + k$ system of equations (n equations, $n + k$ variables). Also introduce these same n slack variables into the objective function.

Step 2 Make k of the variables at a time nonbasic, creating n-by-n systems. Solve all of these systems.

Step 3 Of the solutions in Step 2, select only the feasible ones and evaluate the objective function at these solutions.

Step 4 Examine the values of the objective function found in Step 3 for the maximum or minimum.

Example 5.16 Use the slack variable algorithm to find the minimum of
$$P(x_1, x_2, x_3) = 80x_1 + 25x_2 + 60x_3$$
subject to the constraints
$$2x_1 - 3x_2 + 5x_3 \leq 100$$
$$x_1 + 3x_2 + x_3 \geq 75.$$

Solution Here, there are two inequalities and three variables; so, $n = 2$ and $k = 3$.

Step 1 Introduce two slack variables, x_4 and x_5, into the constraints
$$2x_1 - 3x_2 + 5x_3 + x_4 \quad\quad = 100$$
$$x_1 + 3x_2 + x_3 \quad\quad - x_5 = 75.$$
The objective function, with slack variables, is
$$P(x_1, x_2, x_3, x_4, x_5) = 80x_1 + 25x_2 + 60x_3 + 0 \cdot x_4 + 0 \cdot x_5.$$

Step 2 Make three of the variables at a time nonbasic. This creates ten systems whose solutions are presented in the following table.

System	x_1	x_2	x_3	x_4	x_5	Feasible?
I	175/3	50/9	0	0	0	Yes
II	275/3	0	-50/3	0	0	No
III	75	0	0	-50	0	No
IV	50	0	0	0	-25	No
V	0	275/18	175/6	0	0	Yes
VI	0	25	0	175	0	Yes
VII	0	-100/3	0	0	-175	No
VIII	0	0	75	-275	0	No
IX	0	0	20	0	-55	No
X	0	0	0	100	-75	No

5.4 The Slack Variable Linear Programming Method

Step 3 The feasible solutions are I, V, and VI. The values of the objective function at these three points are

At I: $P\left(\dfrac{175}{3}, \dfrac{50}{9}, 0, 0, 0\right) = (80)\left(\dfrac{175}{3}\right) + 25\left(\dfrac{50}{9}\right) + 60 \cdot 0 + 0 \cdot 0 + 0 \cdot 0$

$\qquad = 4805.55$ (rounded off).

At V: $P\left(0, \dfrac{275}{18}, \dfrac{175}{6}, 0, 0\right) = 2131.94$ (rounded off).

At VI: $P(0, 25, 0, 175, 0) = 625$.

Step 4 The minimum occurs at VI.

EXERCISE 5.4

Use the slack variable algorithm to solve these problems.

1. Find the maximum of

$$P = 13x_1 + 4x_2$$

subject to the constraints

$$x_1 + 2x_2 \leq 10$$
$$-3x_1 + x_2 \leq 2.$$

2. Find the minimum of

$$P = 3x_1 + 4x_2$$

subject to

$$-x_1 + x_2 \leq 5$$
$$x_1 + x_2 \leq 20.$$

▲ 3. In one month, a new car salesman can sell x_1 cars and x_2 trucks according to the constraints

$$2x_1 + 3x_2 \leq 18$$
$$x_2 \leq 4.$$

He has a profit equation of $P = 200x_1 + 145x_2$. Find the maximum profit.

4. An advertising manager buys x_1 hours per week on television and x_2 hours per week on radio. These hours meet the following constraints:

$$3x_1 + 4x_2 \leq 12$$
$$8x_1 + 5x_2 \geq 20.$$

The cost equation for these advertisements is

$$C = 1000x_1 + 550x_2.$$

Find the minimum cost.

5. Suppose a living ecosystem has two consuming organisms, A and B, and two producing resources, I and II. (For example, A and B could be two types of animals that live in a desert and I and II could be two sources of food.) Say there are x_1 A's and x_2 B's. Assume that each A consumes three units of resource I and 15 units of resource II. Let B consume two units of resource I and 28 units of resource II. See the table below. The entry $3x_1$ in the A column, row I, means that the x_1 type A's consume a total of $3x_1$ of resource I, with similar meanings for each of the other entries. A system such as this is called an *input-output* model (the resources produce the input and the consumers use up the output). Suppose that no more than 20 units of resource I are available and no more than 105 units of resource II are available.

	Type A Organism	Type B Organism
Resource I	$3x_1$	$2x_2$
Resource II	$15x_1$	$28x_2$

   a. Write a system of constraints describing this model.

   b. Assume that organisms A and B are themselves producers of some product z, which is to be consumed (by some other organism higher up in the food chain). Suppose each A produces one unit of z and each B produces six units of z; that is, $z = 1 \cdot x_1 + 6 \cdot x_2$. Let this function be the objective function and find the maximum, subject to the constraints in part **a**.

6. A certain input-output system has three consumers, A, B, and C, and two producers, RI and RII. There are x_1 A's, x_2 B's, and x_3 C's. Each A consumes six units of RI and two units of RII. Each B consumes five units of RI and three units of RII. Each C consumes one unit each of RI and RII. Resource RI does not exceed 50, and resource RII, which is actually a *requirement*,

5.4 The Slack Variable Linear Programming Method

cannot be less than 25. Assume that the consumers themselves are in the domain of an objective function, $z = z(x_1, x_2, x_3)$, where

$$z = 10x_1 + 15x_2 + 20x_3.$$

Find the maximum and minimum of z, subject to the constraints of the system. ▲

7. Minimize

$$z = 3x_1 + 7x_2 + 2x_3$$

subject to

$$x_1 + x_2 + x_3 \leq 150$$
$$3x_1 + x_2 \geq 60.$$

8. Maximize

$$z = 150x_1 + 60x_2$$

subject to

$$x_1 + 3x_2 \leq 3$$
$$x_1 + x_2 \leq 7$$
$$x_1 \geq 2.$$

▲ 9. A chain of discount stores has two resources for its retail goods: I—its own brand name products and II—nationally advertised brands. Each store has four departments:
A—household goods, B—automotive supplies, C—sports equipment, and D—clothing.
Let x_1, x_2, x_3, and x_4 represent the sales volume (in $1000) in the four departments, A, B, C, and D, respectively. Assume the following constraints:

$$3x_1 + 5x_2 + x_3 - x_4 \leq 100$$
$$2x_1 - x_2 + 4x_3 + x_4 \leq 80.$$

a. Write out the solutions of the fifteen slack variable systems and identify the feasible ones.

b. Let the revenue function based on sales volume be

$$R(x_1, x_2, x_3, x_4) = 2x_1 + 1.5x_2 + 3x_3 + 1.2x_4,$$

and maximize revenue, subject to the given constraints.

10. Refer to Problem 10 in the previous section (Exercise 5.3). The constraints in the petroleum refinery were

$$x + y + z \leq 100$$
$$x - y - z \leq 20$$

(where x, y, and z are in millions of barrels).

a. Find all the feasible solutions among the ten possible slack variable systems.

b. Assume the objective function is the following profit function:
$$P = 18x + 31y + 23z.$$

Maximize the profit. ▲

6

DETERMINANTS AND VECTORS

6.1

INTRODUCTION TO DETERMINANTS

In the preceding chapters, two-by-two and three-by-three systems of linear equations arose in various practical problems. Some of those systems (such as the basic slack variable constraint equations) were extremely simple since every equation did not always include all the variables. This made the equations easy to solve by *substitution*. For example, the three-by-three system

$$
\begin{aligned}
3x &= 7 \\
5x - s_2 &= 2 \\
2x + s_3 &= 10
\end{aligned}
$$

can be solved with no trouble by substitution, since we know from the first equation that $x = 7/3$. In contrast, the following system is much more difficult to solve by substitution:

$$
\left.\begin{aligned}
4x + 5s_2 + 6s_3 &= 10 \\
6x - s_2 + 7s_3 &= 9 \\
5x + 3s_2 + 4s_3 &= 13.
\end{aligned}\right\} \quad (6.1)
$$

In this chapter, we will discuss some general methods for solving systems such as (6.1). These methods introduce the idea of solving simultaneous equations by working with only the coefficients. The symbols x, y, ..., etc., will be removed from the process and will serve only as headings for columns of their coefficients. The system of equations will thereby be converted to a *table of numbers*, with each row of the table corresponding to an equation, and each column corresponding to one of the variables.

The mathematical operations used to get solutions from such *tabulated* systems are quite simple. Essentially, they involve multiplication and subtraction of numbers. Although the process *is* easy for most systems, the calculations, especially for the larger systems, can become tedious, and the reader is warned to be extremely careful to avoid errors in arithmetic. The numerical calculations are so routine and stupefying that they make excellent fodder for electronic computers.

We begin our approach to the general methods with a special expression, called a **determinant,** which involves certain combinations of numbers, as shown below:

Definition (**Two-by-two determinant**)
If A, B, C, and D are numbers, then the square array

$$\begin{vmatrix} A & B \\ C & D \end{vmatrix} \tag{6.2}$$

is called a *two-by-two determinant* whose value is

$$A \cdot D - C \cdot B. \tag{6.3}$$

In other words, the two-by-two determinant of (6.2) is a function of four variables, A, B, C, and D, and its value is given by the expression in (6.3). Of course, this function could have been written many other ways, for example as $F(A,B,C,D) = A \cdot D - C \cdot B$. The reason for writing it in the form of a square array such as (6.2) will become apparent when we see how it is applied to solving two-by-two systems of equations.

Example 6.1

a. $\begin{vmatrix} 3 & 5 \\ 7 & 10 \end{vmatrix} = 3 \cdot 10 - 7 \cdot 5 = 30 - 35 = -5.$

b. $\begin{vmatrix} 4 & -3 \\ 6 & 2 \end{vmatrix} = 4 \cdot 2 - 6 \cdot (-3) = 8 - (-18) = 8 + 18 = 26.$

c. $\begin{vmatrix} -4 & 8 \\ -5 & -1 \end{vmatrix} = (-4)(-1) - (-5)(8) = 4 + 40 = 44.$

d. $\begin{vmatrix} 3x & 5 \\ 4y & 2 \end{vmatrix} = 6x - 20y.$

6.1 Introduction to Determinants

The rule for writing this combination of four numbers is: Multiply the number in the upper left-hand corner by the number in the lower right, and subtract the product of the number in the lower left-hand corner and the number in the upper

$$\left.\begin{array}{c} \begin{vmatrix} a_1 & b_1 \\ a_2 & b_2 \end{vmatrix} \quad \ominus a_2 \cdot b_1 \\ \oplus a_1 \cdot b_2 \end{array}\right\} = a_1 b_2 - a_2 b_1$$

right. The usual notation for the variables in a determinant is the double subscripted a_{ij}, where the first subscript represents the row and the second represents the column:

$$\begin{vmatrix} a_{11} & a_{12} \\ a_{21} & a_{22} \end{vmatrix} = a_{11} \cdot a_{22} - a_{21} \cdot a_{12}.$$

Note that a_{12} is in the first row, second column; and a_{21} is in the second row, first column. This notation is very convenient, especially in determinants with more than two rows and two columns.

The four numbers in a two-by-two determinant are called **elements** of the determinant. A two-by-two determinant is said to be *evaluated*, or its *value* is found, when the numbers are combined according to the rule given above for the multiplication and subtraction of the numbers in the four corners. We shall see later how to evaluate larger determinants in terms of two-by-two determinants.

Every determinant has a unique value, but the same number may be the value of several (even infinitely many) determinants. For example,

$$845 = \begin{vmatrix} 8 & -15 \\ 3 & 100 \end{vmatrix}, \quad \text{but} \quad 845 = \begin{vmatrix} 80 & -1 \\ 45 & 10 \end{vmatrix}, \quad \text{or} \quad 845 = \begin{vmatrix} 845 & 0 \\ 0 & 1 \end{vmatrix}$$

And, for any number x,

$$845 = \begin{vmatrix} 845 + x & 1 \\ x & 1 \end{vmatrix}$$

Therefore, all these determinants have the same value, even if they contain different elements.

This is a good place to note that certain general statements can be made relating the rows and columns of a determinant to its value. Two examples are: a. If any two rows (or columns) are interchanged, then the value of the determinant changes sign, and b. If any two rows (or columns) are identical, then the value of the determinant is zero.

Example 6.2

a. $\begin{vmatrix} 3 & 7 \\ 5 & 2 \end{vmatrix} = -29$, but $\begin{vmatrix} 5 & 2 \\ 3 & 7 \end{vmatrix} = 29$, and $\begin{vmatrix} 7 & 3 \\ 2 & 5 \end{vmatrix} = 29$.

b. $\begin{vmatrix} x & y \\ x & y \end{vmatrix} = \begin{vmatrix} a & a \\ b & b \end{vmatrix} = 0$.

We will encounter several other properties, but for now our primary reason for studying determinants is to use them in solving systems of equations.

To show that determinants can be helpful in solving simultaneous equations, we will first obtain the solution to the general system of two equations by using a known method and then examine the solution, hoping to find some connection to determinants, as no doubt we will. The general two-by-two system is

$$\left. \begin{array}{ll} \text{(I)} & a_1 x + b_1 y = c_1 \\ \text{(II)} & a_2 x + b_2 y = c_2. \end{array} \right\} \quad (6.4)$$

Multiply Equation (I) by b_2 and Equation (II) by b_1:

$$a_1 b_2 x + b_1 b_2 y = c_1 b_2$$
$$a_2 b_1 x + b_1 b_2 y = c_2 b_1.$$

Subtract the second equation from the first; this eliminates y:

$$a_1 b_2 x - a_2 b_1 x = c_1 b_2 - c_2 b_1.$$

Solving this equation for x, we get

$$x = \frac{c_1 b_2 - c_2 b_1}{a_1 b_2 - a_2 b_1}.$$

By a similar process, we can eliminate x and solve for y. The solution to the system (6.4) is

$$x = \frac{c_1 b_2 - c_2 b_1}{a_1 b_2 - a_2 b_1} \quad \text{and} \quad y = \frac{a_1 c_2 - a_2 c_1}{a_1 b_2 - a_2 b_1}. \quad (6.5)$$

The student may already suspect that there are some two-by-two determinants in the number combinations $c_1 b_2 - c_2 b_1$, $a_1 c_2 - a_2 c_1$, and $a_1 b_2 - a_2 b_1$; those suspicions will be verified.

The equations in (6.5) constitute a *formula* for the solution of the system (6.4), but it is not easy to remember. Fortunately, we have a *mnemonic device* (a device used in memorizing) to help us learn the formula. The mnemonic device is the determinant; this is probably why the determinant is defined as it is. We may write the solution in (6.5) as follows:

6.1 Introduction to Determinants

$$x = \frac{\begin{vmatrix} c_1 & b_1 \\ c_2 & b_2 \end{vmatrix}}{\begin{vmatrix} a_1 & b_1 \\ a_2 & b_2 \end{vmatrix}} \quad \text{and} \quad y = \frac{\begin{vmatrix} a_1 & c_1 \\ a_2 & c_2 \end{vmatrix}}{\begin{vmatrix} a_1 & b_1 \\ a_2 & b_2 \end{vmatrix}}$$

Notice that each denominator is the determinant of the coefficients of the original system (6.4) exactly as they appear on the left in the two equations. The method described below explains how to get the solution directly from the system of equations (6.4); this method is called *Cramer's rule* (Gabriel Cramer, 1704-52).

Definition (Cramer's rule)

For both denominators, use the coefficients of x and y as they appear in the system of equations. For the numerator of x, replace the coefficients of x (a_1 and a_2) with the constants c_1 and c_2. For the numerator of y, leave that a's as they are, but replace the coefficients of y (b_1 and b_2) with the constants c_1 and c_2. (See Figure 6.1.)

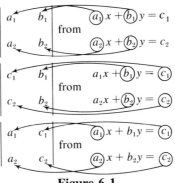

Figure 6.1

Example 6.3 Use Cramer's rule to write the solution for

$$3x - 7y = 1$$
$$5x + 2y = 13.$$

Solution

$$x = \frac{\begin{vmatrix} 1 & -7 \\ 13 & 2 \end{vmatrix}}{\begin{vmatrix} 3 & -7 \\ 5 & 2 \end{vmatrix}}, \quad y = \frac{\begin{vmatrix} 3 & 1 \\ 5 & 13 \end{vmatrix}}{\begin{vmatrix} 3 & -7 \\ 5 & 2 \end{vmatrix}}.$$

If we wish to convert this solution to a more conventional form, we simply evaluate the determinants and get

$$x = \frac{93}{41}, \; y = \frac{34}{41}.$$

We shall eventually be writing our answers to simultaneous linear equations in *vector* form. It is appropriate to introduce here the *column vector notation*. For example, the solutions $93/41$ and $34/41$ above may be written

$$\begin{bmatrix} x \\ y \end{bmatrix} = \begin{bmatrix} 93/41 \\ 34/41 \end{bmatrix}$$

This ordered pair written in a vertical form is called a two-dimensional **column vector**.

Column vectors and, as we shall see later, row vectors have certain computational rules that make them useful for expressing systems of equations and solutions to such systems. The individual numbers in the vectors are called the **components** of the vectors. In the above example, $93/41$ is the x component and $34/41$ is the y component. If we multiply a column vector by a number, we must multiply all of its components by that number. This means that we can bring a number into the vector or factor a number out of the vector. In the above example, we can write the solution as

$$\begin{bmatrix} x \\ y \end{bmatrix} = \frac{1}{41} \begin{bmatrix} 93 \\ 34 \end{bmatrix} = \begin{bmatrix} 93/41 \\ 34/41 \end{bmatrix}.$$

In general,

$$a \cdot \begin{bmatrix} b \\ c \end{bmatrix} = \begin{bmatrix} ab \\ ac \end{bmatrix}.$$

Example 6.4

$$3 \begin{bmatrix} 2 \\ 3 \end{bmatrix} = \begin{bmatrix} 6 \\ 9 \end{bmatrix}, \; -\frac{1}{5} \begin{bmatrix} -13 \\ -22 \end{bmatrix} = \begin{bmatrix} 13/5 \\ 22/5 \end{bmatrix}, \; 7 \begin{bmatrix} -1/7 \\ 3 \end{bmatrix} = \begin{bmatrix} -1 \\ 21 \end{bmatrix}.$$

Note that the equation

$$\begin{bmatrix} x \\ y \end{bmatrix} = \begin{bmatrix} a \\ b \end{bmatrix}$$

means $x = a$ and $y = b$.

6.1 Introduction to Determinants

Example 6.5

$$\begin{bmatrix} x \\ y \end{bmatrix} = \begin{bmatrix} 8 \\ 2 \end{bmatrix}$$

means $x = 8$ and $y = 2$.

$$\begin{bmatrix} u - 1 \\ 3v \end{bmatrix} = \begin{bmatrix} 53 \\ 99 \end{bmatrix}$$

means $u = 54$ and $v = 33$.

EXERCISE 6.1

1. Evaluate the determinants

   a. $\begin{vmatrix} 3 & 2 \\ 1 & 7 \end{vmatrix}$ b. $\begin{vmatrix} -3 & 2 \\ 1 & 7 \end{vmatrix}$ c. $\begin{vmatrix} 3 & -2 \\ 1 & 7 \end{vmatrix}$ d. $\begin{vmatrix} 5 \cdot 1 + 3 & 5 \cdot 7 + 2 \\ 1 & 7 \end{vmatrix}$

2. Evaluate the determinants

   a. $\begin{vmatrix} 5 & 3 \\ 4 & 6 \end{vmatrix}$ b. $\begin{vmatrix} 4 & 4 \\ 2 & 1 \end{vmatrix}$ c. $\begin{vmatrix} 4 & 5 \\ x & y \end{vmatrix}$ d. $\begin{vmatrix} 3x + 4 & 3y + 5 \\ x & y \end{vmatrix}$

3. Show that

   a. $\begin{vmatrix} a_{11} & a_{12} \\ a_{21} & a_{22} \end{vmatrix} = \begin{vmatrix} a_{11} & a_{21} \\ a_{12} & a_{22} \end{vmatrix}$ b. $\begin{vmatrix} a_{11} & a_{12} \\ a_{21} & a_{22} \end{vmatrix} = - \begin{vmatrix} a_{12} & a_{11} \\ a_{22} & a_{21} \end{vmatrix}$

4. Show that if $\begin{vmatrix} ma & nb \\ a & b \end{vmatrix} = 0$, then either $m = n$, $a = 0$, or $b = 0$, and conversely.

5. Write three determinants whose values are 157.

6. Write three determinants whose values are 1.

7. Solve by Cramer's rule:

   a. $3x + 8y = 7$
      $5x - 7y = 10$.

   b. $8x + 3y = 7$
      $-7x + 5y = 10$.

   Convert the answers to conventional notation.

8. Solve by Cramer's rule:

   a. $2x + 3y = 4$
   $5x + 1y = 7.$

   b. $3x + 2y = 4$
   $1x + 5y = 7.$

   Convert the answers to conventional notation.

9. Write the determinant solution to
$$a_{11}x + a_{12}y = b_1$$
$$a_{21}x + a_{22}y = b_2.$$

10. Write the determinant solution to
$$Ax + By = C$$
$$Dx + Ey = F.$$

11. Show that the column vector $\begin{bmatrix} x \\ y \end{bmatrix} = -1/5 \begin{bmatrix} -13 \\ -22 \end{bmatrix}$ is the solution to
$$x + y = 7$$
$$2x - 3y = -8.$$

12. Show that the column vector $\begin{bmatrix} x \\ y \end{bmatrix} = \begin{bmatrix} 1/2 \\ 1/4 \end{bmatrix}$ is the solution to
$$4x + 8y = 4$$
$$20x - 12y = 7.$$

13. Solve for x and y if
$$\begin{bmatrix} x - 1 \\ y + 3 \end{bmatrix} = \begin{bmatrix} 7 \\ 5 \end{bmatrix}.$$

14. Solve for x and y if
$$\begin{bmatrix} 2x + 1 \\ 5y \end{bmatrix} = \begin{bmatrix} 6 \\ 9 \end{bmatrix}.$$

▲ 15. If $p + q = 1$ and $21p + 49q = 33$ are equations for a certain gene frequency distribution, find p and q.

16. A statistical procedure correlating glacial ice wedge depth with width yields the equations $10x + 4y = 3$ and $4x + 3y = 2$. Find x and y. ▲

6.2
THIRD-ORDER DETERMINANTS

A third-order determinant is a determinant with three rows and three columns:

$$\begin{vmatrix} a_1 & b_1 & c_1 \\ a_2 & b_2 & c_2 \\ a_3 & b_3 & c_3 \end{vmatrix}.$$

It will serve as a *mnemonic device* for solving a given system of three linear equations in three unknowns.

The general system

$$\left. \begin{array}{l} a_1 x + b_1 y + c_1 z = d_1 \\ a_2 x + b_2 y + c_2 z = d_2 \\ a_3 x + b_3 y + c_3 z = d_3 \end{array} \right\} \quad (6.6)$$

has the solution

$$\left. \begin{array}{l} x = \dfrac{d_1 b_2 c_3 - d_1 b_3 c_2 - d_2 b_1 c_3 + d_2 b_3 c_1 + d_3 b_1 c_2 - d_3 b_2 c_1}{a_1 b_2 c_3 - a_1 b_3 c_2 - a_2 b_1 c_3 + a_2 b_3 c_1 + a_3 b_1 c_2 - a_3 b_2 c_1} \\[6pt] y = \dfrac{a_1 d_2 c_3 - a_1 d_3 c_2 - a_2 d_1 c_3 + a_2 d_3 c_1 + a_3 d_1 c_2 - a_3 d_2 c_1}{a_1 b_2 c_3 - a_1 b_3 c_2 - a_2 b_1 c_3 + a_2 b_3 c_1 + a_3 b_1 c_2 - a_3 b_2 c_1} \\[6pt] z = \dfrac{a_1 b_2 d_3 - a_1 b_3 d_2 - a_2 b_1 d_3 + a_2 b_3 d_1 + a_3 b_1 d_2 - a_3 b_2 d_1}{a_1 b_2 c_3 - a_1 b_3 c_2 - a_2 b_1 c_3 + a_2 b_3 c_1 + a_3 b_1 c_2 - a_3 b_2 c_1} \end{array} \right\} \quad (6.7)$$

But who would want to memorize this formula? Fortunately, *determinants* have volunteered to do the memorizing for us! That is, we are able to write this formula in a more easily memorized form, using determinants.

When written in determinants, the solution in (6.7) appears as

$$x = \frac{\begin{vmatrix} d_1 & b_1 & c_1 \\ d_2 & b_2 & c_2 \\ d_3 & b_3 & c_3 \end{vmatrix}}{\begin{vmatrix} a_1 & b_1 & c_1 \\ a_2 & b_2 & c_2 \\ a_3 & b_3 & c_3 \end{vmatrix}}, \quad y = \frac{\begin{vmatrix} a_1 & d_1 & c_1 \\ a_2 & d_2 & c_2 \\ a_3 & d_3 & c_3 \end{vmatrix}}{\begin{vmatrix} a_1 & b_1 & c_1 \\ a_2 & b_2 & c_2 \\ a_3 & b_3 & c_3 \end{vmatrix}}, \quad z = \frac{\begin{vmatrix} a_1 & b_1 & d_1 \\ a_2 & b_2 & d_2 \\ a_3 & b_3 & d_3 \end{vmatrix}}{\begin{vmatrix} a_1 & b_1 & c_1 \\ a_2 & b_2 & c_2 \\ a_3 & b_3 & c_3 \end{vmatrix}}. \quad (6.8)$$

The rules for writing the solution in *determinant form* are the same for a system of three linear equations in three unknowns as they are for a system of two linear equations in two unknowns. The denominator in each fraction of (6.8) is the determinant of the coefficients of the variables in system (6.6), and the determinants in the numerators are obtained for each variable by replacing the coefficients of that particular variable with the constant terms

$$\begin{bmatrix} d_1 \\ d_2 \\ d_3 \end{bmatrix}.$$

Suppose someone had just invented the equations in (6.8) as the way to write the solution to system (6.6) without actually knowing it would work and even without knowing what these three-by-three determinants are. How should this person *evaluate* these third-order determinants? The answer is: by looking at the known solution in (6.7) for a clue.

Each denominator in the formula (6.7)

$$a_1 b_2 c_3 - a_1 b_3 c_2 - a_2 b_1 c_3 + a_2 b_3 c_1 + a_3 b_1 c_2 - a_3 b_2 c_1, \qquad (6.9)$$

may be written

$$a_1 \cdot (b_2 c_3 - b_3 c_2) - a_2 \cdot (b_1 c_3 - b_3 c_1) + a_3 \cdot (b_1 c_2 - b_2 c_1). \qquad (6.10)$$

In (6.10), the terms in the parentheses can be expressed as the second-order (or two-by-two) determinants:

$$a_1 \begin{vmatrix} b_2 & c_2 \\ b_3 & c_3 \end{vmatrix} - a_2 \begin{vmatrix} b_1 & c_1 \\ b_3 & c_3 \end{vmatrix} + a_3 \begin{vmatrix} b_1 & c_1 \\ b_2 & c_2 \end{vmatrix}. \qquad (6.11)$$

Now we return to the formulas in (6.8), where each denominator is

$$\begin{vmatrix} a_1 & b_1 & c_1 \\ a_2 & b_2 & c_2 \\ a_3 & b_3 & c_3 \end{vmatrix}. \qquad (6.12)$$

This suggests that a useful way to evaluate the third-order determinant in (6.12) would be to set it equal to the expression in (6.11), which is the same as (6.10) and finally (6.9), the denominator in the formula of (6.7). Similarly, we could evaluate each of the other determinants in (6.8) by the same rules and get (perhaps) the formulas in (6.7). [The student should verify that the determinants in the numerators of (6.8), when expanded, yield the respective numerators in (6.7).]

6.2 Third-Order Determinants

Therefore, the inventor of the determinant formula *defines* the third-order determinant as follows:

$$\begin{vmatrix} a_1 & b_1 & c_1 \\ a_2 & b_2 & c_2 \\ a_3 & b_3 & c_3 \end{vmatrix} = a_1 \begin{vmatrix} b_2 & c_2 \\ b_3 & c_3 \end{vmatrix} - a_2 \begin{vmatrix} b_1 & c_1 \\ b_3 & c_3 \end{vmatrix} + a_3 \begin{vmatrix} b_1 & c_1 \\ b_2 & c_2 \end{vmatrix}, \qquad (6.13)$$

and this is a perfectly correct definition.

Now we can evaluate any third-order determinant since we know how to evaluate the smaller, second-order determinants in the definition. These smaller determinants are called *minors* of the larger determinant. All the minors of (6.12) are found in the larger determinant by the following very mechanical procedure: The minor of a_1 is what is left after taking out the row and column containing a_1. See Figure 6.2a. The minor of a_2 is what is left after deleting the row and column containing a_2. See Figure 6.2b. The minor of a_3 is obtained by deleting the row and column of a_3 as shown in Figure 6.2c.

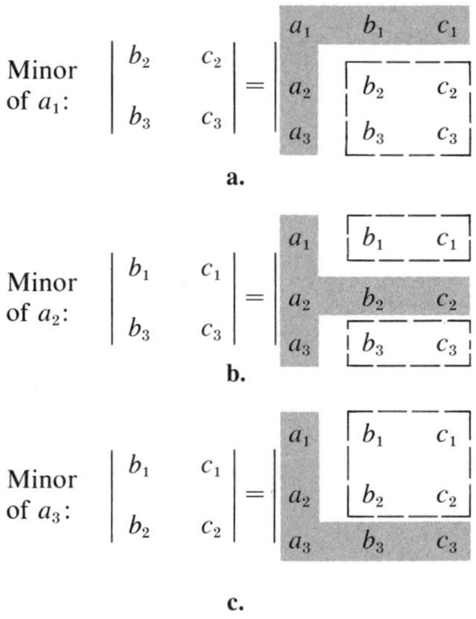

Minor of a_1: $\begin{vmatrix} b_2 & c_2 \\ b_3 & c_3 \end{vmatrix} = \begin{vmatrix} a_1 & b_1 & c_1 \\ a_2 & b_2 & c_2 \\ a_3 & b_3 & c_3 \end{vmatrix}$

a.

Minor of a_2: $\begin{vmatrix} b_1 & c_1 \\ b_3 & c_3 \end{vmatrix} = \begin{vmatrix} a_1 & b_1 & c_1 \\ a_2 & b_2 & c_2 \\ a_3 & b_3 & c_3 \end{vmatrix}$

b.

Minor of a_3: $\begin{vmatrix} b_1 & c_1 \\ b_2 & c_2 \end{vmatrix} = \begin{vmatrix} a_1 & b_1 & c_1 \\ a_2 & b_2 & c_2 \\ a_3 & b_3 & c_3 \end{vmatrix}$

c.

Figure 6.2

The definition in Equation (6.13) may be interpreted as a rule for evaluating third-order determinants. The rule is to multiply the elements of the first column by their respective minors, then add and subtract these products in an alternating pattern. This is called **expansion by minors along the first column.**

Example 6.6

$$\begin{vmatrix} 3 & 2 & 0 \\ 7 & -1 & 1 \\ 8 & 5 & 6 \end{vmatrix} = 3 \cdot \begin{vmatrix} -1 & 1 \\ 5 & 6 \end{vmatrix} - 7 \cdot \begin{vmatrix} 2 & 0 \\ 5 & 6 \end{vmatrix} + 8 \cdot \begin{vmatrix} 2 & 0 \\ -1 & 1 \end{vmatrix}$$

$$= 3 \cdot (-11) - 7 \cdot (12) + 8 \cdot 2 = -101.$$

In double subscript notation, we can write the expansion along the first column as follows:

$$\begin{vmatrix} a_{11} & a_{12} & a_{13} \\ a_{21} & a_{22} & a_{23} \\ a_{31} & a_{32} & a_{33} \end{vmatrix} = a_{11} \begin{vmatrix} a_{22} & a_{23} \\ a_{32} & a_{33} \end{vmatrix} - a_{21} \begin{vmatrix} a_{12} & a_{13} \\ a_{32} & a_{33} \end{vmatrix} + a_{31} \begin{vmatrix} a_{12} & a_{13} \\ a_{22} & a_{23} \end{vmatrix}.$$

Determinants can also be expanded by minors along any other row or column, but the alternation of signs must conform to the following pattern: Multiply the minor of a_{ij} by $(-1)^{i+j} \cdot a_{ij}$, then add these products. The factor $(-1)^{i+j}$ will be -1 when $i + j$ is odd and $+1$ when $i + j$ is even. (See Figure 6.3.) For example, if we expand by minors along the *second row*, we will find the minors of a_{21}, a_{22}, and a_{23}, then we multiply these minors by $-a_{21}$, $+a_{22}$, and $-a_{23}$, and add the products. This expansion is as follows:

$$\begin{vmatrix} a_{11} & a_{12} & a_{13} \\ a_{21} & a_{22} & a_{23} \\ a_{31} & a_{32} & a_{33} \end{vmatrix} = -a_{21} \begin{vmatrix} a_{12} & a_{13} \\ a_{32} & a_{33} \end{vmatrix} + a_{22} \begin{vmatrix} a_{11} & a_{13} \\ a_{31} & a_{33} \end{vmatrix} + -a_{23} \begin{vmatrix} a_{11} & a_{12} \\ a_{31} & a_{32} \end{vmatrix}.$$

$$\begin{vmatrix} + & - & + \\ - & + & - \\ + & - & + \end{vmatrix}$$

Figure 6.3
An indication of the pattern of plus and minus signs used in expanding a determinant by minors.

This allows us to choose the most convenient way to find the value of the determinant. For example, we can select a row or column with the most zeros in it along which to expand.

6.2 Third-Order Determinants

Example 6.7 Given the determinant

$$\begin{vmatrix} 9 & 0 & -7 \\ 10 & 0 & 5 \\ 6 & 8 & 15 \end{vmatrix},$$

evaluate it by expanding along the second column.

Solution

$$-0 \cdot \begin{vmatrix} 10 & 5 \\ 6 & 15 \end{vmatrix} + 0 \cdot \begin{vmatrix} 9 & -7 \\ 6 & 15 \end{vmatrix} - 8 \cdot \begin{vmatrix} 9 & -7 \\ 10 & 5 \end{vmatrix} = 0 + 0 - 8 \cdot (115) = -920.$$

Let us use third-order determinants to solve a system of three simultaneous equations.

Example 6.8 Find the determinant solution and evaluate the determinants for

$$x + 2y - z = 8$$
$$4y + 2z = 5$$
$$2x - 3y + z = 0.$$

Solution First, we note that the coefficient of x in the second equation is zero; that is, $a_2 = 0$. Now, by the determinant formula (6.8):

$$x = \frac{\begin{vmatrix} 8 & 2 & -1 \\ 5 & 4 & 2 \\ 0 & -3 & 1 \end{vmatrix}}{\begin{vmatrix} 1 & 2 & -1 \\ 0 & 4 & 2 \\ 2 & -3 & 1 \end{vmatrix}}, \quad y = \frac{\begin{vmatrix} 1 & 8 & -1 \\ 0 & 5 & 2 \\ 2 & 0 & 1 \end{vmatrix}}{\begin{vmatrix} 1 & 2 & -1 \\ 0 & 4 & 2 \\ 2 & -3 & 1 \end{vmatrix}}, \quad z = \frac{\begin{vmatrix} 1 & 2 & 8 \\ 0 & 4 & 5 \\ 2 & -3 & 0 \end{vmatrix}}{\begin{vmatrix} 1 & 2 & -1 \\ 0 & 4 & 2 \\ 2 & -3 & 1 \end{vmatrix}}.$$

Evaluating these determinants, we get

$$x = \frac{85}{26}, \quad y = \frac{47}{26}, \quad z = \frac{-29}{26}.$$

This determinant method is Cramer's rule applied to a three-by-three system of equations. Determinants and Cramer's rule can be generalized for use in solving four-by-four systems, five-by-five systems, and in general, n-by-n systems, where n is any positive integer. In the next chapter we will discuss another method, the matrix method, for solving systems of equations. As will be seen, a matrix has a close relationship to determinants and vectors.

We now return to a short discussion of column vectors.

COLUMN VECTOR NOTATION

The answer to the problem in Example 6.8, written in column vector notation, is

$$\begin{bmatrix} x \\ y \\ z \end{bmatrix} = \begin{bmatrix} 85/26 \\ 47/26 \\ -29/26 \end{bmatrix} = \frac{1}{26} \begin{bmatrix} 85 \\ 47 \\ -29 \end{bmatrix}.$$

As stated before, the multiplication of a (column) vector by a number is equivalent to the multiplication of each element in the vector by that number.

Equality of two column vectors means component-by-component (element-by-element) equality. That is, the top element of the first vector is equal to the top element of the second, the next component of the first vector is equal to the next component of the second, and so forth.

Both of these properties (multiplication of a vector by a single number is the same as multiplication of all the components by that number, and equal vectors are component-by-component equal) will be assumed for vectors with two, three, four, or any number of elements in them. Of course, a column vector with three components cannot equal one with just two components, and so on. We state these properties in general:

$$n \begin{bmatrix} a \\ b \\ c \end{bmatrix} = \begin{bmatrix} na \\ nb \\ nc \end{bmatrix} = \begin{bmatrix} a \\ b \\ c \end{bmatrix} n, \quad x \begin{bmatrix} a_1 \\ a_2 \\ \vdots \\ a_n \end{bmatrix} = \begin{bmatrix} x \cdot a_1 \\ x \cdot a_2 \\ \vdots \\ x \cdot a_n \end{bmatrix}. \tag{6.14}$$

$$\begin{bmatrix} x \\ y \\ z \end{bmatrix} = \begin{bmatrix} a \\ b \\ c \end{bmatrix} \text{ means } x = a, y = b, \text{ and } z = c.$$

For any positive integer, n,

6.2 Third-Order Determinants

$$\begin{bmatrix} a_1 \\ a_2 \\ \vdots \\ a_n \end{bmatrix} = \begin{bmatrix} b_1 \\ b_2 \\ \vdots \\ b_n \end{bmatrix} \text{ if and only if } a_i = b_i \text{ for all } i = 1,2,3, \ldots, n. \qquad (6.15)$$

In addition to the above basic properties of vectors, we need the rule for adding two vectors (addition is also a component-by-component property). For vectors with three components,

$$\begin{bmatrix} a \\ b \\ c \end{bmatrix} + \begin{bmatrix} d \\ e \\ f \end{bmatrix} = \begin{bmatrix} a+d \\ b+e \\ c+f \end{bmatrix}$$

and, in general,

$$\begin{bmatrix} u_1 \\ u_2 \\ \vdots \\ u_n \end{bmatrix} + \begin{bmatrix} w_1 \\ w_2 \\ \vdots \\ w_n \end{bmatrix} = \begin{bmatrix} u_1 + w_1 \\ u_2 + w_2 \\ \vdots \\ u_n + w_n \end{bmatrix}. \qquad (6.16)$$

Example 6.9   **a.** Combine the following vectors into one vector:

$$3 \cdot \begin{bmatrix} 5 \\ 2 \\ 1 \end{bmatrix} + \begin{bmatrix} 0 \\ 7 \\ 6 \end{bmatrix} \cdot \left(\frac{1}{2}\right).$$

b. Write the following vector equation as a system of linear equations:

$$\begin{bmatrix} a_1 \\ a_2 \end{bmatrix} \cdot x + \begin{bmatrix} b_1 \\ b_2 \end{bmatrix} \cdot y = \begin{bmatrix} c_1 \\ c_2 \end{bmatrix}.$$

Solution

a. $\begin{bmatrix} 15 \\ 19/2 \\ 6 \end{bmatrix}$

b. $\begin{bmatrix} a_1 x + b_1 y \\ a_2 x + b_2 y \end{bmatrix} = \begin{bmatrix} c_1 \\ c_2 \end{bmatrix},$

and by the definition of equality, $a_1 x + b_1 y = c_1$ and $a_2 x + b_2 y = c_2$.

EXERCISE 6.2

1. Evaluate the determinants

$$\begin{vmatrix} 3 & 2 & -1 \\ 1 & -1 & 3 \\ 2 & 0 & 5 \end{vmatrix}, \quad \begin{vmatrix} 0 & 2 & -1 \\ 7 & -1 & 3 \\ -1 & 0 & 5 \end{vmatrix}, \quad \begin{vmatrix} 3 & 0 & -1 \\ 1 & 7 & 3 \\ 2 & -1 & 5 \end{vmatrix}, \quad \begin{vmatrix} 3 & 2 & 0 \\ 1 & -1 & 7 \\ 2 & 0 & -1 \end{vmatrix}.$$

2. Evaluate the determinants

$$\begin{vmatrix} 2 & 9 & 0 \\ -1 & 6 & -7 \\ 1 & 1 & 1 \end{vmatrix}, \quad \begin{vmatrix} 1 & 9 & 0 \\ 0 & 6 & -7 \\ 4 & 1 & 1 \end{vmatrix}, \quad \begin{vmatrix} 2 & 1 & 0 \\ -1 & 0 & -7 \\ 1 & 4 & 1 \end{vmatrix}, \quad \begin{vmatrix} 2 & 9 & 1 \\ -1 & 6 & 0 \\ 1 & 1 & 4 \end{vmatrix}.$$

3. Write the determinant solution (using Cramer's rule) to

$$\begin{aligned} 3x + 2y - z &= 0 \\ x - y + 3z &= 7 \\ 2x \phantom{{}-y} + 5z &= -1. \end{aligned}$$

4. Write the determinant solution to

$$\begin{aligned} 2x + 9y \phantom{{}-7z} &= 1 \\ -x + 6y - 7z &= 0 \\ x + y + z &= 4. \end{aligned}$$

5. Change the determinant solution in Problem 3 to numerical values. (Hint: Use the results in Problem 1.)

6. Change the determinant solution in Problem 4 to numerical values. (See Problem 2.)

7. Write the solution to Problem 5 in column vector notation.

8. Write the solution to Problem 6 in column vector notation.

9. Write the vector equation below as a system of three linear equations:

$$\begin{bmatrix} a_{11} \\ a_{21} \\ a_{31} \end{bmatrix} x_1 + \begin{bmatrix} a_{12} \\ a_{22} \\ a_{32} \end{bmatrix} x_2 + \begin{bmatrix} a_{13} \\ a_{23} \\ a_{33} \end{bmatrix} x_3 = \begin{bmatrix} b_1 \\ b_2 \\ b_3 \end{bmatrix}.$$

6.2 Third-Order Determinants

10. Write the vector equation given below as a system of three linear equations:
$$\begin{bmatrix} 7 \\ 5 \\ 6 \end{bmatrix} x_1 + \begin{bmatrix} 2 \\ -1 \\ 0 \end{bmatrix} x_2 + \begin{bmatrix} -3 \\ 4 \\ -1 \end{bmatrix} x_3 = \begin{bmatrix} 3 \\ -5 \\ 2 \end{bmatrix}.$$

*11. The determinant equation
$$\begin{vmatrix} 3 & 2 & 1 \\ 4 & 1-x & 2 \\ 1 & 0 & 2-x \end{vmatrix} = 0$$
is a quadratic equation in x; solve it.

*12. Solve the following quadratic equation in x:
$$\begin{vmatrix} 2-x & 0 & 2 \\ 3 & 3-x & 6 \\ 5 & 5 & 5 \end{vmatrix} = 0.$$

▲ 13. The supply and use of water in the United States can be grouped into three categories: irrigation, public utilities, and self-supplied amounts. Let $x =$ the irrigation amount, $y =$ the public utilities amount, and $z =$ the self-supplied amount. These amounts are the average daily uses measured in billions of gallons. The estimated 1980 total supply and use is 444 billion gallons per day (that is, $x + y + z = 444$). Suppose the public utilities supply 119 billion gallons less one-third of the self-supplied amount [that is, $y = 119 - (1/3)z$]. Also suppose that twice the irrigation amount plus the public utility amount is 357 billion gallons (that is, $2x + y = 357$). Find the estimated daily average in billions of gallons per day for each category.*

14. In a biological process, let $x =$ amount of lactic acid produced, $y =$ amount of CO_2 produced, $z =$ amount of H_2O produced, and $q =$ amount of glucose consumed in the process. All amounts are in units of $g \cdot cm^{-3} \cdot sec^{-1}$. From the equations
$$x + y + z = 1 - q$$
$$-x + 4y + 10z = 12 + q$$
$$2x - y - \frac{3}{2}z = -4 - 2q,$$
find the amounts x, y, and z. (Assume q is a known constant.) ▲

*Data modified from the U.S. Bureau of the Census *Statistical Abstract of the United States*, 1975, 96th ed. (Washington, D.C.: GPO).

6.3

VECTORS

Let n denote a positive integer. Then an n-dimensional vector $\mathbf{v}$ is an ordered set of n real numbers that satisfies the rules for equality, addition, and multiplication by a number, as stated in the previous section. The vector $\mathbf{v}$ can be written as either a *column vector*,

$$\mathbf{v} = \begin{bmatrix} v_1 \\ v_2 \\ \vdots \\ v_n \end{bmatrix},$$

or a *row vector*,

$$\mathbf{v} = (v_1, v_2, \ldots, v_n).$$

We are free to write vectors as either columns or rows, but this freedom will be decreased considerably in the next chapter, where vectors are regarded as special types of matrices.

Example 6.10 Let $\mathbf{u}$ and $\mathbf{v}$ be the seven-dimensional vectors $\mathbf{u} = (3, 2, -1/\pi, 0, \sqrt{7}, 0, 150)$ and $\mathbf{v} = (-1, 4, 3, 2.5, -6, -12, 0)$.

a. Find $\mathbf{u} + \mathbf{v}$ and $\mathbf{v} + \mathbf{u}$.

b. If $\mathbf{0} = (0,0,0,0,0,0,0)$, find $\mathbf{u} + \mathbf{0}$.

Solution a. $\mathbf{u} + \mathbf{v} = (2, 6, (3\pi - 1)/\pi, 2.5, \sqrt{7}, -6, -12, 150)$. $\mathbf{v} + \mathbf{u}$ is the same as $\mathbf{u} + \mathbf{v}$.

b. $\mathbf{u} + \mathbf{0} = \mathbf{u}$.

Vectors provide a convenient way to represent data in scientific measurements.

▲ **Example 6.11** **Biological vector** In a study of biological membranes that form the walls of a sac, the five-dimensional vector $\mathbf{p}$, defined below, is used to describe certain functions:

$$\mathbf{p} = (T_1, T_2, \sigma, \epsilon, \phi)$$

where T_1 and T_2 = the tension on the biological membrane,

6.3 Vectors

σ = the thickness of the charged fluid layer on both sides of the membrane,
ϵ = the dielectric constant *in situ*, and
ϕ = the electric potential across the membrane.

Example 6.12 **Durable goods vector** To study the weekly earnings in the manufacture of durable goods, an economist collects data from various industries as follows:

x_1 = primary metals industry earnings,
x_2 = fabricated metal products earnings,
x_3 = electrical machinery earnings,
x_4 = nonelectrical machinery earnings,
x_5 = transportation equipment earnings,
x_6 = lumber and wood products earnings,
x_7 = furniture and fixtures earnings, and
x_8 = stone, clay, and glass earnings.

His information is then recorded as a set of eight-dimensional vectors $\{\mathbf{d}\}$:

$$\mathbf{d} = (x_1, x_2, x_3, x_4, x_5, x_6, x_7, x_8).$$

The eight quantities are variables and a functional relation might be stated that relates them to some ninth quantity, such as a corporate earnings index E. For example, $E = E(\mathbf{d}) = \sum_{i=1}^{8} x_i$. ▲

In the previous section, Equations (6.14), (6.15), and (6.16) defined equality, multiplication by a number, and addition of column vectors. These same definitions hold for any vectors and they are summarized here for convenience.

Definition **(Vector equality)**
Let $\mathbf{u} = (u_1, u_2, \ldots, u_n)$ and $\mathbf{v} = (v_1, v_2, \ldots, v_n)$ be n-dimensional vectors. Then $\mathbf{u} = \mathbf{v}$ means that $u_i = v_i$ for all $i = 1, 2, \ldots, n$.

Definition **(Multiplication of a vector by a number)**
If x is any real number and $\mathbf{v} = (v_1, v_2, \ldots, v_n)$ is any n-dimensional vector, then the vector $x \cdot \mathbf{v}$ is defined by $x\mathbf{v} = (xv_1, xv_2, \ldots, xv_n)$. This is called multiplication of a vector by a **scalar** (a nonvector, real number).

Definition **(Addition of vectors)**
Let $\mathbf{u} = (u_1, u_2, \ldots, u_n)$ and $\mathbf{v} = (v_1, v_2, \ldots, v_n)$ be n-dimensional vectors. Then $\mathbf{u} + \mathbf{v}$ is the vector defined by $\mathbf{u} + \mathbf{v} = (u_1 + v_1, u_2 + v_2, \ldots, u_n + v_n)$.

We give a series of examples using these three definitions, starting with some examples of two vectors that are not equal.

Example 6.13   **Examples of unequal vectors**

a. $(3,7,0,4) \neq (3,7,4,0)$.

b. $(0,0,0,1,0,0) \neq (0,0,1,0,0,0)$.

c. $(6,6,6,6,6,6) \neq (6,6,6,6,6,6,6)$.

d. $(a,b,c,d,x) \neq (a,b,c,d,y)$ only if $x \neq y$.

Example 6.14   **Multiplication by a scalar**

a. $10^3 \cdot (0.031, 2.100, 11.020, 0.005) = (31, 2100, 11{,}020, 5)$.

b. $\tfrac{1}{3} \cdot (3, -1, 24, 9, -10) = (1, -\tfrac{1}{3}, 8, 3, -\tfrac{10}{3})$.

▲ **Example 6.15**   Let $\mathbf{v} = (x_1, x_2, x_3, x_4)$, where (all amounts are in 1000 metric tons):

x_1 = copper ore produced by a nation yearly,
x_2 = iron ore produced by a nation yearly,
x_3 = lead ore produced by a nation yearly, and
x_4 = nickel ore produced by a nation yearly.

For the United States, the vector $\mathbf{v}_{us}$ is

$$\mathbf{v}_{us} = (1092, 50{,}000, 326, 18).$$

For the Soviet Union, the vector $\mathbf{v}_{ussr}$ is

$$\mathbf{v}_{ussr} = (800, 92{,}000, 400, 95).$$

The total amount of each of these ores produced by both the United States and the Soviet Union is the sum of the two vectors:

$$\mathbf{v}_{us} + \mathbf{v}_{ussr} = (1892, 142{,}000, 726, 113).$$

The Canadian vector $\mathbf{v}_{can} = (563, 27{,}000, 328, 239)$.
Then

$$\mathbf{v}_{us} + \mathbf{v}_{can} - \mathbf{v}_{ussr} = (855, -15{,}000, 254, 162). \quad ▲$$

The next definition, unlike the first three, was *not* introduced in the previous section.

6.3 Vectors

Definition (**Multiplication of vectors**)
If $\mathbf{u} = (u_1, u_2, \ldots, u_n)$ and $\mathbf{v} = (v_1, v_2, \ldots, v_n)$ are n-dimensional vectors, then the *scalar product*, $\mathbf{u} \cdot \mathbf{v}$, of $\mathbf{u}$ and $\mathbf{v}$ is the *nonvector, real number* $u_1 v_1 + u_2 v_2 + \ldots + u_n v_n$; that is,

$$\mathbf{u} \cdot \mathbf{v} = \sum_{i=1}^{n} u_i \cdot v_i.$$

The use of the multiplication dot makes this called the *dot product*, to distinguish it from another product for vectors (the vector product) which uses a different notation and is a subject for a more advanced treatment of vectors.

NOTES ON THE SCALAR PRODUCT

The dot product is *not* a vector. Scalar multiplication is not a closed operation in the space of vectors; that is, the multiplication of two vectors does not produce a vector, but rather a scalar.

Example 6.16
a. If $\mathbf{u} = (3,7,5,3,1)$ and $\mathbf{v} = (0,8,2,4,6)$, find $\mathbf{u} \cdot \mathbf{v}$.
b. If $\mathbf{a} = (0,7,2,-1)$ and $\mathbf{b} = (2,-3,9,2)$, find $\mathbf{a} \cdot \mathbf{b}$.
c. If $\mathbf{x} = (3,2,4)$, $\mathbf{y} = (1,0,7)$, and $\mathbf{z} = (6,-4,2)$, find $\mathbf{x} \cdot (\mathbf{y} + \mathbf{z})$ and $(\mathbf{x} \cdot \mathbf{y}) + (\mathbf{x} \cdot \mathbf{z})$.
d. If $\mathbf{k} = (1,0,0,0)$ and $\mathbf{h} = (3,6,5,8)$, find $\mathbf{k} \cdot \mathbf{h}$.

Solution
a. $\mathbf{u} \cdot \mathbf{v} = 3 \cdot 0 + 7 \cdot 8 + 5 \cdot 2 + 3 \cdot 4 + 1 \cdot 6 = 84$.
b. $\mathbf{a} \cdot \mathbf{b} = 0 \cdot 2 + 7 \cdot (-3) + 2 \cdot 9 + (-1) \cdot 2 = -5$.
c. $\mathbf{x} \cdot (\mathbf{y} + \mathbf{z}) = (3,2,4) \cdot (7,-4,9) = 21 - 8 + 36 = 49$.
$(\mathbf{x} \cdot \mathbf{y}) + (\mathbf{x} \cdot \mathbf{z}) = (3 + 0 + 28) + (18 - 8 + 8) = 49$.
d. $\mathbf{k} \cdot \mathbf{h} = 3$.

The dot product of two vectors is quite useful in practical applications, especially when one of the vectors represents a rate, such as price per unit, number of births per population unit, or another such marginal quantity.

Example 6.17 **North American birth rate** In 1971, the number of births per 1000 population in North America was the vector $\mathbf{v} = (18,38,42,42,42,38,24,18)$, where the components of **v** correspond to the countries (Canada, Costa Rica, El Salvador, Mexico, Nicaragua, Panama, Puerto Rico, United States). If the population in 1971 in these countries was the vector $\mathbf{p} = 10^6(22.0, 1.4, 3.6, 48.9, 1.9, 1.4, 3.0, 204.0)$, then the total number of births in North America during 1971 is the scalar product, **v·p**, divided by 1000. This is because the scalar product multiplies each country's birth rate (per 1000) times its population and adds up all of these products for all of the countries. Therefore, we write that the total births in 1971 is $^1\!/_{1000}$ **v·p**. From the given vectors, we calculate this total as

$$10^{-3}(18,38,42,42,42,38,24,18)10^6(22.0,1.4,3.6,48.9,1.9,1.4,3.0,204.0).$$

This is an easy job for any of the pocket calculators. The total is 6,531,200 births in North America in 1971. ▲

Some of the examples above suggest that vectors act very much like real numbers, and they do to a certain extent. For example, for the operation of addition, vectors obey the arithmetic rules (called the *group axioms*) for numbers. Specifically, vector addition is commutative and associative; that is, $\mathbf{u} + \mathbf{v} = \mathbf{v} + \mathbf{u}$ (commutative law) and $(\mathbf{u} + \mathbf{v}) + \mathbf{w} = \mathbf{u} + (\mathbf{v} + \mathbf{w})$ (associative law). There is also a zero vector, and each vector has its negative.

On the other hand, since the scalar product of two vectors is not even a vector (multiplication of two quantities does not even produce the same type of quantity in this case), then no one should be surprised to learn that vectors *do not* satisfy all the ordinary arithmetic rules for multiplication. For example, scalar multiplication is not associative. An expression such as $(\mathbf{x}\cdot\mathbf{y})\cdot\mathbf{z}$ is not even defined if we interpret the multiplication dot to mean the scalar product of vectors in both cases. The expressions $(\mathbf{x}\cdot\mathbf{y})\mathbf{z}$ and $\mathbf{x}(\mathbf{y}\cdot\mathbf{z})$ *are* both defined (each is the product of a scalar times a vector), but they are not equal to each other; that is, $(\mathbf{x}\cdot\mathbf{y})\mathbf{z} \neq \mathbf{x}(\mathbf{y}\cdot\mathbf{z})$ in general. This means that we cannot do any parenthesis shifting when multiplying vectors.

Although scalar multiplication is not associative, it is commutative and distributive. See the summary below for the vector arithmetic that we have covered.

LAWS OF VECTORS

Let n be any positive integer, $n \geq 2$. We assume the following algebraic facts for all n-dimensional vectors. In each of these laws, the vectors **u**, **v**, and **w** are arbitrary n-dimensional vectors.

1. Closure law for addition: $\mathbf{u} + \mathbf{v}$ is an n-dimensional vector.

2. Commutative law for addition: $\mathbf{u} + \mathbf{v} = \mathbf{v} + \mathbf{u}$.

6.3 Vectors

3. Associative law for addition: $(\mathbf{u} + \mathbf{v}) + \mathbf{w} = \mathbf{u} + (\mathbf{v} + \mathbf{w})$.
4. Identity law for addition: The n-dimensional zero vector $\mathbf{0} = (0,0,\ldots,0)$ is the vector such that for any n-dimensional vector $\mathbf{v}$, $\mathbf{v} + \mathbf{0} = \mathbf{v}$.
5. Inverse law for addition: For any given vector $\mathbf{v}$, $-1\cdot\mathbf{v}$ is the vector $-\mathbf{v}$ such that $\mathbf{v} + (-\mathbf{v}) = \mathbf{0}$.
6. Scalar multiplication is not closed: The product $\mathbf{u}\cdot\mathbf{v} = \sum_{i=1}^{n} u_i v_i$ is not an n-dimensional vector.
7. The scalar product is commutative: $\mathbf{u}\cdot\mathbf{v} = \mathbf{v}\cdot\mathbf{u}$.
8. Scalar multiplication is distributive over the vector sum: $\mathbf{u}\cdot(\mathbf{v} + \mathbf{w}) = (\mathbf{u}\cdot\mathbf{v}) + (\mathbf{u}\cdot\mathbf{w})$.
9. Multiplication of a vector by a scalar is commutative and distributive: For any scalars a and b, $a\mathbf{v} = \mathbf{v}a$, $(a + b)\mathbf{v} = (a\mathbf{v}) + (b\mathbf{v})$, and $a\cdot(\mathbf{v} + \mathbf{w}) = (a\cdot\mathbf{v}) + (a\cdot\mathbf{w})$.

EXERCISE 6.3

1. Given the vectors
$$\mathbf{u} = (3,7,4,-1),\ \mathbf{v} = (5,8,0,2),\ \text{and } \mathbf{w} = (-6,-10,1,5),$$
find the components of

   a. $\mathbf{u} + \mathbf{v}$

   b. $3\mathbf{w} - 2(\mathbf{u} + \mathbf{v})$

   c. $3\mathbf{u} + \tfrac{1}{2}\mathbf{v} - 2\mathbf{u} + \mathbf{w}$.

2. Given the vectors
$$\mathbf{x} = (7,4,8,6),\ \mathbf{y} = (1,4,6,9),\ \text{and } \mathbf{z} = (3,0,5,0),$$
find the components of

   a. $\mathbf{x} - \mathbf{y}$

   b. $2\mathbf{x} + 3(\mathbf{y} - \mathbf{z})$

   c. $5\mathbf{x} + 3\mathbf{y} - \tfrac{1}{2}\mathbf{z} - \tfrac{1}{2}\mathbf{y}$.

3. For the vectors $\mathbf{u}$, $\mathbf{v}$, and $\mathbf{w}$ in Problem 1:

   a. Let $3\mathbf{p} = \mathbf{u} + \mathbf{v} - \tfrac{1}{2}\mathbf{w}$. Find the components of $\mathbf{p}$.

b. Let **q** be the vector that satisfies the equation
$$5\mathbf{q} + 3\mathbf{u} = 7\mathbf{v} - \mathbf{w}.$$
Find the components of **q**.

4. For the vectors **x**, **y**, and **z** given in Problem 2:

   a. Let **t** be the vector, $\mathbf{t} = \mathbf{x} + \mathbf{y} + \mathbf{z}$. Find the components of **t**.

   b. Let **r** be the vector satisfying the equation
   $$\mathbf{r} - 2\mathbf{x} = 3\mathbf{y} - \mathbf{z}.$$
   Find the components of **r**.

   c. Let **s** be the vector satisfying the equation
   $$2\mathbf{s} + 3(\mathbf{x} - \mathbf{y}) = 2\mathbf{z} - \mathbf{s}.$$
   Find the components of **s**.

5. For the vectors in Problem 1, find the scalar products

   a. $\mathbf{u}\cdot\mathbf{v}$ **b.** $\mathbf{w}\cdot(\mathbf{u} + \mathbf{v})$ **c.** $3\mathbf{w}\cdot(2\mathbf{u} + \mathbf{v})$.

6. For the vectors in Problem 4, find the scalar products

   a. $\mathbf{x}\cdot\mathbf{y}$ **b.** $\mathbf{r}\cdot\mathbf{s}$ **c.** $\mathbf{z}\cdot\mathbf{t}$.

7. Show that, if $\mathbf{e}_1 = (1,0)$ and $\mathbf{e}_2 = (0,1)$, then the vector (a,b) is $a\cdot\mathbf{e}_1 + b\cdot\mathbf{e}_2$.

8. Find three vectors **i**, **j**, and **k** such that the vector (a,b,c) is $a\cdot\mathbf{i} + b\cdot\mathbf{j} + c\cdot\mathbf{k}$.

▲ 9. **Employment vectors** Let $\mathbf{y} = (h,c,t,s,j,a,r,d)$ be a vector whose components represent the number of people employed in the United States as h = hospital personnel, c = cashiers, t = teachers, s = secretaries, j = janitors, a = accountants and bookkeepers, r = retail salesmen, and d = truck drivers.* If in 1968 the employment vector had the values
$$\mathbf{y}_1 = 10^5\,(14.6, 7.3, 22.0, 33.5, 11.0, 17.0, 28.0, 18.4),$$
and in 1969, the vector was
$$\mathbf{y}_2 = 10^5\,(16.2, 8.0, 24.0, 36.5, 11.8, 18.1, 29.5, 19.0):$$

   a. Find the vector **z**, such that $\mathbf{y}_1 + \mathbf{z} = \mathbf{y}_2$.

   b. What is the increase in janitors from 1968 to 1969?

---

*These figures are modified from the 1971 *Information Please Almanac*. New York: Simon & Schuster. Original source: U.S. Bureau of Labor Statistics.

6.3 Vectors

    c. What does the vector **z** represent?

    d. Which job had the largest numerical increase in employment?

    e. If **w** is the eight-dimensional vector of average wages in 1969 for each of the above jobs, what would be scalar product $\mathbf{y}_2 \cdot \mathbf{w}$ represent?

10. Vector of hotel services The components of the vector $\mathbf{x} = (x_h, x_m, x_t, x_c)$ represent quantities, introduced below, in the hotels, motels, trailer parks, and recreational camps business. In the United States in 1967, the vector whose components are the numbers* of each such business was

$$\mathbf{v} = 10^3 (23.6, 42.0, 12.4, 9.0);$$

in other words, there were 23,600 hotels, 42,000 motels, etc. The receipts in each of the units in each component are expressed by the vector

$$\mathbf{r} = 10^3 (163, 65, 22, 26);$$

in other words, each hotel had a yearly receipt (on the average) of \$163,000 in 1969; each motel had a yearly receipt of \$65,000, etc.

    a. What is the value of the scalar product $\mathbf{v} \cdot \mathbf{r}$?

    b. What does the scalar product mean?

    c. How much money did all U.S. hotels take in during the year? How does this compare with the total receipts by all U.S. motels? ▲

---

*Source of figures: U.S. Bureau of the Census *Statistical Abstract of the United States*, 1974, 95th ed. (Washington, D.C.: GPO).

7
MATRICES

7.1
INTRODUCTION TO MATRICES

In the previous chapter, vectors were used as a means for handling several variables at the same time. An even more effective way to do this is by the use of matrix algebra. Matrices obey the same rules as vectors (see Chapter 6, Section 3), and they have certain additional properties. For example, multiplication of square matrices is a closed operation, and under the proper conditions, such matrices have multiplicative inverses. This allows us to apply matrix equations to a larger class of problems than we could do with vector equations.

Let m and n be positive integers. Then an m-by-n *matrix* is a set of $m \cdot n$ real numbers arranged in m *rows* and n *columns* that satisfies certain rules for equality, addition, and multiplication (to be stated later). In general subscript notation, we can write an m-by-n matrix as follows:

$$\begin{bmatrix} a_{11} & a_{12} & a_{13} & \cdots & a_{1j} & \cdots & a_{1n} \\ a_{21} & a_{22} & a_{23} & \cdots & a_{2j} & \cdots & a_{2n} \\ \vdots & \vdots & \vdots & & \vdots & & \vdots \\ a_{i1} & a_{i2} & a_{i3} & \cdots & a_{ij} & \cdots & a_{in} \\ \vdots & \vdots & \vdots & & \vdots & & \vdots \\ a_{m1} & a_{m2} & a_{m3} & \cdots & a_{mj} & \cdots & a_{mn} \end{bmatrix}. \qquad (7.1)$$

The numbers $a_{11}, a_{12}, \ldots,$ are called **elements** of the matrix. a_{ij} is the element in the ith row and the jth column. In matrix (7.1), m = the number of rows and n = the number of columns.

Example 7.1 The array of twelve numbers in three rows and four columns,

$$\begin{array}{c c} & \begin{array}{cccc} \text{Copper} & \text{Iron} & \text{Lead} & \text{Nickel} \end{array} \\ \begin{array}{c} \text{United States} \\ \text{Soviet Union} \\ \text{Canada} \end{array} & \begin{bmatrix} 1093 & 50{,}000 & 326 & 18 \\ 800 & 92{,}000 & 400 & 95 \\ 563 & 27{,}000 & 328 & 239 \end{bmatrix} \end{array},$$

is a three-by-four matrix (but not a four-by-three matrix). In general, an m-by-n matrix is not the same as an n-by-m matrix. Matrices may be thought of either as a column of row vectors or a row of column vectors:

$$\begin{bmatrix} (a_{11}, & a_{12}, & \ldots, & a_{1n}) \\ (a_{21}, & a_{22}, & \ldots, & a_{2n}) \\ & & \vdots & \\ (a_{m1}, & a_{m2}, & \ldots, & a_{mn}) \end{bmatrix} \quad \text{or} \quad \begin{bmatrix} \begin{bmatrix} a_{11} \\ a_{21} \\ \vdots \\ a_{m1} \end{bmatrix}, & \begin{bmatrix} a_{12} \\ a_{22} \\ \vdots \\ a_{m2} \end{bmatrix}, & \ldots & , & \begin{bmatrix} a_{1n} \\ a_{2n} \\ \vdots \\ a_{mn} \end{bmatrix} \end{bmatrix}.$$

We now state the definitions for matrix equality, multiplication of a matrix by a scalar, and addition of two matrices. These are very similar to the corresponding definitions for vectors.

Definition **(Matrix equality)**
The matrices **A** and **B** are equal if they are both m-by-n matrices and have exactly the same numbers in exactly the same positions.

That is, if for given positive integers m and n,

$$\mathbf{A} = \begin{bmatrix} a_{11} & a_{12} & \ldots & a_{1n} \\ a_{21} & a_{22} & \ldots & a_{2n} \\ \vdots & \vdots & & \vdots \\ a_{m1} & a_{m2} & \ldots & a_{mn} \end{bmatrix} \quad \text{and} \quad \mathbf{B} = \begin{bmatrix} b_{11} & b_{12} & \ldots & b_{1n} \\ b_{21} & b_{22} & \ldots & b_{2n} \\ \vdots & \vdots & & \vdots \\ b_{m1} & b_{m2} & \ldots & b_{mn} \end{bmatrix}, \tag{7.2}$$

then $\mathbf{A} = \mathbf{B}$ if and only if $a_{11} = b_{11}$, $a_{12} = b_{12}$, ..., $a_{1n} = b_{1n}$, $a_{21} = b_{21}$, ..., $a_{ij} = b_{ij}$, ..., and $a_{mn} = b_{mn}$. In shorter notation, $\mathbf{A} = \mathbf{B}$ if and only if $a_{ij} = b_{ij}$ for all $i = 1, 2, \ldots, m$, and $j = 1, 2, \ldots, n$.

We will often use the notation $\mathbf{A} = (a_{ij})$ and $\mathbf{B} = (b_{ij})$ to stand for the m-by-n matrices in Equations (7.2).

7.1 Introduction to Matrices

Definition **(Multiplication of a matrix by a scalar)**
If $\mathbf{A}$ is the matrix (a_{ij}) and x is any real number, then $x\mathbf{A} = (xa_{ij})$.

That is,

$$x\mathbf{A} = (xa_{ij}) = \begin{bmatrix} xa_{11} & \cdots & xa_{1n} \\ \vdots & & \vdots \\ xa_{m1} & \cdots & xa_{mn} \end{bmatrix}.$$

Definition **(Addition of matrices)**
Addition is *element-by-element* addition. If $\mathbf{A} = (a_{ij})$ and $\mathbf{B} = (b_{ij})$ are m-by-n matrices, then $\mathbf{A} + \mathbf{B} = (a_{ij} + b_{ij})$.

That is,

$$\mathbf{A} + \mathbf{B} = (a_{ij} + b_{ij}) = \begin{bmatrix} a_{11} + b_{11} & \cdots & a_{1n} + b_{1n} \\ \vdots & & \vdots \\ a_{m1} + b_{m1} & \cdots & a_{mn} + b_{mn} \end{bmatrix}.$$

We will illustrate these three definitions by the following examples.

Example 7.2   **a.**

$$\begin{bmatrix} 0 & 14 & 10 \\ 2 & x & 6 \end{bmatrix} = \begin{bmatrix} 0 & 14 & y \\ 2 & -8 & 6 \end{bmatrix}$$

only if $x = -8$ and $y = 10$.

b. Some unequal matrices:

$$\begin{bmatrix} 3 & 2 & 1 & 7 & 0 \\ 0 & 3 & 4 & 5 & 1 \end{bmatrix} \neq \begin{bmatrix} 3 & 0 \\ 2 & 3 \\ 1 & 4 \\ 7 & 5 \\ 0 & 1 \end{bmatrix}$$

$$\begin{bmatrix} 3 & 2 & 1 & 7 & 0 \\ 0 & 3 & 4 & 5 & 1 \\ 1 & 1 & 1 & 0 & 0 \end{bmatrix} \neq \begin{bmatrix} 3 & 2 & 1 & 7 & 0 \\ 0 & 3 & 4 & 1 & 5 \\ 1 & 1 & 1 & 0 & 0 \end{bmatrix}$$

Example 7.3 Let

$$M = \begin{bmatrix} 0.01 & 2.10 & 3.00 \\ 4.43 & 0.02 & 0.67 \end{bmatrix}.$$

a. Find $10M$.
b. Find $\frac{1}{2}M$.

Solution a.

$$10M = \begin{bmatrix} 0.1 & 21.0 & 30.0 \\ 44.3 & 0.2 & 6.7 \end{bmatrix}.$$

b.

$$\frac{1}{2}M = \begin{bmatrix} 0.005 & 1.050 & 1.500 \\ 2.215 & 0.010 & 0.335 \end{bmatrix}.$$

Example 7.4 Let

$$A = \begin{bmatrix} 3 & 2 & 1 & 0 \\ 7 & 1 & 5 & 6 \end{bmatrix} \quad \text{and} \quad B = \begin{bmatrix} -1 & 5 & 6 & 2 \\ 8 & 16 & -8 & 5 \end{bmatrix}.$$

a. Find $A + B$.
b. Find $3A + 2B$.

Solution a.

$$A + B = \begin{bmatrix} 2 & 7 & 7 & 2 \\ 15 & 17 & -3 & 11 \end{bmatrix}.$$

b.

$$3A + 2B = \begin{bmatrix} 7 & 16 & 15 & 4 \\ 37 & 35 & -1 & 28 \end{bmatrix}.$$

7.1 Introduction to Matrices

Example 7.5*

$$\begin{array}{c} \text{Army} \\ \text{Navy} \\ \text{Marines} \end{array} \begin{bmatrix} \text{Battle Deaths} & \text{Other Deaths} \\ 369 & 2{,}061 \\ 10 & 0 \\ 6 & 0 \end{bmatrix} = \mathbf{W}_S \text{ (Spanish-American War)}$$

$$\begin{array}{c} \text{Army} \\ \text{Navy} \\ \text{Marines} \end{array} \begin{bmatrix} 50{,}510 & 55{,}868 \\ 431 & 6{,}858 \\ 2{,}461 & 390 \end{bmatrix} = \mathbf{W}_I \text{ (World War I)}$$

$$\begin{array}{c} \text{Army} \\ \text{Navy} \\ \text{Marines} \end{array} \begin{bmatrix} 234{,}874 & 83{,}400 \\ 36{,}950 & 25{,}664 \\ 19{,}733 & 4{,}778 \end{bmatrix} = \mathbf{W}_{II} \text{ (World War II)}$$

$$\mathbf{W}_S + \mathbf{W}_I = \begin{bmatrix} \text{Battle Deaths} & \text{Other Deaths} \\ 50{,}879 & 57{,}929 \\ 441 & 6{,}858 \\ 2{,}467 & 390 \end{bmatrix} \begin{array}{l} \text{Army} \\ \text{Navy} \\ \text{Marines} \end{array}$$

$$\mathbf{W}_S + \mathbf{W}_I + \mathbf{W}_{II} = \begin{bmatrix} 285{,}753 & 141{,}329 \\ 37{,}391 & 32{,}522 \\ 22{,}200 & 5{,}168 \end{bmatrix} \begin{array}{l} \text{Army} \\ \text{Navy} \\ \text{Marines} \end{array} \quad \blacktriangle$$

We now present the definition for multiplication of a matrix by a vector. This definition naturally breaks down into two cases, one for column vectors and another for row vectors. The type of vector being multiplied by the matrix makes a difference in the final product.

To make it easier to understand the definitions involving the multiplication of matrices, let us recall the dot product for vectors. If $\mathbf{u}$ is a vector whose components are u_i, where $i = 1,2,3, \ldots, n$, and $\mathbf{v}$ is a vector with components v_i, where $i = 1,2,3, \ldots, n$, then we can denote $\mathbf{u}$ by (u_i) and $\mathbf{v}$ by (v_i), and the dot product $\mathbf{u} \cdot \mathbf{v}$ by

$$\mathbf{u} \cdot \mathbf{v} = \sum_{i=1}^{n} u_i v_i.$$

*Data from the 1971 *Information Please Almanac*. New York: Simon & Schuster; and the U.S. Bureau of the Census *Statistical Abstract of the United States*, 1969 (Washington, D.C.: GPO).

If either **u** or **v** is written as a *row* vector and the other is written as a *column* vector, then when computing their dot product, we will *always write the row vector first and the column vector second*. (The reason for doing this will become clear when we define the product of two matrices.) Thus, the dot product, **u·v**, will appear as follows:

$$(u_1, u_2, \ldots, u_n) \cdot \begin{bmatrix} v_1 \\ v_2 \\ \vdots \\ v_n \end{bmatrix} = u_1 v_1 + u_2 v_2 + \ldots + u_n v_n; \quad (7.3)$$

that is,

$$(u_1, u_2, \ldots, u_n) \cdot \begin{bmatrix} v_1 \\ v_2 \\ \vdots \\ v_n \end{bmatrix} = \sum_{j=1}^{n} u_j v_j. \quad (7.4)$$

The row vector in Equations (7.3) and (7.4) could be a row from a matrix. Some matrix operations are conveniently described by treating each row of the matrix as a row vector and each column as a column vector.

Suppose a row vector **r** is a row from an m-by-n matrix (a_{ij}); say, for example, that **r** is the third row. Then $\mathbf{r} = (a_{3j}) = (a_{31}, a_{32}, \ldots, a_{3n})$ is an n-dimensional vector. If the dot product of **r** with the column vector $\mathbf{v} = (v_i)$ is to be computed, then

$$\mathbf{r} \cdot \mathbf{v} = (a_{31}, a_{32}, \ldots, a_{3n}) \cdot \begin{bmatrix} v_1 \\ v_2 \\ \vdots \\ v_n \end{bmatrix} = \sum_{j=1}^{n} a_{3j} v_j.$$

Or, if the row vector $\mathbf{u} = (u_i)$ is an m-dimensional vector and **c** is a column from an m-by-n matrix, then **c** can be treated as an m-dimensional column vector. Say, for example, **c** is the second column, then

$$\mathbf{c} = (a_{i2}) = \begin{bmatrix} a_{12} \\ a_{22} \\ \vdots \\ a_{m2} \end{bmatrix}.$$

7.1 Introduction to Matrices

What is $\mathbf{u} \cdot \mathbf{c}$? From Equation (7.4), with the row vector first, we get

$$\mathbf{u} \cdot \mathbf{c} = (u_1, u_2, \ldots, u_m) \cdot \begin{bmatrix} a_{12} \\ a_{22} \\ \vdots \\ a_{m2} \end{bmatrix} = \sum_{i=1}^{m} u_i a_{i2}.$$

Example 7.6

If $\mathbf{r}_1 = (a_{11}, a_{12}, a_{13}, a_{14})$
$\mathbf{r}_2 = (a_{21}, a_{22}, a_{23}, a_{24})$ and if $\mathbf{v} = \begin{bmatrix} v_1 \\ v_2 \\ v_3 \\ v_4 \end{bmatrix}$,
$\mathbf{r}_3 = (a_{31}, a_{32}, a_{33}, a_{34})$

find the dot products $\mathbf{r}_1 \cdot \mathbf{v}$, $\mathbf{r}_2 \cdot \mathbf{v}$, and $\mathbf{r}_3 \cdot \mathbf{v}$.

Solution

$$\mathbf{r}_1 \cdot \mathbf{v} = \sum_{j=1}^{4} a_{1j} v_j.$$

$$\mathbf{r}_2 \cdot \mathbf{v} = \sum_{j=1}^{4} a_{2j} v_j.$$

$$\mathbf{r}_3 \cdot \mathbf{v} = \sum_{j=1}^{4} a_{3j} v_j.$$

Definition (Multiplication of a matrix times a column vector)

If $\mathbf{A} = (a_{ij})$ is an m-by-n matrix and $\mathbf{v}$ is an n-dimensional column vector, then the product $\mathbf{A}\mathbf{v}$ is defined as follows:

$$\begin{bmatrix} a_{11} & a_{12} & \cdots & a_{1n} \\ a_{21} & a_{22} & \cdots & a_{2n} \\ \vdots & \vdots & & \vdots \\ a_{m1} & a_{m2} & \cdots & a_{mn} \end{bmatrix} \begin{bmatrix} v_1 \\ v_2 \\ \vdots \\ v_n \end{bmatrix} = \begin{bmatrix} \mathbf{r}_1 \cdot \mathbf{v} \\ \mathbf{r}_2 \cdot \mathbf{v} \\ \vdots \\ \mathbf{r}_m \cdot \mathbf{v} \end{bmatrix}, \quad (7.5)$$

where $\mathbf{r}_1$ is the row vector $(a_{11}, a_{12}, \ldots, a_{1n})$; that is, $\mathbf{r}_1$ is the first row of the matrix, $\mathbf{r}_2$ is the second row $(a_{21}, a_{22}, a_{23}, \ldots, a_{2n})$, etc.

In summation notation the definition is

$$\mathbf{Av} = \begin{bmatrix} \sum_{j=1}^{n} a_{1j} v_j \\ \vdots \\ \sum_{j=1}^{n} a_{mj} v_j \end{bmatrix}. \tag{7.6}$$

Note that $\mathbf{Av}$ in Equation (7.6) is an m-dimensional column vector. The element in the ith row (the ith component) is

$$\sum_{j=1}^{n} a_{ij} v_j.$$

Example 7.7 Let $\mathbf{A} = (a_{ij})$ be a three-by-four matrix and $\mathbf{v} = (v_i)$ be a four-dimensional column vector. Write all the components in the product $\mathbf{Av}$.

Solution
$$\begin{bmatrix} a_{11} & a_{12} & a_{13} & a_{14} \\ a_{21} & a_{22} & a_{23} & a_{24} \\ a_{31} & a_{32} & a_{33} & a_{34} \end{bmatrix} \begin{bmatrix} v_1 \\ v_2 \\ v_3 \\ v_4 \end{bmatrix} = \begin{bmatrix} a_{11} v_1 + a_{12} v_2 + a_{13} v_3 + a_{14} v_4 \\ a_{21} v_1 + a_{22} v_2 + a_{23} v_3 + a_{24} v_4 \\ a_{31} v_1 + a_{32} v_2 + a_{33} v_3 + a_{34} v_4 \end{bmatrix}.$$

Example 7.8 If $\mathbf{A}$ is the matrix $\begin{bmatrix} 3 & 4 & -1 & 0 \\ -8 & 0 & 1 & -5 \\ 3 & 2 & 0 & 8 \end{bmatrix}$ and $\mathbf{v}$ is the vector $\begin{bmatrix} 1 \\ 4 \\ 2 \\ 0 \end{bmatrix}$,

find $\mathbf{Av}$.

Solution
$$\mathbf{Av} = \begin{bmatrix} \mathbf{r}_1 \cdot \mathbf{v} \\ \mathbf{r}_2 \cdot \mathbf{v} \\ \mathbf{r}_3 \cdot \mathbf{v} \end{bmatrix} = \begin{bmatrix} 3 \cdot 1 + 4 \cdot 4 + -1 \cdot 2 + 0 \cdot 0 \\ -8 \cdot 1 + 0 \cdot 4 + 1 \cdot 2 + -5 \cdot 0 \\ 3 \cdot 1 + 2 \cdot 4 + 0 \cdot 2 + 8 \cdot 0 \end{bmatrix} = \begin{bmatrix} 17 \\ -6 \\ 11 \end{bmatrix}.$$

7.1 Introduction to Matrices

Example 7.9 Air pollution is of several types: carbon monoxide, sulfur oxides, etc., and it comes from various sources: transportation, fuel combustion, etc. The matrix showing the fraction* of each type of pollution produced by each source is

$$\begin{array}{c} \\ \text{Transportation} \\ \text{Fuel Combustion} \\ \text{Industrial Processes} \\ \text{Refuse Disposal} \\ \text{Miscellaneous} \end{array} \begin{array}{ccccc} \text{Carbon} & \text{Sulfur} & \text{Hydro-} & \text{Partic-} & \text{Nitrogen} \\ \text{Monoxide} & \text{Oxides} & \text{carbons} & \text{ulates} & \text{Oxides} \end{array}$$

$$\begin{bmatrix} 0.637 & 0.024 & 0.519 & 0.042 & 0.393 \\ 0.019 & 0.732 & 0.022 & 0.315 & 0.485 \\ 0.097 & 0.223 & 0.144 & 0.265 & 0.010 \\ 0.078 & 0.003 & 0.050 & 0.039 & 0.029 \\ 0.169 & 0.018 & 0.265 & 0.339 & 0.083 \end{bmatrix} = \mathbf{P}.$$

In 1968, the tonnage vector (in millions of tons per year) for each type of pollution was

$$\begin{array}{c} \text{Carbon Monoxide} \\ \text{Sulfur Oxides} \\ \text{Hydrocarbons} \\ \text{Particulates} \\ \text{Nitrogen Oxides} \end{array} \begin{bmatrix} 100 \\ 33 \\ 32 \\ 28 \\ 21 \end{bmatrix} = \mathbf{T}.$$

The product **PT** is a column vector whose rows are the sources and whose single column is the total pollution tonnage (in million tons) for each source:

$$\mathbf{PT} = \begin{bmatrix} 90.5 \\ 45.5 \\ 29.3 \\ 11.2 \\ 37.3 \end{bmatrix} \begin{array}{l} \text{Transportation} \\ \text{Fuel combustion} \\ \text{Industrial processes.} \\ \text{Refuse disposal} \\ \text{Miscellaneous} \end{array}$$

Pollution Total for All Types

The first row (90.5) of **PT** comes from the first row of the matrix **P** times the column vector **T**, multiplied in the sense of the *dot product* of two vectors: (0.637)(100) + (0.024)(33) + (0.519)(32) + (0.042)(28) + (0.393)(21). It means that since 0.637 of the 100 million tons of carbon monoxide comes from transportation, and 0.024 of the 33 million tons of sulfur oxides comes from transportation, etc., then 90.5, the sum of all the products, is the total pollution from transportation. ▲

---

*Source: U.S. Bureau of the Census *Statistical Abstract of the United States*, 1970, 91st ed. (Washington, D.C.: GPO).

Example 7.10 Write the following matrix equation as a system of linear equations:

$$\begin{bmatrix} 3 & 0 & 1 & 1 \\ 7 & 1 & -1 & 2 \\ 5 & 0 & 2 & 2 \\ 2 & 1 & 0 & -3 \end{bmatrix} \begin{bmatrix} x_1 \\ x_2 \\ x_3 \\ x_4 \end{bmatrix} = \begin{bmatrix} 2 \\ 1 \\ 6 \\ -5 \end{bmatrix}.$$

Solution

$$3x_1 + 0x_2 + 1x_3 + 1x_4 = 2$$
$$7x_1 + 1x_2 - 1x_3 + 2x_4 = 1$$
$$5x_1 + 0x_2 + 2x_3 + 2x_4 = 6$$
$$2x_1 + 1x_2 + 0x_3 - 3x_4 = -5$$

Definition (**Multiplication of a row vector times a matrix**)
Suppose $\mathbf{u} = (u_i)$ is an m-dimensional row vector, $\mathbf{A} = (a_{ij})$ is an m-by-n matrix, and $\mathbf{c}_1, \mathbf{c}_2$, etc., are the column vectors:

$$\mathbf{c}_1 = \begin{bmatrix} a_{11} \\ a_{21} \\ \vdots \\ a_{m1} \end{bmatrix}, \mathbf{c}_2 = \begin{bmatrix} a_{12} \\ a_{22} \\ \vdots \\ a_{m2} \end{bmatrix}, \ldots, \mathbf{c}_n = \begin{bmatrix} a_{1n} \\ a_{2n} \\ \vdots \\ a_{mn} \end{bmatrix}.$$

Then $\mathbf{uA}$ is defined as follows:

$$(u_1, u_2, \ldots, u_m) \begin{bmatrix} a_{11} & a_{12} & \cdots & a_{1n} \\ a_{21} & a_{22} & \cdots & a_{2n} \\ \vdots & \vdots & & \vdots \\ a_{m1} & a_{m2} & \cdots & a_{mn} \end{bmatrix} = (\mathbf{u} \cdot \mathbf{c}_1, \mathbf{u} \cdot \mathbf{c}_2, \ldots, \mathbf{u} \cdot \mathbf{c}_n).$$

In summation notation,

$$\mathbf{uA} = \left(\sum_{i=1}^{m} u_i a_{i1}, \sum_{i=1}^{m} u_i a_{i2}, \ldots, \sum_{i=1}^{m} u_i a_{in} \right),$$

an n-dimensional row vector, with jth component $= \sum_{i=1}^{m} u_i a_{ij}$.

7.1 Introduction to Matrices

Example 7.11 Find

$$\mathbf{u}\,\mathbf{A} = (0,3,7,5,6,2) \begin{bmatrix} 2 & 7 \\ 8 & 3 \\ 0 & 5 \\ 1 & 7 \\ 5 & -2 \\ 4 & -80 \end{bmatrix}.$$

Solution $\mathbf{uA} = (0\cdot 2 + 3\cdot 8 + 7\cdot 0 + 5\cdot 1 + 6\cdot 5 + 2\cdot 4,\ 0\cdot 7 + 3\cdot 3 + 7\cdot 5 + 5\cdot 7 + 6\cdot -2 + 2\cdot -80) = 67,\ -93)$.

Example 7.12 Write the following matrix equation as a system of linear equations:

$$(x,y,z) \begin{bmatrix} 3 & 4 & 5 \\ 2 & 1 & 0 \\ -6 & 0 & 1 \end{bmatrix} = (2,-1,10).$$

Solution The product is $(3x + 2y - 6z,\ 4x + y,\ 5x + z)$, so the system of equations is

$$3x + 2y - 6z = 2$$
$$4x + y + 0z = -1$$
$$5x + 0y + z = 10.$$

EXERCISE 7.1

1. Let

$$\mathbf{A} = \begin{bmatrix} 3 & 2 & 1 \\ 8 & 16 & -8 \end{bmatrix} \text{ and } \mathbf{B} = \begin{bmatrix} 2 & 7 & 7 \\ 5 & 6 & -2 \end{bmatrix}.$$

Find **a.** $\mathbf{A} + \mathbf{B}$, **b.** $3\mathbf{A} - 2\mathbf{B}$, **c.** $\tfrac{1}{2}\mathbf{A} + 4\mathbf{B}$.

2. Let

$$X = \begin{bmatrix} 1 & 0 & 1 \\ 2 & 1 & 0 \\ 3 & 1 & 1 \end{bmatrix} \quad \text{and} \quad Y = \begin{bmatrix} 9 & 9 & 2 \\ 1 & 2 & 4 \\ 8 & 6 & 9 \end{bmatrix}.$$

Find a. $X + (1/3)Y$, b. $6X - Y$, c. $X + Y$.

3. If **A** and **B** are the matrices in Problem 1, find the elements of a matrix **C** such that

   a. $\tfrac{1}{2}A + 2C = 3B - (\tfrac{3}{4})C$

   b. $2(A + C) = 5B + C$.

4. If **X** and **Y** are the matrices in Problem 2, solve the following matrix equations for **Z** (that is, find the elements of **Z**):

   a. $2X - Y + Z = 3(X - Z)$

   b. $(\tfrac{2}{3})X + (\tfrac{1}{4})Z = 3Y - (\tfrac{1}{3})Z$.

5. Find the products

   a. $\begin{bmatrix} 3 & 2 & 1 & 4 \\ 6 & 8 & 0 & 1 \\ 0 & 2 & 1 & 3 \end{bmatrix} \begin{bmatrix} 5 \\ 4 \\ 7 \\ 1 \end{bmatrix}$ b. $(3,8,1) \begin{bmatrix} 1 & -3 & 6 & 1 & 7 \\ 0 & 2 & 4 & 0 & 1 \\ 0 & 1 & -1 & 8 & 0 \end{bmatrix}.$

6. Find the products

   a. $\begin{bmatrix} 3 & 0 & 1 & 0 & 2 \\ -6 & 1 & 7 & 1 & 4 \end{bmatrix} \begin{bmatrix} 1 \\ 6 \\ -3 \\ 11 \\ 4 \end{bmatrix}$ b. $(0,7,5,7,1) \begin{bmatrix} 3 & -1 \\ -8 & -1 \\ 4 & 0 \\ 9 & 0 \\ 4 & 2 \end{bmatrix}.$

7. a. Write the matrix equation below as a system of linear equations:

   $$\begin{bmatrix} 2 & 1 \\ 3 & 5 \end{bmatrix} \begin{bmatrix} x \\ y \end{bmatrix} = \begin{bmatrix} 3 \\ 7 \end{bmatrix}.$$

   b. Solve the system in part **a** for x and y; check by multiplying by the matrix.

8. Write the following matrix equation as a system of equations and solve it for p and q; check your answer by multiplying the solution vector by the matrix

7.1 Introduction to Matrices

$$\begin{bmatrix} 7 & 3 \\ 4 & 6 \end{bmatrix} \begin{bmatrix} p \\ q \end{bmatrix} = \begin{bmatrix} 1 \\ 6 \end{bmatrix}.$$

9. Find

$$\frac{1}{7} \cdot \begin{bmatrix} 5 & -1 \\ -3 & 2 \end{bmatrix} \begin{bmatrix} 3 \\ 7 \end{bmatrix}.$$

Compare this to Problem 7.

10. Find

$$\frac{1}{30} \cdot \begin{bmatrix} 6 & -3 \\ -4 & 7 \end{bmatrix} \begin{bmatrix} 1 \\ 6 \end{bmatrix}.$$

Compare this to Problem 8.

11. Let

$$\mathbf{I} = \begin{bmatrix} 1 & 0 \\ 0 & 1 \end{bmatrix} \quad \text{and} \quad \mathbf{v} = \begin{bmatrix} x \\ y \end{bmatrix}.$$

a. Show that $\mathbf{Iv} = \mathbf{v}$.

b. If $\mathbf{A} = (a_{ij})$ is a general two-by-two matrix, show that $\mathbf{v} - \mathbf{Av} = (\mathbf{I} - \mathbf{A})\mathbf{v}$.

12. Let

$$\mathbf{I} = \begin{bmatrix} 1 & 0 & 0 \\ 0 & 1 & 0 \\ 0 & 0 & 1 \end{bmatrix} \quad \text{and} \quad \mathbf{v} = \begin{bmatrix} x \\ y \\ z \end{bmatrix}.$$

a. Show that $\mathbf{Iv} = \mathbf{v}$.

b. If $\mathbf{A} = (a_{ij})$ is any three-by-three matrix, show that $\mathbf{v} - \mathbf{Av} = (\mathbf{I} - \mathbf{A})\mathbf{v}$.

▲ *13. The data in this problem are based on a study of lambs born to ewes on two consecutive years, 1952 and 1953, in Tallis, G. M. "The Maximum Likelihood Estimation of Correlation from Contingency Tables." *Biometrics* 18 (1962). Let **A** be the matrix

			1953		
		None	Single	Twins	
	None	58	26	8	
1952	Single	52	58	12	= **A**.
	Twins	1	3	9	

The row headings tell how many lambs were born to the ewes in 1952 and the column headings tell how many were born to ewes in 1953. For example, the number 12 in the second row (headed "single") and the third column (headed "twin") means that 12 ewes had a single birth in 1952 and had twins in 1953.

a. Let the vector

$$\mathbf{v} = \begin{bmatrix} 0 \\ 1 \\ 2 \end{bmatrix}$$

represent the number of lambs produced in each type of birth, and find $\mathbf{Av}$.

b. Interpret the components in the product of $\mathbf{A}$ and $\mathbf{v}$.

c. How many lambs were born in 1953 in this study?

d. Find the product $(0,1,2)\mathbf{A}$.

e. How many lambs were born in 1952 in this study?

*14. The matrix given below represents the fraction of "loyalty" shown toward a previously purchased product. The number in the ith row and jth column is the fraction of people who previously bought brand i but will switch to brand j in the next purchase. (If $i = j$, then the switch is from brand i to brand i; that is, there is no switch at all.)

For example, $a_{23} = 0.02$ means that two percent of the customers who previously bought brand 2 will switch to brand 3, and $a_{11} = 0.83$ means that 83 percent of those who bought brand 1 will buy it again.

$$\text{Previous Purchase} \quad \begin{array}{c} \text{Brand 1} \\ \text{Brand 2} \\ \text{Brand 3} \end{array} \begin{bmatrix} 0.83 & 0.17 & 0.00 \\ 0.06 & 0.92 & 0.02 \\ 0.05 & 0.00 & 0.95 \end{bmatrix} = \mathbf{A}.$$

with column headers "Next Purchase": Brand 1, Brand 2, Brand 3.

Suppose the vector $\mathbf{v}_0$,

$$\mathbf{v}_0 = \begin{bmatrix} 3000 \\ 2000 \\ 1000 \end{bmatrix} \begin{array}{l} \text{Brand 1} \\ \text{Brand 2,} \\ \text{Brand 3} \end{array}$$

represents the state of the market at some given time T_0. That is, at time T_0, 3000 people bought brand 1, 2000 bought brand 2, and 1000 bought brand 3. The product $\mathbf{Av}_0$ represents the state of the market at the next purchase.

a. Find the market state Av_0.

b. Use the vector Av_0 in part **a** to compute the next market state by finding $A(Av_0)$. Hint: Av_0 is also a column vector; call it **u**, then multiply **u** by **A** to find **Au**. ▲

7.2

MATRIX MULTIPLICATION

In the previous section, we defined the product of a matrix and a column vector as the multiplication of the rows of the matrix by the column vector. (See the definition on page 191.)

We will now extend this definition to compute the product of two matrices by repeatedly multiplying the rows of the first matrix by the columns of the second. Multiplication of matrices will have meaning only if the first matrix is as "wide" as the second one is "tall." We want each row in the first matrix to have as many components as each column of the second matrix.

The definition for the product **AB** of two matrices **A** and **B** is stated below; note that the rows of **A** are labeled as row vectors $r_1, r_2, \ldots r_m$, and the columns of **B** are labeled as column vectors $c_1, c_2, \ldots, c_p$. This will make the definition easier to understand.

Definition (**Multiplication of a matrix by a matrix**)

Let **A** be an m-by-n matrix and **B** be an n-by-p matrix as follows:

$$A = (a_{ij}) = \begin{bmatrix} a_{11} & a_{12} & \cdots & a_{1n} \\ a_{21} & a_{22} & \cdots & a_{2n} \\ \vdots & \vdots & & \vdots \\ a_{m1} & a_{m2} & \cdots & a_{mn} \end{bmatrix} \begin{matrix} r_1 \\ r_2 \\ \vdots \\ r_m \end{matrix}$$

$$B = (b_{jk}) = \begin{matrix} c_1 & c_2 & \cdots & c_p \\ \begin{bmatrix} b_{11} & b_{12} & \cdots & b_{1p} \\ b_{21} & b_{22} & \cdots & b_{2p} \\ \vdots & \vdots & & \vdots \\ b_{n1} & b_{n2} & \cdots & b_{np} \end{bmatrix} \end{matrix}.$$

Then the product **A B** is the *m*-by-*p* matrix

$$\mathbf{AB} = \begin{bmatrix} \mathbf{r}_1 \cdot \mathbf{c}_1 & \mathbf{r}_1 \cdot \mathbf{c}_2 & \cdots & \mathbf{r}_1 \cdot \mathbf{c}_p \\ \mathbf{r}_2 \cdot \mathbf{c}_1 & \mathbf{r}_2 \cdot \mathbf{c}_2 & \cdots & \mathbf{r}_2 \cdot \mathbf{c}_p \\ \vdots & \vdots & & \vdots \\ \mathbf{r}_m \cdot \mathbf{c}_1 & \mathbf{r}_m \cdot \mathbf{c}_2 & \cdots & \mathbf{r}_m \cdot \mathbf{c}_p \end{bmatrix}. \qquad (7.7)$$

The result is a matrix of dot products of the rows and columns, $\mathbf{r}_i$ and $\mathbf{c}_k$, with $i = 1, 2, \ldots, m$, and $k = 1, 2, \ldots, p$. The subscripts on the *r*'s and *c*'s identify the positions of the dot products in the new matrix. In carrying out this process, the student can repeat to himself or herself, "First row dot first column" for the entry in the 1,1 position; "first row dot second column" for the entry in the 1,2 position, etc. In summation notation, the product in Equation (7.7) is

$$\mathbf{AB} = \begin{bmatrix} \sum_{j=1}^{n} a_{1j} b_{j1} & \sum_{j=1}^{n} a_{1j} b_{j2} & \cdots & \sum_{j=1}^{n} a_{1j} b_{jp} \\ \sum_{j=1}^{n} a_{2j} b_{j1} & \sum_{j=1}^{n} a_{2j} b_{j2} & \cdots & \sum_{j=1}^{n} a_{2j} b_{jp} \\ \vdots & \vdots & & \vdots \\ \sum_{j=1}^{n} a_{mj} b_{j1} & \sum_{j=1}^{n} a_{mj} b_{j2} & \cdots & \sum_{j=1}^{n} a_{mj} b_{jp} \end{bmatrix}. \qquad (7.8)$$

Example 7.13

a. $\begin{bmatrix} a & b & c \\ d & e & f \end{bmatrix} \begin{bmatrix} p & q \\ r & s \\ t & u \end{bmatrix} = \begin{bmatrix} ap + br + ct & aq + bs + cu \\ dp + er + ft & dq + es + fu \end{bmatrix}.$

b. $\begin{bmatrix} 3 & 2 & 5 \\ 7 & 0 & 8 \end{bmatrix} \begin{bmatrix} 1 & 4 \\ 2 & 7 \\ 6 & 5 \end{bmatrix} = \begin{bmatrix} 3 + 4 + 30 & 12 + 14 + 25 \\ 7 + 0 + 48 & 28 + 0 + 40 \end{bmatrix} = \begin{bmatrix} 37 & 51 \\ 55 & 68 \end{bmatrix}.$

c. The product of the two matrices in part **b** was a two-by-two matrix, but if these same two matrices are multiplied in reverse order, the result is a three-by-three matrix:

7.2 Matrix Multiplication

$$\begin{bmatrix} 1 & 4 \\ 2 & 7 \\ 6 & 5 \end{bmatrix} \begin{bmatrix} 3 & 2 & 5 \\ 7 & 0 & 8 \end{bmatrix} = \begin{bmatrix} 1 \cdot 3 + 4 \cdot 7 & 1 \cdot 2 + 4 \cdot 0 & 1 \cdot 5 + 4 \cdot 8 \\ 2 \cdot 3 + 7 \cdot 7 & 2 \cdot 2 + 7 \cdot 0 & 2 \cdot 5 + 7 \cdot 8 \\ 6 \cdot 3 + 5 \cdot 7 & 6 \cdot 2 + 5 \cdot 0 & 6 \cdot 5 + 5 \cdot 8 \end{bmatrix}$$

$$= \begin{bmatrix} 31 & 2 & 37 \\ 55 & 4 & 66 \\ 53 & 12 & 70 \end{bmatrix}.$$

The dimensions of the two products in parts **b** and **c** occur as follows:

(2 by 3) times (3 by 2) = (2 by 2),

(3 by 2) times (2 by 3) = (3 by 3).

This comes from the general dimensions in the definition of matrix multiplication, which are

(m by n) times (n by p) = (m by p).

An even more dramatic illustration of the difference in the order of multiplication is as follows:

d. Let **A** be the one-by-three matrix (2,4,6) and let **B** be the three-by-one matrix

$$\begin{bmatrix} 1 \\ 3 \\ 5 \end{bmatrix}.$$

Then the product **A B** is the number 44, but the product **B A** is the three-by-three matrix:

$$\mathbf{B A} = \begin{bmatrix} 2 & 4 & 6 \\ 6 & 12 & 18 \\ 10 & 20 & 30 \end{bmatrix}.$$

Notice that in the multiplication of an m-by-n matrix by an n-by-p matrix, the "inner" dimensions n, must be equal for the matrix product to be defined.

Applications of matrix multiplication are quite varied and widespread. Almost any data that can be classified into m-by-n cells are subject to expression in matrix form, and many theoretical statements concerning such data call for sums and products of matrices.

▲ **Example 7.14** Let the matrix **A** below be the employment distribution in a study of certain population samples in 1949, 1953, and 1957:

$$\begin{array}{c} \text{Construction, Mining,} \\ \text{and Manufacturing} \\ \text{Sales and Services} \end{array} \begin{array}{ccc} 1949 & 1953 & 1957 \end{array} \\ \begin{bmatrix} 0.50 & 0.39 & 0.34 \\ 0.48 & 0.60 & 0.62 \end{bmatrix} = \mathbf{A}.$$

Let **B** be the number of people in these samples classified by sex:

$$\begin{array}{c} \\ 1949 \\ 1953 \\ 1957 \end{array} \begin{array}{cc} \text{Men} & \text{Women} \end{array} \\ \begin{bmatrix} 7{,}800 & 6{,}700 \\ 10{,}900 & 9{,}400 \\ 12{,}300 & 10{,}100 \end{bmatrix} = \mathbf{B}.$$

Let **C** be the matrix representing the average number of hours that the men and women were employed each week in the two categories of jobs:

$$\begin{array}{c} \\ \text{Men} \\ \text{Women} \end{array} \begin{array}{cc} \text{Construction} & \text{Sales} \\ \text{(Hours/week)} & \text{(Hours/week)} \end{array} \\ \begin{bmatrix} 32 & 27 \\ 17 & 22 \end{bmatrix} = \mathbf{C}.$$

a. Find **A B** and tell what the elements represent.

b. Find **B C** and tell what the elements represent.

Solution **a.**

$$\mathbf{A B} = \begin{array}{c} \text{Men} \quad \text{Women} \\ \begin{bmatrix} 12{,}333 & 10{,}450 \\ 17{,}910 & 15{,}118 \end{bmatrix} \begin{array}{l} \text{Construction} \\ \text{Sales} \end{array}$$

The elements represent the three-year total number of men and women in the two job categories.

7.2 Matrix Multiplication

b.

$$\mathbf{B\,C} = \begin{array}{c} \text{Construction} \quad \text{Sales} \\ \text{(Hours/Week)} \; \text{(Hours/Week)} \\ \begin{bmatrix} 363{,}500 & 358{,}000 \\ 508{,}600 & 501{,}100 \\ 565{,}300 & 554{,}300 \end{bmatrix} \begin{array}{c} 1949 \\ 1953 \\ 1957 \end{array} \end{array}$$

The elements represent the total weekly hours worked by all of the men and women in the sample in each job category for each of the three years. ▲

The above examples were concerned only with problems in which the matrices were *known*; in the next example we are required to use the rules for matrix algebra to solve problems with an unknown matrix.

Example 7.15 Find the values of x, y, z, and w, such that

$$\begin{bmatrix} 2 & 1 \\ 3 & 0 \end{bmatrix} \begin{bmatrix} x & y \\ z & w \end{bmatrix} = \begin{bmatrix} 1 & 4 \\ 3 & 6 \end{bmatrix} \qquad (7.9)$$

and check your answer.

Solution Here

$$\begin{bmatrix} x & y \\ z & w \end{bmatrix}$$

is the unknown matrix, and the problem is to determine the four quantities, x, y, z, and w, so that Equation (7.9) is true. We present the solution in two parts. First, we will simply give the answer for the unknown matrix and show that it does satisfy the equation. Second, we will show how to get this answer.

The answer is

$$\begin{bmatrix} x & y \\ z & w \end{bmatrix} = \begin{bmatrix} 1 & 2 \\ -1 & 0 \end{bmatrix}.$$

Check:

$$\begin{bmatrix} 2 & 1 \\ 3 & 0 \end{bmatrix} \begin{bmatrix} x & y \\ z & w \end{bmatrix} = \begin{bmatrix} 2 & 1 \\ 3 & 0 \end{bmatrix} \begin{bmatrix} 1 & 2 \\ -1 & 0 \end{bmatrix} = \begin{bmatrix} 1 & 4 \\ 3 & 6 \end{bmatrix}.$$

The method is to start with the original problem, Equation (7.9), and multiply the two matrices on the left; this yields the matrix equation

$$\begin{bmatrix} 2 \cdot x + 1 \cdot z & 2 \cdot y + 1 \cdot w \\ 3 \cdot x + 0 \cdot z & 3 \cdot y + 0 \cdot w \end{bmatrix} = \begin{bmatrix} 1 & 4 \\ 3 & 6 \end{bmatrix}. \tag{7.10}$$

Now, Equation (7.10) is equivalent to Equation (7.9) (that is, it has the same solution). So, if we can find the values for x, y, z, and w that make Equation (7.10) true, then we will have found the solution to Equation (7.9). To solve Equation (7.10), we use the definition for equality of matrices.

Two matrices are equal if and only if their corresponding elements are equal. In Equation (7.10), we set the four elements of the matrix on the left equal to the corresponding four elements of the matrix on the right:

$$2x + z = 1$$
$$2y + w = 4$$
$$3x + 0z = 3$$
$$3y + 0w = 6.$$

This is a system* of four equations in four unknowns and its solution is $x = 1$, $y = 2$, $z = -1$, and $w = 0$, from which we get the matrix

$$\begin{bmatrix} x & y \\ z & w \end{bmatrix} = \begin{bmatrix} 1 & 2 \\ -1 & 0 \end{bmatrix}.$$

This was shown to be the solution to Equation (7.9)

Example 7.16 If $\mathbf{A}$ and $\mathbf{I}$ are the known matrices

$$\mathbf{A} = \begin{bmatrix} 3 & 7 \\ 2 & 9 \end{bmatrix} \quad \text{and} \quad \mathbf{I} = \begin{bmatrix} 1 & 0 \\ 0 & 1 \end{bmatrix},$$

find the unknown two-by-two matrix $\mathbf{X} = (x_{ij})$ such that

$$\mathbf{A}\mathbf{X} = \mathbf{I}. \tag{7.11}$$

Solution Start by writing Equation (7.11) as

$$\begin{bmatrix} 3 & 7 \\ 2 & 9 \end{bmatrix} \begin{bmatrix} x_{11} & x_{12} \\ x_{21} & x_{22} \end{bmatrix} = \begin{bmatrix} 1 & 0 \\ 0 & 1 \end{bmatrix}. \tag{7.12}$$

---

*Actually, there are two distinct two-by-two systems here—two equations in x and z and two equations in y and w.

7.2 Matrix Multiplication

Multiply together the two matrices on the left in Equation (7.12), getting

$$\begin{bmatrix} 3x_{11} + 7x_{21} & 3x_{12} + 7x_{22} \\ 2x_{11} + 9x_{21} & 2x_{12} + 9x_{22} \end{bmatrix} = \begin{bmatrix} 1 & 0 \\ 0 & 1 \end{bmatrix}. \tag{7.13}$$

Set corresponding elements equal in Equation (7.13):

$$3x_{11} + 7x_{21} = 1$$
$$2x_{11} + 9x_{21} = 0$$
$$3x_{12} + 7x_{22} = 0$$
$$2x_{12} + 9x_{22} = 1.$$

Solving these two systems of two equations for x_{11}, x_{21} and x_{12}, x_{22}, we get $x_{11} = 9/13$, $x_{12} = -7/13$, $x_{21} = -2/13$, and $x_{22} = 3/13$. So, the unknown matrix X in Equation (7.11) is

$$X = \begin{bmatrix} 9/13 & -7/13 \\ -2/13 & 3/13 \end{bmatrix}.$$

Check:

$$AX = \begin{bmatrix} 3 & 7 \\ 2 & 9 \end{bmatrix} \begin{bmatrix} 9/13 & -7/13 \\ -2/13 & 3/13 \end{bmatrix} = \begin{bmatrix} (27-14)/13 & (-21+21)/13 \\ (18-18)/13 & (-14+27)/13 \end{bmatrix} = \begin{bmatrix} 1 & 0 \\ 0 & 1 \end{bmatrix}.$$

The matrix I,

$$I = \begin{bmatrix} 1 & 0 \\ 0 & 1 \end{bmatrix},$$

in Equation (7.11) is called the **multiplicative identity matrix** for two-by-two matrices. (See problems 9 and 10 in Exercise 7.2.) The matrix X such that $AX = I$ is called the **inverse** of A. (See Problems 11 and 12 in Exercise 7.2.)

EXERCISE 7.2

1. Find the products

   a. $\begin{bmatrix} 1 & 4 & 5 \\ 6 & -7 & 1 \\ 0 & 1 & 2 \end{bmatrix} \begin{bmatrix} 2 & 1 \\ 0 & 6 \\ -3 & 2 \end{bmatrix}$ b. $\begin{bmatrix} 2 & 4 \\ 3 & 6 \end{bmatrix}^2$

c. $\begin{bmatrix} 1 & 0 & 0 \\ 0 & 1 & 0 \\ 0 & 0 & 1 \end{bmatrix} \begin{bmatrix} a & b & c \\ d & e & f \\ g & h & i \end{bmatrix}$

2. Find the products

a. $\begin{bmatrix} 2 & 4 & 5 & 2 \\ 1 & 5 & 8 & 7 \end{bmatrix} \begin{bmatrix} -1 & 0 \\ 7 & 2 \\ 7 & 4 \\ 3 & 5 \end{bmatrix}$
b. $\begin{bmatrix} 0.6 & 0.4 \\ 0.3 & 0.7 \end{bmatrix}^2$

c. $\begin{bmatrix} 1 & 0 \\ 0 & 1 \end{bmatrix} \begin{bmatrix} x & y \\ u & v \end{bmatrix}$

▲ 3. Let **A** be a matrix in which each row represents one of a company's retail stores and each column represents an item sold in the stores. The number a_{ij} = the number of items of the jth type sold in the ith store.

$$\begin{array}{c} \\ \text{Store 1} \\ \text{Store 2} \\ \text{Store 3} \end{array} \begin{array}{cccc} \text{Item 1} & \text{Item 2} & \text{Item 3} & \text{Item 4} \end{array} \\ \begin{bmatrix} 7 & 8 & 10 & 3 \\ 4 & 11 & 15 & 6 \\ 2 & 8 & 9 & 10 \end{bmatrix} = \mathbf{A}.$$

Let **B** be the matrix in which the columns represent revenue and profit per item (in dollars); the rows represent the four items in matrix **A**:

$$\begin{array}{c} \\ \text{Item 1} \\ \text{Item 2} \\ \text{Item 3} \\ \text{Item 4} \end{array} \begin{array}{cc} \text{Revenue} & \text{Profit} \\ \text{Per Item} & \text{Per Item} \end{array} \\ \begin{bmatrix} 2 & 0.75 \\ 4 & 1.30 \\ 3.50 & 2.00 \\ 7.50 & 4.75 \end{bmatrix} = \mathbf{B}.$$

Find the product **A B**, and interpret the meaning of an entry in it.

4. Let the matrices **A** and **B** be as given below. Find the product **A B**.

$$\begin{array}{c} \\ \text{Men} \\ \text{Women} \end{array} \begin{array}{ccc} \text{Tribe 1} & \text{Tribe 2} & \text{Tribe 3} \end{array} \\ \begin{bmatrix} 100 & 50 & 80 \\ 110 & 60 & 100 \end{bmatrix} = \mathbf{A}$$

$$\begin{array}{c} \\ \text{Tribe 1} \\ \text{Tribe 2} \\ \text{Tribe 3} \end{array} \begin{array}{ccc} \text{Fishing} & \text{Hunting} & \text{Farming} \end{array} \\ \begin{bmatrix} 0.60 & 0.30 & 0.10 \\ 0.20 & 0.50 & 0.30 \\ 0.15 & 0.10 & 0.75 \end{bmatrix} = \mathbf{B}$$

7.2 Matrix Multiplication

5. Find the values of x_{11}, x_{12}, x_{21}, and x_{22} that make the matrix **X**,

$$\mathbf{X} = \begin{bmatrix} x_{11} & x_{12} \\ x_{21} & x_{22} \end{bmatrix},$$

satisfy the matrix equation $\mathbf{AX} = \mathbf{B}$, where **A** and **B** are the known matrices

$$\mathbf{A} = \begin{bmatrix} 2 & 6 \\ -1 & 0 \end{bmatrix} \quad \text{and} \quad \mathbf{B} = \begin{bmatrix} 4 & 7 \\ 3 & 1 \end{bmatrix}.$$

6. Find x, y, z, and w such that

$$\begin{bmatrix} x & y \\ z & w \end{bmatrix} \begin{bmatrix} 5 & 3 \\ 8 & 5 \end{bmatrix} = \begin{bmatrix} 3 & -4 \\ 1 & 5 \end{bmatrix}.$$

7. For the matrices **A** and **B** in Problem 5 and for an unknown two-by-two matrix **Y** such that $\mathbf{YA} = \mathbf{B}$, find the elements of **Y**.

8. Find p, q, r, and s such that

$$\begin{bmatrix} 5 & 3 \\ 8 & 5 \end{bmatrix} \begin{bmatrix} p & q \\ r & s \end{bmatrix} = \begin{bmatrix} 3 & -4 \\ 1 & 5 \end{bmatrix}.$$

9. The matrix

$$\mathbf{I} = \begin{bmatrix} 1 & 0 \\ 0 & 1 \end{bmatrix}$$

is the *identity* matrix for two-by-two matrices.

   a. Find $\begin{bmatrix} 1 & 0 \\ 0 & 1 \end{bmatrix} \begin{bmatrix} 3 & 7 \\ 5 & 2 \end{bmatrix}$.

   b. Find $\begin{bmatrix} 1 & 0 \\ 0 & 1 \end{bmatrix} \begin{bmatrix} x & y \\ u & w \end{bmatrix}$ and $\begin{bmatrix} x & y \\ u & w \end{bmatrix} \begin{bmatrix} 1 & 0 \\ 0 & 1 \end{bmatrix}$.

   c. Is it true, in general, that for any two-by-two matrix **A**, $\mathbf{IA} = \mathbf{A}$?

   d. If $\mathbf{A} = \begin{bmatrix} 3 & 4 \\ 5 & 6 \end{bmatrix}$ and $\mathbf{B} = \begin{bmatrix} 7 & 2 \\ 1 & 8 \end{bmatrix}$ find $\mathbf{A} - \mathbf{AB}$ and $\mathbf{A}(\mathbf{I} - \mathbf{B})$.

   e. Is it true that $\mathbf{A} - \mathbf{AB} = \mathbf{A}(\mathbf{I} - \mathbf{B})$?

   f. Find $\mathbf{I} \begin{bmatrix} 3 \\ 7 \end{bmatrix}$ and $\mathbf{I} \begin{bmatrix} u \\ v \end{bmatrix}$.

10. Refer to Problem 9 for the definition of **I**.

    a. Find $\begin{bmatrix} 1 & 0 \\ 0 & 1 \end{bmatrix} \begin{bmatrix} 5 & 6 \\ 1 & 3 \end{bmatrix}$.

b. If $M = \begin{bmatrix} a_{11} & a_{12} \\ a_{21} & a_{22} \end{bmatrix}$, find **IM** and **MI**.

c. Is it true, in general, that for any two-by-two matrix **A**, $AI = A$?

d. If $X = \begin{bmatrix} x_1 & x_2 \\ x_3 & x_4 \end{bmatrix}$ and $Y = \begin{bmatrix} y_1 & y_2 \\ y_3 & y_4 \end{bmatrix}$, find $X - XY$ and $X(I - Y)$.

e. Find $I \begin{bmatrix} a \\ b \end{bmatrix}$.

11. Let $A = \begin{bmatrix} a & b \\ c & d \end{bmatrix}$, and let $D = $ the determinant $\begin{vmatrix} a & b \\ c & d \end{vmatrix} = ad - bc$.

(D is called the determinant of the matrix **A**). The matrix A^{-1} constructed by interchanging a and d; changing the signs of b and c; and dividing everything by $D(= ad - bc)$ is called the **inverse of A**:

$$A^{-1} = \begin{bmatrix} d/D & -b/D \\ -c/D & a/D \end{bmatrix}.$$

Any matrix whose determinant is not zero has an inverse and this inverse is unique. **A** times its inverse, $A^{-1} = I$, identity matrix.

Let $A = \begin{bmatrix} 6 & 2 \\ 5 & 3 \end{bmatrix}$.

a. Find the value of the determinant of **A**.

b. Find A^{-1}.

c. Show that $\begin{bmatrix} 3/8 & -2/8 \\ -5/8 & 6/8 \end{bmatrix} \begin{bmatrix} 6 & 2 \\ 5 & 3 \end{bmatrix} = \begin{bmatrix} 1 & 0 \\ 0 & 1 \end{bmatrix}$.

d. Find $A \begin{bmatrix} 7 & 2 \\ 5 & 0 \end{bmatrix}$.

e. Find $A^{-1} \begin{bmatrix} 52 & 12 \\ 50 & 10 \end{bmatrix}$.

f. Let $X = \begin{bmatrix} 1 & 3 \\ 2 & 1 \end{bmatrix}$ and let $B = AX$. Compute the elements of **B**.

Compute $A^{-1}B$ and show that it is the same as **X**. (**A** transforms **X** into **B** and A^{-1} transforms **B** back into **X**.)

7.3 Inverse of a Matrix

12. Refer to Problem 11 for the definition of A^{-1}, the inverse of A. Let $A = \begin{bmatrix} 3 & 7 \\ 2 & 5 \end{bmatrix}$.

    a. Find the determinant D of A.

    b. Find A^{-1}.

    c. Show that $A^{-1} \cdot A = A \cdot A^{-1} = \begin{bmatrix} 1 & 0 \\ 0 & 1 \end{bmatrix}$.

    d. Find $A \begin{bmatrix} 3 & -2 \\ -4 & 2 \end{bmatrix}$.

    e. Find $A^{-1} \begin{bmatrix} -19 & 8 \\ -14 & 6 \end{bmatrix}$.

    f. Let $Y = \begin{bmatrix} 1 & 0 \\ 3 & 4 \end{bmatrix}$; transform Y into another matrix B by the multiplication $AY = B$. Find the elements of B. Transform B back into Y by multiplying $A^{-1}B$.

13. Write your height (nearest inch) and weight (nearest pound) as a column vector:

    $$V = \begin{bmatrix} \text{Height} \\ \text{Weight} \end{bmatrix}.$$

    "Encode" this information by multiplying V by the matrix

    $$C = \begin{bmatrix} -2 & 1 \\ -3 & 1 \end{bmatrix}.$$

    That is, let $U = C \cdot V$. U will be the *coded vector*. Exchange coded vectors (U) with a classmate. Can you decode his or her coded vector? (Hint: Try using C^{-1}. See the definition of the inverse of a matrix in Problem 11).

7.3

INVERSE OF A MATRIX

In Problem 11 of Exercise 7.2, the multiplicative inverse of the two-by-two matrix A,

$$A = \begin{bmatrix} a & b \\ c & d \end{bmatrix},$$

was written as the two-by-two matrix

$$\mathbf{A}^{-1} = \begin{bmatrix} \dfrac{d}{D} & \dfrac{-b}{D} \\ \dfrac{-c}{D} & \dfrac{a}{D} \end{bmatrix} \tag{7.14}$$

where D is the determinant

$$\begin{vmatrix} a & b \\ c & d \end{vmatrix} = ad - bc.$$

It can readily be shown that the matrix $\mathbf{A}^{-1}$ given in Equation (7.14) is the inverse by multiplying $\mathbf{A}$ times $\mathbf{A}^{-1}$ and seeing if the product is the identity matrix:

$$\mathbf{A} \cdot \mathbf{A}^{-1} = \begin{bmatrix} a & b \\ c & d \end{bmatrix} \begin{bmatrix} d/D & -b/D \\ -c/D & a/D \end{bmatrix} = \begin{bmatrix} (ad-bc)/D & (-ab+ab)/D \\ (cd-dc)/D & (-bc+ad)/D \end{bmatrix} = \begin{bmatrix} 1 & 0 \\ 0 & 1 \end{bmatrix}.$$

Equation (7.14) shows that it is easy to compute the inverse of a two-by-two matrix.

This is not true, however, for three-by-three, four-by-four, or larger matrices. Computation of their inverses is a much more involved process, but it is a process that is *mechanical*; therefore, it can be performed by a computer. (Most computers have built-in matrix inversion programs. The user need only type in the elements of the matrix in some designated order and tell the computer to print out the inverse.)

The process for finding the inverse of three-by-three matrices is given below. For simplicity we will use letters a, b, c, etc., without subscripts. Let A be the three-by-three matrix

$$A = \begin{bmatrix} a & b & c \\ d & e & f \\ g & h & i \end{bmatrix}.$$

ALGORITHM FOR FINDING THE INVERSE OF A THREE-BY-THREE MATRIX, A.

Step 1 Find D, the determinant of $\mathbf{A}$.

$$D = \det(\mathbf{A}) = \begin{vmatrix} a & b & c \\ d & e & f \\ g & h & i \end{vmatrix}.$$

7.3 Inverse of a Matrix

Now, expanding by minors along the first column,

$$D = a \begin{vmatrix} e & f \\ h & i \end{vmatrix} - d \begin{vmatrix} b & c \\ h & i \end{vmatrix} + g \begin{vmatrix} b & c \\ e & f \end{vmatrix}.$$

a. If the resulting $D = 0$, then **A** has no inverse. STOP.

b. If $D \neq 0$, then **A** has an inverse; proceed to Step 2.

Step 2 Find the minor for each of the nine elements, a, b, c, d, e, f, g, h, and i. This means (for three-by-three matrices) we need to find nine minors. Let M_x denote the minor of x. Then,

$$M_a = \begin{vmatrix} e & f \\ h & i \end{vmatrix} ; \quad M_b = \begin{vmatrix} d & f \\ g & i \end{vmatrix} ; \quad M_c = \begin{vmatrix} d & e \\ g & h \end{vmatrix}.$$

$$M_d = \begin{vmatrix} b & c \\ h & i \end{vmatrix} ; \quad M_e = \begin{vmatrix} a & c \\ g & i \end{vmatrix} ; \quad M_f = \begin{vmatrix} a & b \\ g & h \end{vmatrix}.$$

$$M_g = \begin{vmatrix} b & c \\ e & f \end{vmatrix} ; \quad M_h = \begin{vmatrix} a & c \\ d & f \end{vmatrix} ; \quad M_i = \begin{vmatrix} a & b \\ d & e \end{vmatrix}.$$

Step 3 Change the sign of the minor in each of the following positions: first row, second column; second row, first column; second row, third column; third row, second column (that is, change the sign of the minor in the ith row and jth column only when $i + j =$ an odd number). In this case, we change the signs of the minors M_b, M_d, M_f, and M_h.

Step 4 Write the minors with the above sign changes in a matrix as follows:

$$\begin{bmatrix} M_a & -M_b & M_c \\ -M_d & M_e & -M_f \\ M_g & -M_h & M_i \end{bmatrix}.$$

Step 5 *Transpose* the matrix in Step 4. This means *write every row as a column and every column as a row*. Every element, except those along the diagonal from the upper left to the lower right, will be shifted to a new position. The transposed matrix is

$$\begin{bmatrix} M_a & -M_d & M_g \\ -M_b & M_e & -M_h \\ M_c & -M_f & M_i \end{bmatrix}.$$

Step 6 Divide every element in the matrix of Step 5 by D, the det $(\mathbf{A})$:

$$\begin{bmatrix} M_a/D & -M_d/D & M_g/D \\ -M_b/D & M_e/D & -M_h/D \\ M_c/D & -M_f/D & M_i/D \end{bmatrix}. \quad (7.15)$$

The above matrix (7.15) is $\mathbf{A}^{-1}$, the inverse of $\mathbf{A}$.

Example 7.17 If

$$\mathbf{A} = \begin{bmatrix} a & b & c \\ d & e & f \\ g & h & i \end{bmatrix} = \begin{bmatrix} 1 & 2 & 3 \\ 0 & 1 & 4 \\ 3 & 1 & 1 \end{bmatrix} \quad (7.16)$$

use the algorithm to find $\mathbf{A}^{-1}$ (if it exists) and check your answer by seeing whether

$$\mathbf{A}\mathbf{A}^{-1} = \begin{bmatrix} 1 & 0 & 0 \\ 0 & 1 & 0 \\ 0 & 0 & 1 \end{bmatrix}.$$

Solution **Step 1** The determinant of $\mathbf{A}$ is

$$\text{Det}(\mathbf{A}) = D = 1 \cdot \begin{vmatrix} 1 & 4 \\ 1 & 1 \end{vmatrix} - 0 \cdot \begin{vmatrix} 2 & 3 \\ 1 & 1 \end{vmatrix} + 3 \cdot \begin{vmatrix} 2 & 3 \\ 1 & 4 \end{vmatrix} = -3 + 0 + 15 = 12.$$

$12 \neq 0$; so we proceed to Step 2.

Step 2 The minors of the nine elements are

$$M_a = \begin{vmatrix} 1 & 4 \\ 1 & 1 \end{vmatrix} = -3; \quad M_b = \begin{vmatrix} 0 & 4 \\ 3 & 1 \end{vmatrix} = -12; \quad M_c = \begin{vmatrix} 0 & 1 \\ 3 & 1 \end{vmatrix} = -3.$$

$$M_d = \begin{vmatrix} 2 & 3 \\ 1 & 1 \end{vmatrix} = -1; \quad M_e = \begin{vmatrix} 1 & 3 \\ 3 & 1 \end{vmatrix} = -8; \quad M_f = \begin{vmatrix} 1 & 2 \\ 3 & 1 \end{vmatrix} = -5.$$

$$M_g = \begin{vmatrix} 2 & 3 \\ 1 & 4 \end{vmatrix} = 5; \quad M_h = \begin{vmatrix} 1 & 3 \\ 0 & 4 \end{vmatrix} = 4; \quad M_i = \begin{vmatrix} 1 & 2 \\ 0 & 1 \end{vmatrix} = 1.$$

Step 3 Change the signs of M_b, M_d, M_f, and M_h, getting $-M_b = 12$, $-M_d = 1$, $-M_f = 5$, and $-M_h = -4$.

7.3 Inverse of a Matrix

Step 4 The matrix defined in this step of the algorithm is

$$\begin{bmatrix} M_a & -M_b & M_c \\ -M_d & M_e & -M_f \\ M_g & -M_h & M_i \end{bmatrix} = \begin{bmatrix} -3 & 12 & -3 \\ 1 & -8 & 5 \\ 5 & -4 & 1 \end{bmatrix}.$$

Step 5 To transpose, write the columns as rows and the rows as columns:

$$\begin{bmatrix} -3 & 1 & 5 \\ 12 & -8 & -4 \\ -3 & 5 & 1 \end{bmatrix}.$$

Step 6 Divide every element of the above matrix by $D = \det(\mathbf{A}) = 12$; this yields the inverse of $\mathbf{A}$,

$$\mathbf{A}^{-1} = \begin{bmatrix} -3/12 & 1/12 & 5/12 \\ 12/12 & -8/12 & -4/12 \\ -3/12 & 5/12 & 1/12 \end{bmatrix}. \tag{7.17}$$

The fractions in this matrix could be simplified, but it will be easier to check if we leave all of them as they are, with the same denominator. To check, we multiply matrix $\mathbf{A}$ in Equation (7.16) by the inverse $\mathbf{A}^{-1}$ in Equation (7.17) to see if we get the three-by-three identity matrix as the product:

$$\mathbf{A}\mathbf{A}^{-1} = \begin{bmatrix} 1 & 2 & 3 \\ 0 & 1 & 4 \\ 3 & 1 & 1 \end{bmatrix} \begin{bmatrix} -3/12 & 1/12 & 5/12 \\ 12/12 & -8/12 & -4/12 \\ -3/12 & 5/12 & 1/12 \end{bmatrix}$$

$$= \begin{bmatrix} (-3+24-9)/12 & (1-16+15)/12 & (5-8+3)/12 \\ (0+12-12)/12 & (0-8+20)/12 & (0-4+4)/12 \\ (-9+12-3)/12 & (3-8+5)/12 & (15-4+1)/12 \end{bmatrix} = \begin{bmatrix} 1 & 0 & 0 \\ 0 & 1 & 0 \\ 0 & 0 & 1 \end{bmatrix}.$$

The method outlined in the above algorithm also works (with some modifications) for two-by-two and for higher order matrices, such as four-by-four, etc.

In general, matrix multiplication is not commutative; that is, $\mathbf{AB} \neq \mathbf{BA}$. Two notable exceptions are the identity matrix and the inverse. If n is any positive integer (greater than or equal to two) and $\mathbf{I}$ is the n-by-n identity matrix, then for any n-by-n matrix $\mathbf{A}$, $\mathbf{IA} = \mathbf{AI}$; and if $\mathbf{A}^{-1}$ is the inverse of $\mathbf{A}$, then $\mathbf{AA}^{-1} = \mathbf{A}^{-1}\mathbf{A}$. This property will be useful in the discussion below concerning the use of inverses in solving systems of linear equations.

Example 7.18   **a.** Write the system

$$1 \cdot x + 2 \cdot y + 3 \cdot z = 5$$
$$1 \cdot y + 4 \cdot z = 6 \quad (7.18)$$
$$3 \cdot x + 1 \cdot y + 1 \cdot z = -12$$

as a matrix equation

$$\mathbf{Ax} = \mathbf{b}, \quad (7.19)$$

where $\mathbf{A}$ is a three-by-three matrix and $\mathbf{x}$ and $\mathbf{b}$ are column vectors.

b. Find the inverse $\mathbf{A}^{-1}$ of $\mathbf{A}$.

c. Multiply both sides of Equation (7.19) by $\mathbf{A}^{-1}$ to obtain $\mathbf{x}$.

Solution   **a.** Equation (7.18) is equivalent to

$$\begin{bmatrix} 1 & 2 & 3 \\ 0 & 1 & 4 \\ 3 & 1 & 1 \end{bmatrix} \begin{bmatrix} x \\ y \\ z \end{bmatrix} = \begin{bmatrix} 5 \\ 6 \\ -12 \end{bmatrix}, \quad (7.20)$$

which can be written as Equation (7.19) if we let

$$\mathbf{A} = \begin{bmatrix} 1 & 2 & 3 \\ 0 & 1 & 4 \\ 3 & 1 & 1 \end{bmatrix}, \quad \mathbf{x} = \begin{bmatrix} x \\ y \\ z \end{bmatrix}, \quad \text{and} \quad \mathbf{b} = \begin{bmatrix} 5 \\ 6 \\ -12 \end{bmatrix}.$$

b. Notice that the matrix $\mathbf{A}$ here is the same as the one in Example 7.17, where we found the inverse as given in Equation (7.17).

c. Multiplying both sides of Equation (7.20) by $\mathbf{A}^{-1}$ from Equation (7.17), we get

$$\begin{bmatrix} -3/12 & 1/12 & 5/12 \\ 12/12 & -8/12 & -4/12 \\ -3/12 & 5/12 & 1/12 \end{bmatrix} \begin{bmatrix} 1 & 2 & 3 \\ 0 & 1 & 4 \\ 3 & 1 & 1 \end{bmatrix} \begin{bmatrix} x \\ y \\ z \end{bmatrix} = \begin{bmatrix} -3/12 & 1/12 & 5/12 \\ 12/12 & -8/12 & -4/12 \\ -3/12 & 5/12 & 1/12 \end{bmatrix} \begin{bmatrix} 5 \\ 6 \\ -12 \end{bmatrix}. \quad (7.21)$$

7.3 Inverse of a Matrix

Simplifying Equation (7.21), we get

$$\begin{bmatrix} 1 & 0 & 0 \\ 0 & 1 & 0 \\ 0 & 0 & 1 \end{bmatrix} \begin{bmatrix} x \\ y \\ z \end{bmatrix} = \begin{bmatrix} (-15 + 6 - 60)/12 \\ (60 - 48 + 48)/12 \\ (-15 + 30 - 12)/12 \end{bmatrix}.$$

And, finally,

$$\mathbf{x} = \begin{bmatrix} x \\ y \\ z \end{bmatrix} = \begin{bmatrix} -69/12 \\ 5 \\ 1/4 \end{bmatrix}.$$

This solution satisfies Equation (7.18) by substitution, which can easily be verified.

In general, for any positive integer n, the system of n linear equations

$$\begin{aligned} a_{11}x_1 + a_{12}x_2 + \ldots + a_{1n}x_n &= b_1 \\ a_{21}x_1 + a_{22}x_2 + \ldots + a_{2n}x_n &= b_2 \\ &\vdots \\ a_{n1}x_1 + a_{n2}x_2 + \ldots + a_{nn}x_n &= b_n \end{aligned}$$

is equivalent to the matrix equation

$$\mathbf{A}\mathbf{x} = \mathbf{b}, \tag{7.22}$$

where $\mathbf{A} = (a_{ij})$ is the n-by-n matrix of coefficients and $\mathbf{x}$ and $\mathbf{b}$ are the n-dimensional column vectors (x_i) and (b_i), respectively. In a more complete display, Equation (7.22) is

$$\begin{bmatrix} a_{11} & a_{12} & \ldots & a_{1n} \\ a_{21} & a_{22} & \ldots & a_{2n} \\ \vdots & \vdots & & \vdots \\ a_{n1} & a_{n2} & \ldots & a_{nn} \end{bmatrix} \begin{bmatrix} x_1 \\ x_2 \\ \vdots \\ x_n \end{bmatrix} = \begin{bmatrix} b_1 \\ b_2 \\ \vdots \\ b_n \end{bmatrix}.$$

If the $\det(\mathbf{A})$ is not zero, then Equation (7.22) can be solved by multiplying both sides by $\mathbf{A}^{-1}$, getting $\mathbf{x} = \mathbf{A}^{-1} \mathbf{b}$, a column vector.

A similar method applies to matrix equations with n-by-n matrices in place of the column vectors $\mathbf{x}$ and $\mathbf{b}$. We state this specifically in the theorem below, but first we need a definition.

Definition (Singular matrix)

If for any positive integer n greater than or equal to two, $\mathbf{A}$ is an n-by-n matrix, then $\mathbf{A}$ is said to be *singular* if the determinant of $\mathbf{A}$ is zero. Otherwise [if $\det(\mathbf{A}) \neq 0$], $\mathbf{A}$ is called *nonsingular*.

We state the following theorem without proof.

Theorem 7.1 Given the matrix equation $\mathbf{AB} = \mathbf{C}$, where each of $\mathbf{A}$, $\mathbf{B}$, and $\mathbf{C}$ is an n-by-n matrix, then:

1. The equation can be solved for $\mathbf{B}$ (if $\mathbf{A}$ is nonsingular) with multiplication by $\mathbf{A}^{-1}$ as follows:

$$\mathbf{A}^{-1}\mathbf{AB} = \mathbf{A}^{-1}\mathbf{C}$$
$$\mathbf{IB} = \mathbf{B} = \mathbf{A}^{-1}\mathbf{C}.$$

2. The equation can be solved for $\mathbf{A}$ (if $\mathbf{B}$ is nonsingular) with multiplication by $\mathbf{B}^{-1}$ as follows:

$$\mathbf{ABB}^{-1} = \mathbf{C}\cdot\mathbf{B}^{-1}$$
$$\mathbf{AI} = \mathbf{A} = \mathbf{C}\mathbf{B}^{-1}.$$

Example 7.19 Solve for the matrix with the x_i elements:

$$\underbrace{\begin{bmatrix} 3 & 2 \\ 1 & 5 \end{bmatrix}}_{\mathbf{A}} \underbrace{\begin{bmatrix} x_1 & x_2 \\ x_3 & x_4 \end{bmatrix}}_{\mathbf{B}} = \underbrace{\begin{bmatrix} 5 & 1 \\ 6 & 2 \end{bmatrix}}_{\mathbf{C}}.$$

Solution By the theorem, the solution is $\mathbf{B} = \mathbf{A}^{-1}\mathbf{C}$, which is

$$\begin{bmatrix} x_1 & x_2 \\ x_3 & x_4 \end{bmatrix} = \begin{bmatrix} 5/13 & -2/13 \\ -1/13 & 3/13 \end{bmatrix} \begin{bmatrix} 5 & 1 \\ 6 & 2 \end{bmatrix} = \begin{bmatrix} 1 & 1/13 \\ 1 & 5/13 \end{bmatrix}.$$

Example 7.20 Solve for $\mathbf{A}$ if $\mathbf{A} - \mathbf{AB} = \mathbf{C}$, where

$$\mathbf{B} = \begin{bmatrix} 2 & 1 \\ 5 & -1 \end{bmatrix} \text{ and } \mathbf{C} = \begin{bmatrix} 3 & 4 \\ 1 & 6 \end{bmatrix}.$$

Solution First we write $\mathbf{A} - \mathbf{AB}$ as $\mathbf{A(I - B)}$; so, the equation to be solved is $\mathbf{A(I - B) = C}$. By Theorem 7.1: $\mathbf{A = C(I - B)^{-1}}$. Now we need to find $\mathbf{I - B}$ and its inverse, $\mathbf{(I - B)^{-1}}$:

$$\mathbf{I - B} = \begin{bmatrix} 1 & 0 \\ 0 & 1 \end{bmatrix} - \begin{bmatrix} 2 & 1 \\ 5 & -1 \end{bmatrix} = \begin{bmatrix} -1 & -1 \\ -5 & 2 \end{bmatrix},$$

and

$$\mathbf{(I - B)^{-1}} = \begin{bmatrix} -2/7 & -1/7 \\ -5/7 & 1/7 \end{bmatrix}.$$

Therefore,

$$\mathbf{A} = \begin{bmatrix} 3 & 4 \\ 1 & 6 \end{bmatrix} \begin{bmatrix} -2/7 & -1/7 \\ -5/7 & 1/7 \end{bmatrix} = \begin{bmatrix} -26/7 & 1/7 \\ -32/7 & 5/7 \end{bmatrix}.$$

To check this solution, compute $\mathbf{AB}$ and $\mathbf{A - AB}$, which will turn out to be $\mathbf{C}$.

EXERCISE 7.3

1. Find the inverse $\mathbf{A}^{-1}$ of the matrix
$$\mathbf{A} = \begin{bmatrix} 1 & 0 & 3 \\ 4 & 7 & 2 \\ -1 & 1 & 0 \end{bmatrix}$$
and check by multiplying $\mathbf{A A^{-1}}$.

2. Find the inverse $\mathbf{B}^{-1}$ of the matrix
$$\mathbf{B} = \begin{bmatrix} 1 & 2 & 0 \\ 3 & 2 & 1 \\ 4 & 0 & 1 \end{bmatrix}$$
and check by multiplying $\mathbf{BB^{-1}}$.

3. For the matrix $\mathbf{A}$ in Problem 1, find $\mathbf{I - A}$ and $\mathbf{(I - A)^{-1}}$.
4. For the matrix $\mathbf{B}$ in Problem 2, find $\mathbf{I - B}$ and $\mathbf{(I - B)^{-1}}$.

5. If **Y** is the column vector,

$$Y = \begin{bmatrix} 10 \\ 7 \\ 1 \end{bmatrix},$$

solve the equation $X - AX = Y$ for **X**, where **A** is the matrix of Problem 1.

6. If **B** is the matrix of Problem 2 and **Z** is the column vector,

$$Z = \begin{bmatrix} 5 \\ -1 \\ 3 \end{bmatrix},$$

solve the equation $W - BW = Z$, for **W**.

7. Solve the matrix equation $MP = Q$ for **P**, where **M** and **Q** are the matrices,

$$M = \begin{bmatrix} 3 & 2 \\ 1 & 0 \end{bmatrix} \text{ and } Q = \begin{bmatrix} 1 & 6 \\ 5 & 1 \end{bmatrix}.$$

8. Solve the matrix equation $ST = W$ for **S**, where **T** and **W** are the matrices,

$$T = \begin{bmatrix} 6 & 2 \\ 2 & 1 \end{bmatrix} \text{ and } W = \begin{bmatrix} 1 & 1 \\ 6 & 4 \end{bmatrix}.$$

9. a. Find the elements of the matrix **X**:

$$\begin{bmatrix} 3 & 2 \\ 7 & 5 \end{bmatrix} \begin{bmatrix} 3 & 2 & 0 & 6 & 0 & 1 \\ 1 & 8 & 1 & 2 & 8 & 1 \end{bmatrix} = X.$$

b. Find the inverse of the matrix

$$\begin{bmatrix} 3 & 2 \\ 7 & 5 \end{bmatrix}$$

and multiply it by **X**. Do you get the above two-by-six matrix back again?

10. a. Find the product **AB** for

$$A = \begin{bmatrix} 6 & 5 \\ 5 & 4 \end{bmatrix}, \quad B = \begin{bmatrix} 3 & 6 & 1 & 1 & 4 & 2 \\ 1 & 2 & 8 & 0 & 4 & 1 \end{bmatrix}.$$

b. Let $AB = C$. Find A^{-1} and multiply it by **C**. Is $A^{-1}C = B$?

7.4
LEONTIEF'S PRIZE-WINNING MATRIX

An application of matrix algebra won the Nobel Prize in Economics for Wassily Leontief in 1973. He developed a matrix method, called an **input-output model**, for describing the interaction between various segments of an economic system. We studied a special case of this model in Chapters 2 and 5 when we used systems of inequalities and equations to define resources and the weighted consumptions of these resources.

In his book *The Structure of American Economy, 1919-1939* (New York: Oxford University Press, 1951), Professor Leontief used matrices that represented as many as 46 segments of the economy. Since we don't relish the computation, by hand, of the inverse of a 46-by-46 matrix, we will confine our discussion here to greatly simplified economic systems that can be described as two-by-two and three-by-three matrices. These will be adequate to illustrate the main points.

In the Leontief model, the input-output matrix is a square matrix in which each column and each row represent a segment of the economy. The same segments in the same order are represented by both the rows and the columns. An entry in the ith row and jth column stands for the amount of the product of the ith industry consumed by the jth industry. In other words, a_{ij} = that part of i's output consumed by j. (A different matrix would be needed to represent that part of j's input produced by i.)

Example 7.21 From one of Leontief's charts for 1939, we extract the following table

Input \ Output	Agriculture and Fishing	Ferrous Metals	Nonmetallic Minerals	Construction
Agriculture and Fishing	0.076	0	0	0.008
Ferrous metals	0.006	0.305	0	0.152
Nonmetallic Minerals	0.005	0.010	0.101	0.530
Construction	0.026	0.004	0.002	0

The number 0.010 in the third row and second column (a_{32}) is the fraction of the third industry's output consumed by the second industry (the nonmetallic minerals consumed by the ferrous metals industry). Similarly, entry $a_{41} = 0.026$ means that 2.6 percent of the construction industry's output is consumed by the agriculture and fishing industry, and so on.

In general, a Leontief matrix reveals important information about an economic system; for example, it tells us the *overall cost* to the rest of the system in producing one unit of a given good. Furthermore, if we know the *gross product* (such as GNP) for an economic system, then we can compute the net product by subtracting the overall cost from the gross product.

Example 7.22 We will show how to find the overall cost to the system given in Example 7.21. First, we need to write the table of that system as a matrix:

$$\mathbf{M} = \begin{array}{c} \\ \text{AG} \\ \text{FE} \\ \text{NM} \\ \text{CON} \end{array} \begin{array}{cccc} \text{AG} & \text{FE} & \text{NM} & \text{CON} \\ \begin{bmatrix} 0.076 & 0 & 0 & 0.008 \\ 0.006 & 0.305 & 0 & 0.152 \\ 0.005 & 0.010 & 0.101 & 0.530 \\ 0.026 & 0.004 & 0.002 & 0 \end{bmatrix} \end{array}.$$

We have made the obvious abbreviations (AG for the agriculture and fishing industry, etc.). What is the cost to the NM in the (gross) production of 2000 units of AG? 3000 units of FE? 10,000 units of NM? and 15,000 units of CON? To answer this, we observe that AG consumes 0.005 units of NM for each unit (of AG) produced; so, if AG produces 2000 units, it uses up $(2000)(0.005) = 10$ units of NM. Similarly, FE consumes 0.010 units of NM per unit of FE produced; so, in the production of 3000 units, FE consumes $(3000)(0.010) = 30$ units of NM.

The production of 10,000 units of NM consumes $(10,000)(0.101)$ or 1010 units of NM; and the production of 15,000 units of CON consumes $(15,000)(0.530) = 7950$ units of NM. The total number of units of NM consumed in the gross production is $10 + 30 + 1010 + 7950 = 9000$. This result is actually the dot product of the two vectors

$$(0.005, 0.010, 0.101, 0.530) \begin{bmatrix} 2000 \\ 3000 \\ 10000 \\ 15000 \end{bmatrix} = 9000.$$

$\mathbf{r}_3$ = third row of $\mathbf{M}$, rate of consumption of nonmetals by the other industries.

System's gross production of all items, $\mathbf{X}$.

Cost to the nonmetals in producing the gross product vector $\mathbf{X}$.

7.4 Leontief's Prize-Winning Matrix

Now, if we obtain the product of the entire matrix **M**, not just row r_3, times the gross product vector **X**, then we will have the cost that **X** causes to each of the industries:

$$\begin{bmatrix} 0.076 & 0 & 0 & 0.008 \\ 0.006 & 0.305 & 0 & 0.152 \\ 0.005 & 0.010 & 0.101 & 0.530 \\ 0.026 & 0.004 & 0.002 & 0 \end{bmatrix} \begin{bmatrix} 2000 \\ 3000 \\ 10000 \\ 15000 \end{bmatrix} = \begin{bmatrix} 272 \\ 3207 \\ 9000 \\ 84 \end{bmatrix}$$

M
Input-output matrix model for the system

X
Gross product vector

MX
Cost vector (the cost to the entire system in producing **X**)

To compute the *net production* for the entire economic system, we subtract the cost vector **MX** from the gross product vector **X**. In symbols, let **Y** denote the net production. Then, $\mathbf{Y} = \mathbf{X} - \mathbf{MX}$, or $\mathbf{Y} = (\mathbf{I} - \mathbf{M})\mathbf{X}$.

Example 7.23 What is the net production of the gross product **X** in the model **M**? (Assume the same **X** and **M** in Example 7.22.)

Solution For Example 7.22,

$$\mathbf{MX} = \begin{bmatrix} 272 \\ 3207 \\ 9000 \\ 84 \end{bmatrix},$$

and the net product **Y** is defined by $\mathbf{Y} = \mathbf{X} - \mathbf{MX}$; therefore,

$$\mathbf{Y} = \begin{bmatrix} 2000 \\ 3000 \\ 10000 \\ 15000 \end{bmatrix} - \begin{bmatrix} 272 \\ 3207 \\ 9000 \\ 84 \end{bmatrix} = \begin{bmatrix} 1728 \\ -207 \\ 1000 \\ 14916 \end{bmatrix}.$$

Notice the negative entry -207 in the second row (ferrous metals) of the net product **Y**; it shows that in producing **X** the system **M** consumed more ferrous metals than it produced. It depended upon an outside source for this material (much as certain economies, such as Japan and Holland, depend upon outside sources for their petroleum). An economic society could decide that such dependence is intolerable and it may wish to achieve a gross product that will yield a net product with only positive entries. In this and other cases, it is necessary to know what gross product vector **X** is required in order to maintain a certain net product **Y**. This is equivalent to solving the equation $\mathbf{X} - \mathbf{MX} = \mathbf{Y}$, or $(\mathbf{I} - \mathbf{M})\mathbf{X} = \mathbf{Y}$ for **X**, given a known net product **Y**. Solving the equation for **X**, yields $\mathbf{X} = (\mathbf{I} - \mathbf{M})^{-1} \mathbf{Y}$.

Example 7.24 Assume an economy is based on two products, food and fuel. Let its input-output matrix **M** be

$$\mathbf{M} = \begin{matrix} & \text{Food} & \text{Fuel} \\ & \begin{bmatrix} 0.12 & 0.07 \\ 0.15 & 0.03 \end{bmatrix} & \begin{matrix} \text{Food} \\ \text{Fuel} \end{matrix} \end{matrix}$$

If the system is to survive and be self-sufficient, it must have a net production $\mathbf{Y} = (1800, 1000)$; that is, 1800 units of food and 1000 units of fuel. What gross product **X** must it produce in order to achieve its net production goal?

Solution The formula for net production is $(\mathbf{I} - \mathbf{M}) \mathbf{X} = \mathbf{Y}$; therefore, to find the gross product, we write $\mathbf{X} = (\mathbf{I} - \mathbf{M})^{-1} \mathbf{Y}$. This means we need to find $\mathbf{I} - \mathbf{M}$ and its inverse $(\mathbf{I} - \mathbf{M})^{-1}$:

$$\mathbf{I} - \mathbf{M} = \begin{bmatrix} 1 & 0 \\ 0 & 1 \end{bmatrix} - \begin{bmatrix} 0.12 & 0.07 \\ 0.15 & 0.03 \end{bmatrix} = \begin{bmatrix} 0.88 & -0.07 \\ -0.15 & 0.97 \end{bmatrix}$$

and

$$(\mathbf{I} - \mathbf{M})^{-1} = \begin{bmatrix} 0.97/0.843 & 0.07/0.843 \\ 0.15/0.843 & 0.88/0.843 \end{bmatrix} = \begin{bmatrix} 1.15 & 0.08 \\ 0.18 & 1.04 \end{bmatrix}.$$

The entries are only approximate, since the determinant of $(\mathbf{I} - \mathbf{M})$ is rounded off at 0.843 and the figures for $0.97/0.843$ etc., are also rounded off. We now find **X** to be

$$\mathbf{X} = \begin{bmatrix} 1.15 & 0.08 \\ 0.18 & 1.04 \end{bmatrix} \begin{bmatrix} 1800 \\ 1000 \end{bmatrix} = \begin{bmatrix} 2150 \\ 1364 \end{bmatrix}.$$

The required gross product is 2150 units of food and 1364 units of fuel. This will yield, by the matrix **M**, the survival level net product of approximately 1800 units of food and 1000 units of fuel.

A check is to compute the net product **Y** from the computed gross product **X**:

$$\mathbf{X} - \mathbf{MX} = \begin{bmatrix} 2150 \\ 1364 \end{bmatrix} - \begin{bmatrix} 0.12 & 0.07 \\ 0.15 & 0.03 \end{bmatrix} \begin{bmatrix} 2150 \\ 1364 \end{bmatrix} = \begin{bmatrix} 1796.5 \\ 1000.6 \end{bmatrix} \approx \mathbf{Y}. \quad \blacktriangle$$

7.4 Leontief's Prize-Winning Matrix

EXERCISE 7.4

▲ For Problems 1–4 use the following chart, which depicts a six-segment economy (modified from one of Leontief's examples).

Input \ Output	Food	Minerals and Metals	Energy	Textiles	Foreign Trade	Government
Food	—	—	—	0.11	0.23	0.01
Minerals and Metals	0.05	0.12	0.04	0.25	0.37	0.17
Energy	0.03	0.10	—	0.04	0.25	0.08
Textiles	0.06	0.07	0.12	0.08	0.20	0.35
Foreign Trade	0.06	0.10	0.01	0.06	—	0.01
Government	0.07	0.02	0.02	—	—	0.16

1. What fraction of energy production was consumed by the government?
2. What fraction of the minerals produced was consumed by foreign trade?
3. If the gross product vector (in millions of units) is

$$\mathbf{X} = \begin{bmatrix} 30 \\ 10 \\ 15 \\ 12 \\ 7 \\ 15 \end{bmatrix} \begin{matrix} \text{Food} \\ \text{Minerals} \\ \text{Energy} \\ \text{Textiles} \\ \text{Foreign Trade} \\ \text{Government} \end{matrix}$$

find the cost to the energy industry of producing **X**.

4. For the gross product vector **X** in Problem 3, find the cost of **X** to the food industry.

For Problems 5–9 use the following two-segment economic model:

$$\begin{matrix} & \text{Food} & \text{Clothing} \\ \text{Food} & \begin{bmatrix} 0.35 & 0.15 \\ 0.06 & 0.01 \end{bmatrix} & = \mathbf{M}. \\ \text{Clothing} & & \end{matrix}$$

5. If the gross product is

$$X = \begin{bmatrix} 3500 \\ 2500 \end{bmatrix} \begin{matrix} \text{Food} \\ \text{Clothing} \end{matrix},$$

  a. Find the cost vector **M X**.

  b. Find the net product **X − M X**.

6. If the gross product is

$$U = \begin{bmatrix} 1500 \\ 100 \end{bmatrix} \begin{matrix} \text{Food} \\ \text{Clothing} \end{matrix},$$

  a. Find the cost vector **M U**.

  b. Find the net production vector **U − M U**.

7. Show that

$$\mathbf{I - M} = \begin{bmatrix} 0.65 & -0.15 \\ -0.06 & 0.99 \end{bmatrix} \text{ and } (\mathbf{I - M})^{-1} = \begin{bmatrix} 1.560 & 0.236 \\ 0.095 & 1.024 \end{bmatrix}.$$

8. a. For the net product **Y**,

$$Y = \begin{bmatrix} -300 \\ 2600 \end{bmatrix},$$

  find the gross product **X** required to produce this net product.

  b. Check your answer in part **a** by $(\mathbf{I - M})\mathbf{X} = \mathbf{Y}$.

9. a. For the net product **Y**,

$$Y = \begin{bmatrix} 1000 \\ 500 \end{bmatrix},$$

  find the gross produce **X** required to produce **Y**.

  b. Check your answer by the equation $(\mathbf{I - M})\mathbf{X} = \mathbf{Y}$.

8
GAUSS-JORDAN ELIMINATION

8.1
INTRODUCTION

The determinant and matrix methods discussed in preceding chapters are excellent for solving *square*, or n-by-n (n is a positive integer), systems of linear equations—n equations and n variables. But if we want to solve certain complex problems (such as those from linear programming) that have more variables than equations, we need a method that can be applied to *nonsquare*, n-by-$(n + k)$ systems—n equations and $n + k$ variables.

Recall that the use of slack variables in a linear programming problem creates a system in which there are more variables than equations. In Chapter 5, Introduction to Linear Programming, we simply wrote the n-by-$(n + k)$ system as several n-by-n systems of *basic variables*. That is, for n equations in $n + k$ variables, we let k of the variables be *nonbasic* by setting them equal to zero; then for each basic system, we found the solution and determined the value of the objective function.

The above procedure virtually ignores the nonbasic variables while the basic system is being solved, and it treats each basic system independently. A more efficient method would be one that keeps track of the coefficients of the nonbasic variables while solving the basic system. This method would permit us to compare the basic variables to the nonbasic ones and determine which of the nonbasic variables is likely to increase or decrease the value of the objective function.

We are going to introduce a method that will do this and will reduce the amount of calculation needed to solve the various basic systems.

This new method is called **Gauss-Jordan elimination** (Carl Friedrich Gauss, 1777-1855, and Camille Jordan, 1838-1922). In this section we will study its application to square systems, two-by-two, and three-by-three systems; in a later section we will work with nonsquare systems in preparation for the simplex method of linear programming.

In this process, a matrix equation such as

$$\begin{bmatrix} a_{11} & a_{12} & a_{13} & a_{14} \\ a_{21} & a_{22} & a_{23} & a_{24} \\ a_{31} & a_{32} & a_{33} & a_{34} \\ a_{41} & a_{42} & a_{43} & a_{44} \end{bmatrix} \begin{bmatrix} x_1 \\ x_2 \\ x_3 \\ x_4 \end{bmatrix} = \begin{bmatrix} b_1 \\ b_2 \\ b_3 \\ b_4 \end{bmatrix}$$

is gradually transformed into the equation

$$\begin{bmatrix} 1 & 0 & 0 & 0 \\ 0 & 1 & 0 & 0 \\ 0 & 0 & 1 & 0 \\ 0 & 0 & 0 & 1 \end{bmatrix} \begin{bmatrix} x_1 \\ x_2 \\ x_3 \\ x_4 \end{bmatrix} = \begin{bmatrix} c_1 \\ c_2 \\ c_3 \\ c_4 \end{bmatrix}.$$

The sequence of steps used is called an **elimination** process because each equation in the system is successively reduced until it has only one variable left. (The others have been eliminated.)

A simple two-by-two example will illustrate the basic ideas.

Example 8.1 To solve the system

(I) $\qquad\qquad 3x + 5y = 2$

(II) $\qquad\qquad 2x - y = 4$ $\qquad\qquad$ (8.1)

by Gauss-Jordan elimination, we perform the operation in the two steps given below:

Step 1

a. In the first equation, make the x-coefficient $= 1$. Here, we divide Equation (I) by 3 (the coefficient of x); this gives us a new Equation (I):

(I) $\qquad\qquad x + \dfrac{5}{3} y = \dfrac{2}{3}.$

b. Eliminate the x term in the second equation (reduce the x-coefficient to zero in the second equation). This is accomplished here by subtracting 2 times the (new) Equation (I) from Equation (II); that is,

(I) $\qquad\qquad x + \dfrac{5}{3} y = \dfrac{2}{3}$

8.1 Introduction

(II − 2I) $\qquad 2x - 2x - y - 2 \cdot \left(\dfrac{5}{3}\right) y = 4 - 2 \cdot \left(\dfrac{2}{3}\right).$

Simplifying, we get two new Equations (I) and (II):

$$\left.\begin{array}{l} \text{(I)} \qquad x + \dfrac{5}{3} y = \dfrac{2}{3} \\[2mm] \text{(II)} \qquad -\dfrac{13}{3} y = \dfrac{8}{3}. \end{array}\right\} \quad (8.2)$$

Loosely speaking, we have eliminated x from the second equation. What we actually did, of course, was to replace the original system (8.1) by an equivalent system (8.2), in which the second equation has no x term.

The reason that system (8.2) is said to be equivalent to (8.1) is because both systems have the same solution. One was derived from the other by a set of reversible steps based on the properties of equals (division of both sides of an equation by a constant, and subtraction of two equations).

Step 2

a. In the second equation in system (8.2), make the y-coefficient equal to 1. This is done here by dividing both sides of Equation (II) by $-13/3$, which yields the new Equation (II):

(II) $\qquad y = -\dfrac{8}{13}.$

b. Eliminate the y term in the first equation; that is, reduce the y-coefficient to zero in Equation (I). Here, we subtract $5/3$ of the new Equation (II) from Equation (I).

$(I - \dfrac{5}{3} II) \qquad x + \dfrac{5}{3} y - \dfrac{5}{3} y = \dfrac{2}{3} - \dfrac{5}{3}\left(-\dfrac{8}{13}\right)$

(II) $\qquad y = -\dfrac{8}{13}.$

Simplifying, we get new Equations (I) and (II):

$$\left.\begin{array}{l} \text{(I)} \qquad x = \dfrac{22}{13} \\[2mm] \text{(II)} \qquad y = -\dfrac{8}{13}. \end{array}\right\} \quad (8.3)$$

The system (8.3) is the solution to system (8.1) obtained by elimination. The corresponding matrix equations for (8.1) and (8.3) clearly show the *elimination* that took place:

$$\begin{bmatrix} 3 & 5 \\ 2 & -1 \end{bmatrix} \cdot \begin{bmatrix} x \\ y \end{bmatrix} = \begin{bmatrix} 2 \\ 4 \end{bmatrix} \text{ was transformed to } \begin{bmatrix} 1 & 0 \\ 0 & 1 \end{bmatrix} \cdot \begin{bmatrix} x \\ y \end{bmatrix} = \begin{bmatrix} 22/13 \\ -8/13 \end{bmatrix}.$$

Note that if the system is one in which no solution exists, then during the elimination process, an "equation" will appear in which all the coefficients on the left side will be zero, but the constant on the right will not be zero.

Before stating the general case, we will examine some slightly more complex examples.

Example 8.2 Solve, by Gauss-Jordan elimination, the system

$$\left. \begin{array}{ll} \text{(I)} & x_1 + 2x_2 - x_3 = 5 \\ \text{(II)} & 6x_1 - 5x_2 + 3x_3 = 4 \\ \text{(III)} & -2x_1 + x_2 - x_3 = 3. \end{array} \right\} \quad (8.4)$$

Solution **Step 1**

a. In the first equation, make the x_1-coefficient equal to 1. Here it is already 1.

b. Eliminate the x_1 terms in Equations (II) and (III). This reduction is accomplished by subtracting 6 times Equation (I) from Equation (II) and then adding 2 times Equation (I) to Equation (III):

$$\begin{array}{ll} \text{(I)} & x_1 + 2x_2 - x_3 = 5 \\ \text{(II} - 6\text{I)} & 6x_1 - 6x_1 - 5x_2 - 12x_2 + 3x_3 + 6x_3 = 4 - 30 \\ \text{(III} + 2\text{I)} & -2x_1 + 2x_1 + x_2 + 4x_2 - x_3 - 2x_3 = 3 + 10. \end{array}$$

Simplifying, we get the new system of equations

$$\left. \begin{array}{ll} \text{(I)} & x_1 + 2x_2 - x_3 = 5 \\ \text{(II)} & -17x_2 + 9x_3 = -26 \\ \text{(III)} & 5x_2 - 3x_3 = 13. \end{array} \right\} \quad (8.5)$$

This system has no x_1 terms in Equations (II) and (III); thus, x_1 has been eliminated from the second and third equations. Systems (8.4) and (8.5) are equivalent: they have the same solution.

8.1 Introduction

Step 2

a. In the second equation, make the x_2-coefficient $= 1$. To do this, divide Equation (II) by the x_2-coefficient, -17; this yields the new Equation (II):

(II) $$x_2 - \frac{9}{17}x_3 = \frac{26}{17}.$$

b. Reduce the x_2-coefficients to zero in the first and third equations. Subtract 2 times Equation (II) from Equation (I) and subtract 5 times Equation (II) from Equation (III):

(I − 2II) $$x_1 + (2 - 2)x_2 + \left[-1 - 2\left(-\frac{9}{17}\right)\right]x_3 = 5 - 2\left(\frac{26}{17}\right)$$

(II) $$x_2 \qquad -\frac{9}{17}x_3 = \frac{26}{17}$$

(III − 5II) $$(5 - 5)x_2 + \left[-3 - 5\left(-\frac{9}{17}\right)\right]x_3 = 13 - 5\left(\frac{26}{17}\right).$$

Simplifying, we get the new system

(I) $$x_1 \qquad + \frac{1}{17}x_3 = \frac{33}{17}$$

(II) $$x_2 \qquad -\frac{9}{17}x_3 = \frac{26}{17}$$

(III) $$-\frac{6}{17}x_3 = \frac{91}{17}.$$

Step 3

a. In the third equation, make the x_3-coefficient $= 1$ (divide by $-6/17$):

(III) $$x_3 = -\frac{91}{6}.$$

b. Reduce the x_3-coefficient to zero in the first and second equations; that is, subtract $1/17$ times Equation (III) from Equation (I) and add $9/17$ times Equation (III) to Equation (II).

$\left(I - \dfrac{1}{17}III\right)$ $$x_1 \qquad + \left(\frac{1}{17} - \frac{1}{17}\right)x_3 = \frac{33}{17} - \frac{1}{17}\left(-\frac{91}{6}\right).$$

$$\left(\text{II} + \frac{9}{17}\text{III}\right) \qquad x_2 + \left(\frac{-9}{17} + \frac{9}{17}\right)x_3 = \frac{26}{17} + \frac{9}{17}\left(-\frac{91}{6}\right).$$

$$\text{(III)} \qquad\qquad\qquad\qquad\qquad x_3 = -\frac{91}{6}.$$

Simplifying, we get

$$\left.\begin{aligned}\text{(I)} &\qquad x_1 = \frac{17}{6}\\ \text{(II)} &\qquad x_2 = -\frac{13}{2}\\ \text{(III)} &\qquad x_3 = -\frac{91}{6}.\end{aligned}\right\} \quad (8.6)$$

The system (8.6) is the solution to system (8.4).

We can save some writing in the process if we express the successive systems in tabular form and perform the steps on the numbers in the table without carrying along the x_1's, x_2's, and x_3's.

Example 8.3 We write the system (8.4) of Example 8.2 in the form of a table as follows:

	Solution Vector	Coefficient Matrix			Constant Vector
	v	x_1	x_2	x_3	$\mathbf{x}_0$
(I)	x_1	1	2	-1	5
(II)	x_2	6	-5	3	4
(III)	x_3	-2	1	-1	3

The problem is to transform this to a table with a one in each diagonal position of the coefficient matrix and a zero in every other position of that matrix.

There is already a one in the first row, first column position of the diagonal; we now proceed to clear the rest of the first column. We must reduce the 6

8.1 Introduction

to 0 and reduce the -2 to 0, which means replace all of row (II) by a new row computed from the reduction (II $-$ 6I) and replace all of row (III) by a new row computed from the reduction (III $+$ 2I). This produces the new table:

	v	x_1	x_2	x_3	x_0
(I)	x_1	1	2	-1	5
(II)	x_2	0	-17	9	-26
(III)	x_3	0	5	-3	13

Next we want a one in the second row, second column diagonal position. Divide row (II) in the last table by -17; this gives us a new row (II):

(I)	1	2	-1	5
(II)	0	1	$-9/17$	$26/17$
(III)	0	5	-3	13

Now, we clear the rest of column 2 by the row operations: subtract 2 times row (II) from row (I) and subtract 5 times row (II) from row (III), getting a new table:

(I)	1	0	$1/17$	$33/17$
(II)	0	1	$-9/17$	$26/17$
(III)	0	0	$-6/17$	$91/17$

Continuing in a similar manner,

(I)	1	0	$1/17$	$33/17$
(II)	0	1	$-9/17$	$26/17$
(III)	0	0	1	$-91/6$

	v	x_1	x_2	x_3	x_0
(I − 1/17 III)	x_1	1	0	0	17/6
(II + 9/17 III)	x_2	0	1	0	−13/2
(III)	x_3	0	0	1	−91/6

Sometimes the coefficient of the first variable in the first equation is zero; in such a case, the first equation should be interchanged with one of the other equations in which the x_1-coefficient is not zero. The solution of the system is not affected by interchanging equations.

Example 8.4 Solve the system

$$2x_2 - x_3 = 1$$
$$4x_1 + 6x_3 = 5$$
$$x_1 + 3x_2 + x_3 = 2.$$

Solution Here, the coefficient of x_1 in the first equation is zero. We simply interchange the first and second equations, getting the same system with the equations rearranged:

$$\left. \begin{aligned} 4x_1 + 6x_3 &= 5 \\ 2x_2 - x_3 &= 1 \\ x_1 + 3x_2 + x_3 &= 2. \end{aligned} \right\} \quad (8.7)$$

We set up the table for system (8.7) and solve:

	v	x_1	x_2	x_3	x_0
(I)	x_1	4	0	6	5
(II)	x_2	0	2	−1	1
(III)	x_3	1	3	1	2

Divide (I) by 4:

8.1 Introduction

$$\begin{array}{c} \text{(I)} \\ \text{(II)} \\ \text{(III)} \end{array} \left[\begin{array}{ccc|c} 1 & 0 & 3/2 & 5/4 \\ 0 & 2 & -1 & 1 \\ 1 & 3 & 1 & 2 \end{array}\right]$$

Replace (III) by (III − I):

$$\begin{array}{c} \text{(I)} \\ \text{(II)} \\ \text{(III)} \end{array} \left[\begin{array}{ccc|c} 1 & 0 & 3/2 & 5/4 \\ 0 & 2 & -1 & 1 \\ 0 & 3 & -1/2 & 3/4 \end{array}\right]$$

Divide (II) by 2:

$$\begin{array}{c} \text{(I)} \\ \text{(II)} \\ \text{(III)} \end{array} \left[\begin{array}{ccc|c} 1 & 0 & 3/2 & 5/4 \\ 0 & 1 & -1/2 & 1/2 \\ 0 & 3 & -1/2 & 3/4 \end{array}\right]$$

Replace (III) by (III − 3II):

$$\begin{array}{c} \text{(I)} \\ \text{(II)} \\ \text{(III)} \end{array} \left[\begin{array}{ccc|c} 1 & 0 & 3/2 & 5/4 \\ 0 & 1 & -1/2 & 1/2 \\ 0 & 0 & 1 & -3/4 \end{array}\right]$$

Replace (I) by $[I - (3/2)\,III]$ and (II) by $[II + (1/2)III]$:

$$\begin{array}{c} \text{(I)} \\ \text{(II)} \\ \text{(III)} \end{array} \left[\begin{array}{ccc|c} 1 & 0 & 0 & 19/8 \\ 0 & 1 & 0 & 1/8 \\ 0 & 0 & 1 & -3/4 \end{array}\right]$$

The solution is $v = (x_1, x_2, x_3) = (^{19}/_8, \ ^1/_8, \ -^3/_4)$.

SUMMARY AND GENERAL PROCEDURE

Suppose the n-by-n system

$$a_{11} x_1 + a_{12} x_2 + \ldots + a_{1n} x_n = b_1$$
$$a_{21} x_1 + a_{22} x_2 + \ldots + a_{2n} x_n = b_2$$
$$\vdots \qquad \vdots \qquad \qquad \vdots \qquad \vdots$$
$$a_{n1} x_1 + a_{n2} x_2 + \ldots + a_{nn} x_n = b_n$$

is arranged so that the leading coefficient $a_{11} \neq 0$.

Step 1

a. Divide the first equation by a_{11}. This makes the x_1-coefficient $= 1$.

b. Reduce the x_1-coefficients to zero in all the other equations. Multiply the new first equation (the one obtained in step a) by a_{k1} and subtract from the kth equation; this produces the replacement for the kth equation. This new kth equation will not have an x_1 term.

Step 2

a. In the second equation, make the x_2-coefficient $= 1$.
Note that if the x_2-coefficient is zero in the second equation, find another equation below the second one in which the x_2 coefficient is *not* zero, and interchange it with the second equation. If the only nonzero coefficient is in the first equation, then there will be either no solution or no unique solution. If all the x_2-coefficients are zero, including the one in the first equation, then the system is not n-by-n.

b. Reduce the x_2-coefficients to zero in all other equations.

Step k

a. In the kth equation, make the x_k-coefficient $= 1$. (If the x_k coefficient is zero in the kth and all subsequent equations, see the note in Step 2a.)

b. Reduce the x_k-coefficient to zero in all other equations.

EXERCISE 8.1

1. Solve for x, y, and z by reducing the z column elements in rows (I) and (II) to zero.

	v	x	y	z	b
(I)	x	1	0	2/3	7
(II)	y	0	1	3/4	9
(III)	z	0	0	1	8

2. Solve for x_1, x_2, x_3, and x_4 by reducing the x_4 column elements in rows (I), (II), and (III) to zero.

	v	x_1	x_2	x_3	x_4	x_0
(I)	x_1	1	0	0	2	0
(II)	x_2	0	1	0	−1/5	2
(III)	x_3	0	0	1	2/3	5
(IV)	x_4	0	0	0	1	−2

3. Solve by Gauss-Jordan elimination:

v	x	y	z	b
x	0	2	1	5
y	4	1	3	0
z	−1	6	5	1

4. Solve by Gauss-Jordan elimination:

v	x	y	z	b
x	0	2	5	3
y	2	0	6	9
z	1	2	0	1

5. Use Gauss-Jordan elimination to solve:

v	x_1	x_2	x_3	x_4	x_0
x_1	2	1	8	8	19
x_2	2	4	20	20	40
x_3	4	3	24	22	49
x_4	-1	$1/2$	1	$5/2$	$5/2$

6. Use Gauss-Jordan elimination to solve:

v	x_1	x_2	x_3	x_4	x_0
x_1	3	-2	1	1	7
x_2	2	6	3	0	-1
x_3	2	-6	1	5	14
x_4	1	2	1	1	1

▲ 7. A linear learning model has three *weights* (coefficients relating dependent and independent variables in the model), c_1, c_2, and c_3. Suppose these weights are correlated to a test score T by the linear equation $T = c_1 p + c_2 q + c_3 r$. In three trials, the following data were obtained:

$$T = 3 \text{ when } (p, q, r) = (1, 0, 7)$$
$$T = 5 \text{ when } (p, q, r) = (2, 1, 1)$$
$$T = 4 \text{ when } (p, q, r) = (1, 3, 2).$$

  a. Write the information given above as a system of three equations in the three weights, c_1, c_2, and c_3.
  b. Solve the system for c_1, c_2, and c_3.
  c. Find the test score T when $(p, q, r) = (1, 1, 1)$.

8. Three types of products, A, B, and C, are produced in numbers x_1, x_2, and x_3 (x_1 units of product A, x_2 units of B, and x_3 units of C are produced). Each unit of A produced uses 2 units of resource I, 5 units of resource II, and 3 units of resource III. (These data have been recorded in the A column of the table below.)

x_1 Units of Product A	x_2 Units of Product B	x_3 Units of Product C	
2			Resource I
5			Resource II
3			Resource III

  a. If each unit of product B consumes 4 units of resource I, 1 unit of resource II, and *produces* 1 unit (consumes -1 unit) of resource III, fill in the B column.
  b. If each unit of product C consumes 1 unit of resource I, 3 units of resource II, and 5 units of resource III, fill in the C column.
  c. If there is a total of 125 units of resource I, 145 units of resource II, and 135 units of resource III, write a system of linear equations relating these data to the data in the table.
  d. Solve the resulting system by Gauss-Jordan elimination.

9. A warehouse has 550 cartons stored in one corner. The cartons are of different sizes:

x of them are 1 ft by 2 ft by 3 ft (six cubic feet)

y of them are 2 ft by 2 ft by 2 ft (eight cubic feet)

z of them are 1 ft by 3 ft by 3 ft (nine cubic feet)

The total volume of all the cartons is 4025 cubic feet. After 75 of the six-cubic-feet cartons are removed, the stock clerk notices that there are twice as many of the six-cubic-feet cartons as the eight-cubic-feet cartons left.

a. If $x - 75$ is the number of six-cubic-feet cartons left, write an equation relating this to the number y.

b. Complete the following equations:

$$x + y + z = ?$$
$$6x + 8y + 9z = ?$$
$$x - 2y \quad\quad = ?$$

c. Write the above system of equations in tabular form:

v	x	y	z	b
x				
y				
z				

d. Solve the above system by Gauss-Jordan elimination. ▲

8.2

SOLVING FOR SYSTEMS OF BASIC VARIABLES

If an *m*-by-*n* system of equations has more variables than equations, we may apply the Gauss-Jordan process and solve the system for a set of basic variables.

8.2 Solving for Systems of Basic Variables

Example 8.5 Solve by Gauss-Jordan elimination for some two basic variables:

$$\left. \begin{array}{r} x_1 + 6x_2 + x_3 = 0 \\ 4x_2 + x_3 = -\dfrac{1}{4} \end{array} \right\} \quad (8.8)$$

Solution In the two-equation, three-variable system (8.8), we make one of the variables *nonbasic*, leaving the other two basic, as was done in solving the linear programming problems in Chapter 5. Start by writing the equations of system (8.8) in a tabular form as follows:

Table 8.1

	v	x_1	x_2	x_3	b
(I)		1	6	1	0
(II)		0	4	1	$-1/4$

How should we fill in the column **v**? This column is the vector of the variables in the system, and its components will represent the components of the solution to the system; so, if we decide which two of the three variables to make basic, then we can put those two variables into the **v** column and solve the table for those variables.

But what does it mean to say we can solve the tabulated system for the basic variables? Simply that we transform the coefficient matrix for the two basic variables into the identity matrix, while continuing to carry the nonbasic variables in the table.

Let us choose x_1 and x_3 as the basic variables, then we can fill in the **v** column of Table 8.1 and get

Table 8.2

	v	x_1	x_2	x_3	b
(I)	x_1	1	6	1	0
(II)	x_3	0	4	1	$-1/4$

Now we need to change the coefficient matrix for the basic variables x_1 and x_3. That is, we will transform the matrix by Gauss-Jordan elimination.

$$\begin{array}{c} \\ x_1 \\ x_3 \end{array} \begin{array}{cc} x_1 & x_3 \\ \begin{bmatrix} 1 & 1 \\ 0 & 1 \end{bmatrix} \end{array} \text{ will become the identity matrix } \begin{array}{c} \\ x_1 \\ x_3 \end{array} \begin{array}{cc} x_1 & x_3 \\ \begin{bmatrix} 1 & 0 \\ 0 & 1 \end{bmatrix} \end{array}. \qquad (8.9)$$

The first matrix in (8.9) above comes from the x_1 and x_3 columns of Table 8.2. The variable x_2 keeps its place in the table, but we look to only the basic variables x_1 and x_3 for the solution.

Please note that every calculation (multiplication, addition, etc.) that we do on a row must be done to *every variable* in that row, not just to the basic variables. The nonbasic variables will get the same treatment as the basic ones. This keeps the equations true for the nonbasic variables, so that if we want to make a nonbasic variable basic later, the equations will still be valid for it.

We now apply the steps of the Gauss-Jordan elimination process to Table 8.2. The coefficient of x_1 is already 1 in row (I) and x_1 is already eliminated in row (II). So we proceed to the second step, part of which is to make the coefficient of the second basic variable (here it is x_3) equal to 1, but this is also already the case. Now for the rest of Step 2, we will reduce the x_3-coefficient to zero in row (I). This is accomplished by decreasing row (I) by one times row (II); we get

Table 8.3

	v	x_1	x_2	x_3	b
(I)	x_1	1	2	0	$1/4$
(II)	x_3	0	4	1	$-1/4$

The coefficient matrix for x_1 and x_3 has been changed into the identity matrix. This means that we have solved the system (8.8) for x_1 and x_3 as basic variables and x_2 as nonbasic ($x_2 = 0$). Table 8.3 shows that this solution is $x_1 = 1/4$, $x_2 = 0$, and $x_3 = -1/4$.

If we wish to consider a different system of basic variables, we can now exchange one of them for the nonbasic one.

8.2 Solving for Systems of Basic Variables

Example 8.6 Solve the system in Table 8.3, but with x_2 and x_3 as the basic variables.

Solution We replace the variables in the v column of Table 8.3 by x_2 and x_3:

Table 8.4

	v	x_1	x_2	x_3	b
(I)	x_2	1	2	0	1/4
(II)	x_3	0	4	1	−1/4

Now we need to solve this system for the basic variables x_2 and x_3; that is, we will change the coefficient matrix of Table 8.4 from

$$\begin{array}{c} \\ x_2 \\ x_3 \end{array} \begin{array}{cc} x_2 & x_3 \\ \begin{bmatrix} 2 & 0 \\ 4 & 1 \end{bmatrix} \end{array} \text{ to the identity matrix } \begin{array}{c} \\ x_2 \\ x_3 \end{array} \begin{array}{cc} x_2 & x_3 \\ \begin{bmatrix} 1 & 0 \\ 0 & 1 \end{bmatrix} \end{array}.$$

We bring about this change by the Gauss-Jordan elimination process, as follows. Multiply row (I) by $\frac{1}{2}$ (to make the coefficient of x_2 equal to one):

	v	x_1	x_2	x_3	b
(I)	x_2	1/2	1	0	1/8
(II)	x_3	0	4	1	−1/4

Now replace row (II) by one in which the coefficient of x_2 is zero; that is, eliminate x_2 in row (II). This is accomplished by subtracting four times row (I) from row (II).

	v	x_1	x_2	x_3	b
(I)	x_2	1/2	1	0	1/8
(II)	x_3	−2	0	1	−3/4

This table gives us the solution with x_2 and x_3 as the basic variables and x_1 as nonbasic. (You can tell what the basic variables are by looking at the components in the V column. You can tell that the system is solved for its basic variables because the coefficient matrix for these variables is the identity matrix.) The solution is $x_1 = 0$, $x_2 = \frac{1}{8}$, and $x_3 = -\frac{3}{4}$.

Notice that in all of these solutions, the nonbasic variable had its coefficient treated in the same way as those of the basic variables. This method of keeping track of the nonbasic variables prepares them for entry into the system without having to start all over again.

Let us consider a larger example.

Example 8.7

Table 8.5

	v	x_1	x_2	x_3	x_4	x_5	x_6	b
(I)	x_2	3	1	2	0	0	8	2
(II)	x_4	4	0	-2	1	0	10	1
(III)	x_5	1	0	3	0	1	6	4

a. What are the basic variables in this table?

b. Is the system solved for its basic variables?

c. Write the solution represented by this table.

d. Change the system into one in which the basic variables are x_4, x_5, and x_6.

e. Solve the system for x_4, x_5, and x_6.

Solution

a. From the column v, we see that the basic variables are x_2, x_4, and x_5.

b. The system is solved for these basic variables since their coefficient matrix is the identity matrix

$$\begin{array}{c} \\ x_2 \\ x_4 \\ x_5 \end{array} \begin{array}{ccc} x_2 & x_4 & x_5 \\ \left[\begin{array}{ccc} 1 & 0 & 0 \\ 0 & 1 & 0 \\ 0 & 0 & 1 \end{array}\right] \end{array}.$$

8.2 Solving for Systems of Basic Variables

c. The solution represented by the table is

   Basic variables: $x_2 = 2$, $x_4 = 1$, $x_5 = 4$.

   Nonbasic variables: $x_1 = x_3 = x_6 = 0$.

d. To change the system from one with x_2, x_4, and x_5 as basic variables to one with x_4, x_5, and x_6 as basic means that x_6 replaces x_2 in the set of basic variables.

Therefore, we must relabel row (I) (the former x_2 row) as the x_6 row. Then to keep the subscripts in numerical order, x_4, x_5, x_6, we put that row at the bottom of the table. Now Table 8.5 becomes:

Table 8.6

	v	x_1	x_2	x_3	x_4	x_5	x_6	b
(I)	x_4	4	0	−2	1	0	10	1
(II)	x_5	1	0	3	0	1	6	4
(III)	x_6	3	1	2	0	0	8	2

Call the rows in Table 8.6 "new" and the ones in Table 8.5 "old." Notice that new row (I), labeled x_4, is the old row (II), also labeled x_4. The new row (II), labeled x_5, is old row (III), also x_5; but the new row (III), labeled x_6, is old row (I), labeled x_2.

In general, to exchange a basic variable, we replace the old basic variable in the **v** column by the new basic variable and rearrange the rows to maintain subscript order.

Table 8.6 is a system whose basic variables are x_4, x_5, and x_6, but it is not yet solved for these variables since the coefficient matrix

$$\begin{array}{c} \\ x_4 \\ x_5 \\ x_6 \end{array} \begin{array}{c} x_4 \; x_5 \; x_6 \\ \begin{bmatrix} 1 & 0 & 10 \\ 0 & 1 & 6 \\ 0 & 0 & 8 \end{bmatrix} \end{array} \quad (8.10)$$

is not the identity matrix.

We can apply Gauss-Jordan elimination to change the matrix in expression (8.10) into the identity matrix, and we will thereby solve Table 8.6 for its basic variables. In system (8.10) we need to work on only the x_6 column. Returning to Table 8.6, we divide row (III) by 8 and get

	v	x_1	x_2	x_3	x_4	x_5	x_6	b
(I)	x_4	4	0	-2	1	0	10	1
(II)	x_5	1	0	3	0	1	6	4
(III)	x_6	3/8	1/8	1/4	0	0	1	1/4

Now we need to eliminate the x_6 coefficients 10 and 6 from rows (I) and (II) by replacing these two rows, respectively, by (I $-$ 10III) and (II $-$ 6III). This yields the following table:

Table 8.7

	v	x_1	x_2	x_3	x_4	x_5	x_6	b
(I)	x_4	1/4	$-5/4$	$-9/2$	1	0	0	$-3/2$
(II)	x_5	$-5/4$	$-3/4$	3/2	0	1	0	5/2
(III)	x_6	3/8	1/8	1/4	0	0	1	1/4

In this table the basic variables have the identity matrix as their coefficient matrix; therefore, the system is solved for these basic variables, and the solution is

$$\text{Basic variables: } x_4 = \frac{-3}{2}, \ x_5 = \frac{5}{2}, \ x_6 = \frac{1}{4}.$$

Nonbasic variables: $x_1 = x_2 = x_3 = 0$.

In each of the above examples, the coefficient matrix for the basic variables could be changed into the identity matrix. What happens if this is impossible? For example, the coefficient matrix for the chosen basic variables might be one

8.2 Solving for Systems of Basic Variables

in which some step of the Gauss-Jordan process cannot be performed. In such cases, we must abandon the particular set of basic variables and claim that the solution fails to exist for them.

Example 8.8 Show that the following system does not have a solution with x_1 and x_3 as the basic variables:

$$x_1 + 6x_2 + 2x_3 = 5$$
$$2x_1 + 10x_2 + 4x_3 = 7.$$

Solution We put the system into a table with x_1 and x_3 as the basic variables and try to solve it by Gauss-Jordan elimination.

	v	x_1	x_2	x_3	b
(I)	x_1	1	6	2	5
(II)	x_3	2	10	4	7

Now to make the x_1-coefficient zero in row (II), we replace (II) by (II − 2I), getting the following table:

	v	x_1	x_2	x_3	b
(I)	x_1	1	6	2	5
(II)	x_3	0	−2	0	−3

The next step in the process would have been to divide row (II) by the x_3-coefficient in that row, but since the coefficient is zero, this step is impossible. If we had another row below row (II), we might interchange rows, but this is also impossible. The process cannot be applied to the system for this set of basic variables.

In fact, the coefficient matrix for x_1 and x_3 cannot be transformed into the identity matrix by any method. This matrix is *singular* (its determinant is zero; see Chapter 7). The system has no solution for x_1 and x_3.

EXERCISE 8.2

Use the Gauss-Jordan elimination process in solving these problems.

1. Solve
$$2x_1 - 6x_2 + x_3 = 4$$
$$3x_1 + x_2 - 2x_3 = 7$$
with x_1 and x_2 as the basic variables.

2. Solve
$$x_1 - 2x_2 + 3x_3 = 0$$
$$-2x_1 + x_2 + x_3 = 1$$
with x_1 and x_3 as the basic variables.

3. In Problem 1, let x_2 and x_3 be the basic variables and solve.
4. In Problem 2, let x_1 and x_2 be the basic variables and solve.
5. For the system in the following table:

	v	x_1	x_2	x_3	x_4	b
(I)	x_3	2	1	6	1	2
(II)	x_4	3	0	4	7	1

   a. What are the basic variables?
   b. Is this system solved for its basic variables?
   c. Solve for the basic variables and write out the complete solution.
   d. Change the basic variables to x_2 and x_3 and solve.

8.2 Solving for Systems of Basic Variables

6. For the system in the following table:

	v	x_1	x_2	x_3	x_4	b
(I)	x_1	1	0	2	3	6
(II)	x_2	4	1	6	1	5

a. What are the basic variables?

b. Is this system solved for its basic variables?

c. Solve and write out the complete solution for these basic variables.

d. Change the basic variables to x_2 and x_4 and solve.

7. a. How do you know that the following system is solved for x_1, x_3, and x_4?

b. Change it to a system with basic variables x_1, x_2, and x_4 and solve.

	v	x_1	x_2	x_3	x_4	x_5	b
	x_1	1	4	0	0	3	0
	x_3	0	-2	1	0	7	2
	x_4	0	5	0	1	-2	1

8. Change the system in Table 8.7 to one in which the basic variables are x_1, x_5, and x_6 and solve.

▲ **9.** A department store can choose from five brands of pocket calculators: A, B, C, D, and E. Let x_1 = the number of Brand A that can be sold, x_2 = the number of Brand B that can be sold, x_3 = the number of Brand C that can be sold, x_4 = the number of Brand D that can be sold, and x_5 = the number of Brand E that can be sold. Suppose that the equations describing sales to two markets (over the counter and through the mail) are

$$3x_1 + 2x_2 + x_3 + 2x_4 + 5x_5 = 76$$

$$6x_1 + x_2 - 4x_3 + 5x_4 + x_5 = 80.$$

a. Suppose that the manager wants to sell only two different brands of calculators at the same time and decides to sell brands A and E. Find the number of each sold.

b. Suppose there is a decision to change to brands A and B. How many of each are sold? (Round off to the nearest whole number.)

10. Use the equations in Problem 9.

   a. If only brands C and D are sold, how many of each are sold? (Round off.)

   b. If the brands sold are changed from C and D to A and C, how many of each are sold? (Round off.) ▲

11. Show that the system:

v	x_1	x_2	x_3	x_4	x_5	b
	1	3	6	4	5	10
	2	6	18	1	15	11

   a. Has no solution for the basic variables x_1, and x_2.

   b. Has a solution for the basic variables x_1 and x_3. Solve for these variables.

12. Show that the system in Problem 11:

   a. Has no solution for the basic variables x_3 and x_5

   b. Has a solution for x_4 and x_5 as basic. Find this solution.

9
THE SIMPLEX METHOD FOR LINEAR PROGRAMMING

9.1

THE INCOMING VARIABLE

We are now ready to introduce a systematic method for solving linear programming problems. The initial stage, called **tabulation of the problem,** is to put all the constraint inequalities and the objective function into a table.

Let us also define the **basis** of a linear programming problem.

Definition (Basis)
The basic variables of a system are called the *basis*. Any variable that is basic is said to be in the basis.

Example 9.1 Suppose we wish to find the maximum (or minimum) of the function

$$z = 3x_1 - 2x_2 + x_3 \tag{9.1}$$

subject to the constraints

$$\left. \begin{array}{l} x_1 + x_2 + x_3 \leq 100 \\ 2x_1 + 5x_2 - 4x_3 \leq 15. \end{array} \right\} \tag{9.2}$$

The constraint inequalities in system (9.2) can be converted to slack variable equations as we did in Section 2.2. We also introduce slack variables into the objective function in Equation (9.1) as we did in Section 5.4. Here, if we let x_4 and x_5 be the slack variables, then we can write the objective function as

$$z = 3x_1 - 2x_2 + x_3 + 0x_4 + 0x_5$$

and the constraints as

(I) $\qquad x_1 + x_2 + x_3 + x_4 \qquad = 100$
(II) $\qquad 2x_1 + 5x_2 - 4x_3 \qquad + x_5 = 15.$

Now let us assume that we wish to put this data in a table with x_3 and x_5 as the basis. Then we can write Table 9.1 below. The top row labeled C_i is the

Table 9.1

C_i	3	−2	1	0	0

	V	x_1	x_2	x_3	x_4	x_5	x_0	C_j
(I)	x_3	1	1	1	1	0	100	1
(II)	x_5	2	5	−4	0	1	15	0

set of coefficients of the objective function. The column C_j on the extreme right is the set of the objective function coefficients for those variables in the basis. The V column (as in Chapter 8) contains the basis.

Solving Table 9.1 by Gauss-Jordan elimination, we get Table 9.2.

Notice that Table 9.2 has one additional part, namely the value of z for this basis.

Table 9.2

C_i	3	−2	1	0	0

	V	x_1	x_2	x_3	x_4	x_5	x_0	C_j
(I)	x_3	1	1	1	1	0	100	1
(II)	x_5	6	9	0	4	1	415	0

z

100

9.1 The Incoming Variable

This value can be computed only after the solution for the basis has been obtained. If we think of the x_0 and C_j columns as vectors (letting x_0 be written as a row vector and C_j as a column), then we get z as the dot product of x_0 and C_j:

$$z = (100, 415) \cdot \begin{bmatrix} 1 \\ 0 \end{bmatrix} = 100 \cdot 1 + 415 \cdot 0 = 100.$$

For simplicity, compute the products of the two columns directly and add as follows:

x_0		C_j		
100	·	1	=	100
415	·	0	=	$\dfrac{+0}{100}$ = z.

We will also compute the dot product for the columns x_1, x_2, x_3, x_4, and x_5 with C_j in this same direct manner:

x_1		C_j		
1	·	1	=	1
6	·	0	=	$\dfrac{+0}{1} = z_1$.

x_2		C_j		
1	·	1	=	1
9	·	0	=	$\dfrac{+0}{1} = z_2$.

(9.3)

x_3		C_j		
1	·	1	=	1
0	·	0	=	$\dfrac{+0}{1} = z_3$.

$$\begin{bmatrix} x_4 \\ 1 \\ 4 \end{bmatrix} \cdot \begin{bmatrix} C_j \\ 1 \\ 0 \end{bmatrix} = \frac{1+0}{1} = z_4.$$

$$\begin{bmatrix} x_5 \\ 0 \\ 1 \end{bmatrix} \cdot \begin{bmatrix} C_j \\ 1 \\ 0 \end{bmatrix} = \frac{0+0}{0} = z_5.$$

Each of these dot products, z_1, z_2, z_3, z_4, and z_5, is a number called the **opportunity cost** generated by the variables x_1, x_2, x_3, x_4, and x_5, respectively. If x_i is a given basic variable, z_i is the per-unit contribution that x_i makes to the objective function. For a nonbasic variable, the opportunity cost is also a measure of the per-unit contribution but it is a contribution from which the current system derives no benefit (so long as the variable remains nonbasic). This verbal description is necessarily awkward because the opportunity cost is a mathematical concept. A mathematical explanation that clearly shows the role played by the z_i's for both basic and nonbasic variables will be given in Theorem 9.1 at the end of this section.

For each nonbasic variable x_i, the coefficient C_i would be the per-unit contribution to the objective function were that variable to be brought into the system. If, for some x_i, the objective function's coefficient C_i exceeds its opportunity cost z_i (that is, $C_i > z_i$), then there is more to be gained than lost by bringing x_i into the basis. The difference $C_i - z_i$ is the **potential** (per-unit) **net contribution** of a given variable. Thus, if $C_i - z_i$ is positive for a specific variable x_i, then that variable has a potential to increase the value z of the objective function. If $C_i - z_i$ is negative, then x_i has a potential to decrease the value of z.

The way z_i is computed insures that the basic variables will have a potential net contribution of zero. (For the basic variables $C_i = z_i$.) That is, there is no potential for either increasing or decreasing the value of the objective function by bringing into the basis a variable that is already there.

We will pursue the practical aspects of z_i and $C_i - z_i$ after we complete the tabulation of the linear programming problem by introducing two new rows, z_i and $C_i - z_i$, into the table.

From the dot products in Equations (9.3), we get

9.1 The Incoming Variable

z_i	1	1	1	1	0

From rows C_i in Table 9.2 and z_i above, we get the row

$C_i - z_i$	2	−3	0	−1	0

Adding these two rows to Table 9.2, we get Table 9.3, the complete linear programming table for the systems (9.1) and (9.2) with x_3 and x_5 as the basis.

Table 9.3

C_i	3	−2	1	0	0

V	x_1	x_2	x_3	x_4	x_5	x_0	C_j
x_3	1	1	1	1	0	100	1
x_5	6	9	0	4	1	415	0

						z
z_i	1	1	1	1	0	100
$C_i - z_i$	2	−3	0	−1	0	

What is the significance of $C_i - z_i$ in Table 9.3? It tells us which nonbasic variable should be brought into the basis. It can be used to determine the best incoming variable. In maximizing problems, this will be the variable with the *greatest positive* potential net contribution (largest $C_i - z_i$). In minimizing problems, it will be the variable with the *most negative* potential net contribution.

Example 9.2 Use Table 9.3 to determine the best choice of the incoming variable:

  a. If the problem is to maximize the objective function.

  b. If the problem is to minimize the objective function.

Solution a. Here the best variable to bring in is one that has the greatest (positive) potential net contribution to the objective function, $C_i - z_i$. Since $C_1 - z_1$ is the largest, then x_1 is the best incoming variable for the maximizing problem.

  b. In the minimizing problem we want to bring into the basis the variable for which $C_i - z_i$ is the *most negative*. In this case, $C_2 - z_2 = -3$; so the potential net contribution is most negative for the x_2 column. This means that we want to make x_2 basic since it will most likely decrease the value of the objective function.

The above discussion tells how to select the best incoming variable. In the next section we will introduce a method for determining which basic variable will be the best one to replace (that is, the best outgoing variable). For the rest of the examples in this section and in Exercise 9.1, the outgoing variable will be given in the statement of the problem.

Example 9.3 Suppose we wish to maximize the objective function for the problem in Table 9.3. Let x_1 be the incoming variable and x_5 the outgoing variable.

  a. What is the new basis?

  b. Solve for this new basis.

  c. Show that this solution is feasible.

  d. Show that no other variable can increase the value of the objective function.

Solution a. Since the old basis in Table 9.3 is x_3 and x_5 and since x_1 is incoming and x_5 is outgoing, then this replacement of x_5 by x_1 means that the new basis is x_1, x_3.

  b. Label the old x_5 row as the x_1 row and arrange the rows in numerical order of the subscripts. The new table is shown in Table 9.4.

  In Table 9.4 we dropped the z_i and $C_i - z_i$ rows because we can not compute their entries until the table is solved for the new set of basic variables. Notice that the entries in the C_j column in Table 9.4 correspond to the coefficients (in the objective function) of the new basis x_1, x_3.

  Before solving Table 9.4, and to save some space in determining the solution,

9.1 The Incoming Variable

Table 9.4

C_i	3	−2	1	0	0		

V	x_1	x_2	x_3	x_4	x_5	x_0	C_j
(I) x_1	6	9	0	4	1	415	3
(II) x_3	1	1	1	1	0	100	1

we will introduce an abbreviated table, called a **tableau**. The tableau is the body of the table corresponding to the constraints. The tableau for Table 9.4 is

(I)	6	9	0	4	1	415
(II)	1	1	1	1	0	100

Solving for x_1 and x_3 by Gauss-Jordan elimination, first we divide row (I) by 6 and replace (I) by $[(1/6)I]$:

(I)	1	3/2	0	2/3	1/6	415/6
(II)	1	1	1	1	0	100

Then replace row (II) by (II − I):

(I)	1	3/2	0	2/3	1/6	415/6
(II)	0	−1/2	1	1/3	−1/6	185/6

This last tableau shows that the basic solution is $x_1 = 415/6$ and $x_3 = 185/6$ (the nonbasic variables are $x_2 = x_4 = x_5 = 0$).

c. The solution $(x_1, x_2, x_3, x_4, x_5) = (415/6, 0, 185/6, 0, 0)$ is feasible since none of the components is negative (see Chapter 5, Sections 2 and 3).

d. To show that no other variable can increase the objective function, we need the z_i and $C_i - z_i$ rows; see Table 9.5. Notice that all the nonbasic variables have a negative value of $C_i - z_i$; this means that they can only decrease the objective function by being brought into the basis. (In general, if there is no positive potential net contribution, then there can be no increase by changing the basis.) Since this is the case here, we have found the maximum value of the objective function. It is $z = 715/3$ (or $238 1/3$).

Table 9.5

C_i	3	−2	1	0	0		
V	x_1	x_2	x_3	x_4	x_5	x_0	C_i
x_1	1	3/2	0	2/3	1/6	415/6	3
x_3	0	−1/2	1	1/3	−1/6	185/6	1
						z	
z_i	3	4	1	7/3	1/3	715/3	
$C_i - z_i$	0	−6	0	−7/3	−1/3		

Now we consider a minimizing problem.

Example 9.4 Suppose we want to minimize the objective function for the problem in Table 9.3. Let x_2 be the incoming variable. (This was determined to be the best incoming variable for the minimizing problem in Example 9.2.) Let x_3 be the outgoing variable.

a. What is the new basis?

b. Solve for the new basis.

c. Show that this solution is *not feasible*.

9.1 The Incoming Variable

Solution   a. The new basis is x_2, x_5.

b. The solution for this basis is in the following tableau:

V	x_1	x_2	x_3	x_4	x_5	x_0
x_2	1	1	1	1	0	415
x_5	-3	0	-9	-5	1	-485

That is, the solution is $(x_1, x_2, x_3, x_4, x_5) = (0, 415, 0, 0, -485)$.

c. The above solution is not feasible because it has a negative component, $x_5 = -485$. Recall that for a solution to be feasible (that is, a corner point), none of its components can be negative.

We can still consider x_2 as the candidate for entering into the basis, but we will have to replace the other basic variable, x_5.

Example 9.5   Try again to minimize the objective function in Table 9.3 by bringing in x_2; let x_5 be the outgoing variable.

a. What is the new basis?

b. Solve for this new basis.

c. Is the solution feasible?

d. Show that no other variable can decrease the value of the objective function.

Solution   a. The new basis is x_2, x_3.

b. The solution for this new basis is in the tableau.

V	x_1	x_2	x_3	x_4	x_5	x_0
x_2	2/3	1	0	4/9	1/9	415/9
x_3	1/3	0	1	5/9	-1/9	485/9

c. The solution is feasible: $(x_1, x_2, x_3, x_4, x_5) = (0, {}^{415}/_9, {}^{485}/_9, 0, 0)$. No component is negative.

d. To show that none of the nonbasic variables can decrease the value of the objective function, we construct Table 9.6. The $C_i - z_i$ row (the potential net contribution to the objective) contains no negative values. In fact, for every nonbasic variable x_i, the value of $C_i - z_i$ is positive, which means that any of these variables brought into the basis could only increase the value of z. Therefore, $z = {}^{-345}/_9$ is the minimum.

Table 9.6

C_i	3	−2	1	0	0

V	x_1	x_2	x_3	x_4	x_5	x_0	C_j
x_2	2/3	1	0	4/9	1/9	415/9	−2
x_3	1/3	0	1	5/9	−1/9	485/9	1

						z
z_i	−1	−2	1	−1/3	−1/3	−345/9
$C_i - z_i$	4	0	0	1/3	1/3	

PROOF THAT THE INCOMING VARIABLE IMPROVES THE SOLUTION

We now state a theorem that justifies the method outlined above for choosing the incoming variable. What this theorem says, is that if x_k is the incoming nonbasic variable and z_k is computed for it, then the value of the objective function in the new basis will be increased or decreased according to whether $C_k - z_k$ is positive or negative.

Theorem 9.1 Let m and n be positive integers with $m > n$ and suppose that a tabulated n by m system is solved for its basis with the feasible solution being

9.1 The Incoming Variable

$$\begin{bmatrix} x_1 \\ x_2 \\ \vdots \\ x_n \end{bmatrix} = \begin{bmatrix} b_1 \\ b_2 \\ \vdots \\ b_n \end{bmatrix} \quad \text{each } b_i \geqq 0.$$

Let P_1 be the value of the objective function for this basis. Denote the entries in the column of a nonbasic variable x_k as

$$\begin{bmatrix} a_{k1} \\ a_{k2} \\ \vdots \\ a_{kn} \end{bmatrix}$$

Suppose x_k and x_n are variables such that the replacement of x_n by x_k produces a new system in which the new basis $(x_1, x_2, \ldots, x_{n-1}, x_k)$ is feasible, and let P_2 be the value of the objective function in this new system.
Then,

$$P_2 = P_1 + \left(\frac{b_n}{a_{kn}}\right)(C_k - z_k).$$

Proof First we note that this theorem is general since the basic and nonbasic variables can be rearranged as necessary so that the subscripts will correspond to those in the hypotheses of the theorem.

By definition

$$z_k = \sum_{i=1}^{n} C_i a_{ki}.$$

The value P_1, of the objective function in the first system is

$$P_1 = \sum_{i=1}^{n} C_i b_i.$$

When x_n is replaced by x_k and the new system is solved (by Gauss-Jordan elimination), the new feasible solution is

$$x_1 = b_1 - (a_{k1})\left(\frac{b_n}{a_{kn}}\right)$$

$$x_2 = b_2 - (a_{k2})\left(\frac{b_n}{a_{kn}}\right)$$

$$x_i = b_i - (a_{ki})\left(\frac{b_n}{a_{kn}}\right)$$
$$\vdots$$
$$x_{n-1} = b_{n-1} - (a_{k\,n-1})\left(\frac{b_n}{a_{kn}}\right)$$
$$x_k = \frac{b_n}{a_{kn}}.$$

We use this solution to compute P_2, the value of the objective function in the new system:

$$P_2 = \sum_{i=1}^{n-1} C_i\left(b_i - a_{ki}\left(\frac{b_n}{a_{kn}}\right)\right) + C_k\left(\frac{b_n}{a_{kn}}\right)$$

or

$$P_2 = \sum_{i=1}^{n-1} C_i b_i - \left(\frac{b_n}{a_{kn}}\right)\sum_{i=1}^{n-1} C_i a_{ki} + C_k\left(\frac{b_n}{a_{kn}}\right).$$

By the definitions of z_k and P_1 we can replace the sums in the above equation as follows:

$$\sum_{i=1}^{n-1} C_i b_i = P_1 - C_n b_n \quad \text{and} \quad \sum_{i=1}^{n-1} C_i a_{ki} = z_k - C_n a_{kn}.$$

So now,

$$P_2 = P_1 - C_n b_n - \left(\frac{b_n}{a_{kn}}\right)(z_k - C_n a_{kn}) + C_k\left(\frac{b_n}{a_{kn}}\right).$$

After some simplifying algebraic manipulations, we get

$$P_2 = P_1 + \left(\frac{b_n}{a_{kn}}\right)(C_k - z_k).$$

This completes the proof.

The following corollaries are immediate consequences.

Corollary 1 If $C_k - z_k > 0$, the value of the objective function is increased by replacing x_n by x_k. (That is, $P_2 > P_1$.)

Corollary 2 If $C_k - z_k < 0$, the value of the objective function is decreased by replacing x_n by x_k. (That is, $P_2 < P_1$.)

9.1 The Incoming Variable

Corollary 3 If $C_k - z_k = 0$, the value of the objective function is unchanged by replacing x_n by x_k. (That is, $P_2 = P_1$.)

EXERCISE 9.1

In the following problems, assume that all variables are nonnegative.

1. For the table given here:

C_i	4	10	3	2	2

V	x_1	x_2	x_3	x_4	x_5	x_0	C_j
x_1	1	2	0	1	4	15	
x_3	0	3	1	-1	5	22	

		z
z_i		
$C_i - z_i$		

a. Fill in the C_j column.

b. Compute z, z_i, and $C_i - z_i$; fill in the rows.

c. Assume this is a *maximizing* problem and determine the best nonbasic variable to bring into the basis.

d. Assume this is a *minimizing* problem and determine the best nonbasic variable to bring into the basis.

2. Repeat the instructions in parts **a**, **b**, **c**, and **d** of Problem 1 for the following table.

C_i	3	1	5	4

V	x_1	x_2	x_3	x_4	x_0	C_j
x_1	1	0	1/2	−3	18	
x_2	0	1	1	4	37	

		z
z_i		
$C_i - z_i$		

3. Repeat parts **a, b, c,** and **d** of Problem 1 for the following table.

C_i	15	1	3	12	0	0

V	x_1	x_2	x_3	x_4	x_5	x_6	x_0	C_j
x_2	1	1	0	2	0	6	10	
x_3	3	0	1	2	0	−1	64	
x_5	2	0	0	−3	1	7	23	

		z
z_i		
$C_i - z_i$		

4. Repeat parts **a, b, c,** and **d** of Problem 1 for the following table.

9.1 The Incoming Variable

C_i	2	3	4	0	0	0

V	x_1	x_2	x_3	x_4	x_5	x_6	x_0	C_j
x_4	2	5	4	1	0	0	10	
x_5	3	7	1	0	1	0	15	
x_6	1	2	5	0	0	1	20	

		z
z_i		
$C_i - z_i$		

5. Given the system in the following table:

C_i	4	10	3	2	2

V	x_1	x_2	x_3	x_4	x_5	x_0	C_j
x_3	0	3	1	−1	5	22	3
x_4	1	2	0	1	4	15	2

a. Solve the table for its basis x_3, x_4.

b. Compute z, z_i, and $C_i - z_i$ for the solution.

c. If this is a maximizing problem, show that the maximum is reached with this basis.

6. Given the system:

C_i	4	10	3	2	2		

V	x_1	x_2	x_3	x_4	x_5	x_0	C_j
x_3	0	3	1	−1	5	22	3
x_5	1	2	0	1	4	15	2

a. Solve this table for its basis x_3, x_5.

b. Compute z, z_i, and $C_i - z_i$ for this solution.

c. If this is a minimizing problem, show that the minimum has been reached with this basis.

9.2

THE OUTGOING VARIABLE

In the preceding section, we found the best incoming variable to be one for which the potential net effect on the objective function was to increase it for maximizing problems and to decrease it for minimizing problems. Now we need to find the best outgoing variable.

It would be nice if we could identify one outgoing variable for maximizing problems and a different one for minimizing. Coupled with the technique for getting the incoming variable, this could give us a twofold attack on the maximization or minimization of the objective. Unfortunately, this is not possible. All we can do is determine which variable replaced by the incoming variable will make the (new) basis feasible.

Of course, replacement of the outgoing variable by the incoming variable still may improve the objective function; that was why we picked that particular incoming variable in the first place. It turns out that *the variable exchange that maintains feasibility is independent of whether the problem is one of maximization or minimization.*

The search for the best outgoing variable is actually a search for the variable whose replacement results in a feasible basis. We will state a rule for picking the replacement variable, but first, we need to describe a small calculation.

9.2 The Outgoing Variable

For each *positive* element p in the column corresponding to the incoming variable, compute the quotient b/p, where b is the element in the x_0 column and in the same row as p. All the quotients b/p will be nonnegative. If one is zero, no improvement from this incoming variable is possible. If all the quotients are positive, find the smallest one. If several are equal to the smallest, select any of these.

The row in which the smallest nonnegative quotient b/p ($p > 0$) occurs is headed by a variable in the basis. Replacement of this variable by the incoming variable will guarantee a feasible solution for the next basis.

The quotients (b/p) described above are called **replacement quotients.** p is in the column for the incoming variable, $p > 0$; b is in the x_0 column and in the same row as p. The rule for determining the outgoing variable is as follows.

REPLACEMENT RULE

Given a system solved for a feasible basis, compute the replacement quotients, b/p. If any $b/p = 0$, with $p > 0$, then no improvement from an incoming variable is possible; stop. If all b/p are positive (when $p > 0$), pick the row with the smallest positive value and replace the basic variable heading that row.

Notes on this rule

1. Since $b \geq 0$ and b/p is computed for only $p > 0$, then no quotient b/p can be negative. If you accidentally ignore the signs of the elements in the incoming column and compute b/p and find it is negative, then p is negative. The basic variable of this row *cannot* be replaced by the incoming variable. To do so would make the next basis infeasible.

2. If $b/p = 0$ with $p > 0$, then no improvement is possible, *but if $b/p = 0$ with $p < 0$ and some other quotient b_i/p_i is positive, then replacement of the variable in the row with the smallest such positive quotient will improve the solution.*

3. If two or more rows are tied for the smallest quotient, pick one of these rows at random.

4. If all the entries in the column for the incoming variable are negative, then the incoming variable either cannot improve the objective function or cannot produce a feasible solution by replacing one of the current basic variables.

The examples below illustrate the replacement rule. At the end of this section, we will give a proof that replacement as described above will always produce a new feasible basis.

Example 9.6 Suppose Table 9.7 represents the constraint equations in a maximizing or minimizing problem. Let x_5 be the incoming variable as found in Section 9.1.

Table 9.7

	V	x_1	x_2	x_3	x_4	x_5	x_6	x_0
(I)	x_1	1	0	0	0	5	2	15
(II)	x_2	0	1	0	0	−6	7	12
(III)	x_3	0	0	1	0	1	−9	11
(IV)	x_4	0	0	0	1	−2	5	0

This table is solved for its basis and its solution is feasible. Notice that for the incoming column, x_5, the entry in row (II) is negative (−6); therefore, we need not compute the quotient for this row. The variable of this row x_2 should not be replaced; to do so would result in a table with an infeasible basis. (Verify this for yourself.) In row (I), the x_1 row, the number in the incoming column is 5 and the corresponding number in the x_0 column is 15; therefore the replacement quotient (b/p) is $15/5 = 3$. For the x_3 row, the replacement quotient is $11/1$. The smallest nonnegative* replacement quotient (for a positive entry in the x_5 column) is found in the x_1 row; so x_1 is the variable to be replaced by x_5. Replacement of x_3 would have resulted in an infeasible basis, and replacement of x_4 would result in a feasible basis, but the objective function would remain the same.

The replacement rule requires that the the table be solved for a feasible basis. If a table is not solved for its basis, solve it! It may take several trials to get the first feasible solution, but it is necessary.

We now state another example.

Example 9.7 Suppose that in Table 9.7 the incoming variable had been x_6 (instead of x_5). Which basic variable should be replaced?

Solution Refer back to Table 9.7. The four replacement quotients are

for x_1, $15/2 = 7.5$;
for x_2, $12/7 = 1.714 \ldots$;
for x_3, $11/(-9) =$ negative, do not replace;
for x_4, $0/5 = 0$, smallest nonnegative result.

---

*Although for row x_4, the quotient $0/(-2) = 0$ is nonnegative, the denominator is not positive. Remember, we are selecting the *smallest nonnegative replacement quotient computed from positive entries in the column for the incoming variable.*

9.2 The Outgoing Variable

The quotient for x_4 comes from division by a positive entry in the column for the incoming variable, x_6. The x_4 row is the outgoing row; that is, x_4 is the variable to be replaced by x_6. The result will be a new basis that is feasible, but one that will not improve the value of the objective function. The replacement of any other row will result in an infeasible basis.

Example 9.8 Let Table 9.8 be for a maximizing problem. It is solved for its current basis and the solution is feasible.

Table 9.8

C_i	3	5	1	−2	0	0

V	x_1	x_2	x_3	x_4	x_5	x_6	$\mathbf{x}_0$	C_j
x_3	2	3	1	−1	0	0	22	1
x_5	1	−1	0	7	1	0	15	0
x_6	−1	4	0	3	0	1	6	0

a. Does this solution make the objective function maximum?

b. What variable should be brought into the system to improve the solution?

c. For the incoming variable found in part **b**, what is the outgoing variable?

d. Solve for the new basis.

e. If the new basis does not make the objective function maximum (there is another nonbasic variable with a positive net potential), repeat steps **b**, **c**, and **d** until the maximum is attained.

Solution a. We need to compute z, z_i, and $C_i - z_i$ for this basis. The objective function is not yet at its maximum since two of the nonbasic variables have a positive net potential: $C_1 - z_1 = 1$ and $C_2 - z_2 = 2$.

z_i	2	3	1	−1	0	0	22
$C_i - z_i$	1	2	0	−1	0	0	

b. Select x_2 as the incoming variable since its positive net potential is the largest.

c. For the incoming variable x_2, the three replacement quotients are

for x_3, $22/3 = 7.333 \ldots$;

for x_5, $15/(-1) =$ negative, do not replace x_5;

for x_6, $6/4 = 1.5$, the smallest positive replacement quotient.

Therefore x_6 is the outgoing variable.

d. The new basis is x_2, x_3, x_5. Keep rows x_3 and x_5 as in Table 9.8, relabel row x_6 as x_2, and put it above the other two rows. This gives the following tableau, not yet solved for its basis:

x_2	-1	4	0	3	0	1	6
x_3	2	3	1	-1	0	0	22
x_5	1	-1	0	7	1	0	15

Solving by the Gauss-Jordan elimination process, we get Table 9.9.

Table 9.9

C_i	3	5	1	-2	0	0		
V	x_1	x_2	x_3	x_4	x_5	x_6	x_0	C_i
x_2	$-1/4$	1	0	$3/4$	0	$1/4$	$3/2$	5
x_3	$11/4$	0	1	$-13/4$	0	$-3/4$	$35/2$	1
x_5	$3/4$	0	0	$31/4$	1	$1/4$	$33/2$	0

							z	
z_i	$3/2$	5	1	$1/2$	0	$1/2$	25	
$C_i - z_i$	$3/2$	0	0	$-5/2$	0	$-1/2$		

9.2 The Outgoing Variable

e. The new basis in Table 9.9 does not make the objective function maximum. The positive net potential $C_1 - z_1 = 3/2$ means that x_1 could improve the value of the objective function. Let x_1 be the incoming variable. The positive replacement quotients are

$$\text{for } x_3, \quad \frac{(35/2)}{(11/4)} = 6.36, \text{ approximately;}$$

$$\text{for } x_5, \quad \frac{(33/2)}{(3/4)} = 22.$$

The smaller one of these is for x_3; therefore, x_3 is the outgoing variable (to be replaced by x_1). The new basis is x_1, x_2, x_5; see Table 9.10. No variable has a positive net potential; therefore, the solution is maximum.

Table 9.10

C_i	3	5	1	-2	0	0		

V	x_1	x_2	x_3	x_4	x_5	x_6	x_0	C_j
x_1	1	0	$4/11$	$-13/11$	0	$-3/11$	$70/11$	3
x_2	0	1	$1/11$	$5/11$	0	$2/11$	$34/11$	5
x_5	0	0	$-3/11$	$95/11$	1	$5/11$	$129/11$	0

							z	
z_i	3	5	$17/11$	$-14/11$	0	$1/11$	$380/11$	
$C_i - z_i$	0	0	$-6/11$	$-8/11$	0	$-1/11$		

The remainder of this section is devoted to an explanation of why the replacement rule insures that the new basis will be feasible. (This material can be skipped without loss of continuity.)

Proof of the replacement rule Suppose a system has m equations and n variables (m and n are positive integers with $n > m$), and for convenience suppose the basis consists of the first m variables. (Any basic feasible system can be arranged so that this is true.) Also suppose that the system is solved for the feasible basis and that x_{m+1} is the incoming variable as determined in Section 9.1. (By rearrangement, the incoming variable can always be put into the x_{m+1} column.) Table 9.11 is the general form.

Table 9.11

C_i	c_1	c_2	...	c_m	c_{m+1}	...	c_n		
V	x_1	x_2	...	x_m	x_{m+1}	...	x_n	$\mathbf{x}_0$	C_j
x_1	1	0	...	0	p_1	...	a_{n1}	b_1	c_1
x_2	0	1	...	0	p_2	...	a_{n2}	b_2	c_2
⋮	⋮	⋮		⋮	⋮		⋮	⋮	⋮
x_m	0	0	...	1	p_m	...	a_{nm}	b_m	c_m

First, notice that all the replacement quotients get their *signs* from the x_{m+1} column; this is because all the entries in the $\mathbf{x}_0$ column are nonnegative. Let b_i/p_i be any negative replacement quotient ($b_i/p_i < 0$). Then since $b_i \geq 0$, $b_i/p_i < 0$ only if $p_i < 0$ and $b_i \neq 0$. (If $b_i = 0$, $b_i/p_i = 0$ but is not negative.)

Now we will show that replacing a basic variable that has a negative replacement quotient results in an infeasible solution. Suppose that $b_i/p_i < 0$ and let x_i be selected as the outgoing variable, then x_i would be replaced by x_{m+1} and the x_i row would be relabeled x_{m+1} and moved to the bottom of the table, as shown in Table 9.12.

The old x_i row is now the x_{m+1} row and the entry p_i (circled) is on the diagonal of the basic matrix. We need to convert the coefficient in that diagonal position to a 1. If we divide the entire x_{m+1} row by p_i, we get the 1 on the diagonal, but we get $b_i/p_i < 0$ in the $\mathbf{x}_0$ column. This makes the x_{m+1} component of the next solution negative. So the new basis is not feasible.

We will also use Table 9.12 to show why we need to pick the smallest nonnegative replacement quotient. Consider only the positive values p_k in the incoming column and assume that *none of the corresponding values b_k in the $\mathbf{x}_0$ column is zero.* (If, for some $p_k > 0$, the value $b_k = 0$, then x_k is the outgoing variable, but the new basis will not improve the solution.)

Now, suppose some variable (say, x_i again) has a positive replacement quotient but it is not the smallest. Say, x_j has a smaller one; that is,

9.2 The Outgoing Variable

$$0 < \frac{b_j}{p_j} < \frac{b_i}{p_i},$$

which may be written as

$$b_j < p_j \cdot \frac{b_i}{p_i}. \tag{9.4}$$

Table 9.12

Nonbasic Column ↓

V	x_1	x_2	... x_{i-1}	x_i	x_{i+1}	...	x_j	...	x_m	x_{m+1}	...	x_0
x_1	1	0	... 0	0	0	...	0	...	0	p_1	...	b_1
x_2	0	1	... 0	0	0	...	0	...	0	p_2	...	b_2
⋮	⋮	⋮	⋮	⋮	⋮		⋮		⋮	⋮		⋮
x_{i-1}	0	0	... 1	0	0	...	0	...	0	p_{i-1}	...	b_{i-1}
x_{i+1}	0	0	... 0	0	1	...	0	...	0	p_{i+1}	...	b_{i+1}
⋮	⋮	⋮	⋮	⋮	⋮		⋮		⋮	⋮		⋮
x_j	0	0	... 0	0	0	...	1	...	0	p_j	...	b_j
⋮	⋮	⋮	⋮	⋮	⋮		⋮		⋮	⋮		⋮
x_m	0	0	... 0	0	0	...	0	...	1	p_m	...	b_m
x_{m+1}	0	0	... 0	1	0	...	0	...	0	(p_i)	...	b_i

(Old x_i row) → x_{m+1}

Assume that you attempt to replace x_i by x_{m+1}. To solve Table 9.12, first divide the (new) row labeled x_{m+1} by the element in the diagonal, p_i, getting

V	x_1	x_2	...	x_j	...	x_{m+1}	...	x_0
x_1	1	0	...	0	...	p_1	...	b_1
x_2	0	1	...	0	...	p_2	...	b_2
⋮	⋮	⋮		⋮		⋮		⋮
x_j	0	0	...	1	...	p_j	...	b_j
⋮	⋮	⋮		⋮		⋮		⋮
x_{m+1}	0	0	...	0	...	1	...	b_i/p_i

Now when you try to reduce the value p_j in the x_j row to zero using Gauss-Jordan elimination, you get new x_j row = old x_j row minus p_j times the x_{m+1} row:

V	x_1	x_2	...	x_j	...	x_{m+1}	...	x_0
⋮	⋮	⋮		⋮		⋮		⋮
x_j	0	0	...	1	...	0	...	$b_j - p_j \cdot {}^{b_i}\!/\!_{p_i}$
⋮	⋮	⋮		⋮		⋮		⋮

The value in the x_0 column, $b_j - p_j({}^{b_i}\!/\!_{p_i})$, is negative by Inequality (9.4). This means that the x_j component of the new basic solution is negative, and the solution is therefore not feasible. This shows that x_j cannot be selected as the outgoing variable; that is, removing a variable for which there is a smaller positive replacement quotient results in an infeasible basis. What about variables for which the replacement quotient is zero?

If ${}^{b_i}\!/\!_{p_i} = 0$, then in the elimination of p_j in the x_{m+1} column and x_j row, the corresponding x_0 value $b_j - p_j({}^{b_i}\!/\!_{p_i}) = b_j$, which means the x_j component of the solution is unchanged. This is true for all the basic variables. The new variable will have a value of zero just as the one it replaces; therefore, bringing in a variable for which the replacement quotient is zero will lead to a feasible solution, but it will not improve the value of the objective function.

The only remaining case is to replace the variable for which the quotient ${}^{b_i}\!/\!_{p_i}$ is smallest (and positive). In this case, the x_{m+1} component (the new basic variable) of the solution will be positive (${}^{b_i}\!/\!_{p_i}$) and the other components x_j will be $b_j - p_j({}^{b_i}\!/\!_{p_i}) \geq 0$. This means that the resulting basis is feasible and has been changed in a way that the incoming variable can improve the value of the objective function. This justifies the replacement rule.

EXERCISE 9.2

1. In the following table, assume x_4 is the incoming variable. Find the outgoing variable and solve for the new basis.

9.2 The Outgoing Variable

V	x_1	x_2	x_3	x_4	x_5	x_6	x_0
x_1	1	0	0	0	2	3	5
x_2	0	1	0	4	-7	6	16
x_3	0	0	1	2	-1	9	9

2. In the table in Problem 1, assume x_6 is the incoming variable. Find the outgoing variable and solve for the new basis.

3. Suppose the following table represents a maximization problem.

C_i	8	39	-1	11	0

V	x_1	x_2	x_3	x_4	x_5	x_0	C_j
x_1	1	7	0	2	3	15	8
x_3	0	5	1	6	2	20	-1

    a. Find the value of the objective function for the current basis.

    b. Find the best incoming variable.

    c. Find the outgoing variable.

    d. Solve for the new basis and show that it provides the maximum value of the objective function.

4. Suppose the table in Problem 3 is for a minimization problem.

    a. Find the best incoming variable (to minimize the objective function).

    b. Find the outgoing variable.

    c. Solve for the new basis and show that it provides the minimum value of the objective function.

5. Minimize
$$z = 5x_1 + 2x_2$$
subject to the constraints
$$x_1 + x_2 \geq 15$$
$$5x_1 + x_2 \leq 35$$
$$x_1 + 2x_2 \geq 24$$
$$x_1 + 5x_2 \leq 75.$$

6. Maximize
$$z = 6x_1 + x_2$$
subject to the constraints given in Problem 5.

7. Maximize
$$z = 3x_1 + 5x_2 + 6x_3$$
subject to
$$4x_1 - 2x_2 + 6x_3 \leq 10$$
$$2x_1 + x_2 + 5x_3 \leq 18$$
$$x_1 + x_2 + x_3 \leq 50.$$

8. Minimize
$$z = 10x_1 + 4x_2 + 25x_3$$
subject to
$$x_1 + 5x_2 - x_3 \geq 20$$
$$2x_1 + x_2 - 6x_3 \geq 80$$
$$3x_1 + 2x_3 \leq 150.$$

9.3

OBTAINING AN INITIAL FEASIBLE SOLUTION

In Sections 9.1 and 9.2, we saw how to exchange a basic variable for a nonbasic variable. Such an exchange of variables is called **pivoting.** The nonbasic variable

9.3 Obtaining an Initial Feasible Solution

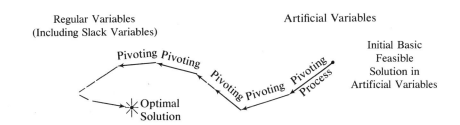

Schematic Idea of the Artificial Variables
Figure 9.1

is pivoted into the system and a basic one is pivoted out. The incoming column is usually called the *pivot column* and the outgoing row is the *pivot row*. The intersection of the pivot column and pivot row is called the *pivot position* (or simply *the pivot*).

Pivoting is the process of moving from one feasible basis to a better one (where "better" means one that increases or decreases the value of the objective function depending upon whether the problem is to maximize or minimize the objective). As we saw in the previous sections, pivoting is very efficient; it sure beats the trial and error methods of Chapter 5. It greatly reduces the number of basic systems we need to consider, but one thing it does *not* do is tell us how to get a feasible solution in the first place. (Pivoting always starts from a feasible solution.)

We will now see an ingenious method for getting a first feasible solution in any system. The method requires the introduction of variables called *artificial variables* into the system. These variables *immediately* provide a feasible basis whose components can be pivoted out of the system one at a time while the regular variables are pivoted into the system. (See Figure 9.1)

Example 9.9 Let the constraints for a given problem be

$$\left. \begin{array}{r} 3x_1 + 2x_2 - x_3 = 7 \\ -5x_2 - 2x_3 = -10. \end{array} \right\} \quad (9.5)$$

First, we *make all of the constant terms* (the right-hand sides of the equations) *nonnegative*. Multiplying the second equation on both sides by -1 gives

$$\left. \begin{array}{r} 3x_1 + 2x_2 - x_3 = 7 \\ 5x_2 + 2x_3 = 10. \end{array} \right\} \quad (9.6)$$

Next, we introduce two more variables, x_4 and x_5 (x_4 into the first equation and x_5 into the second); these are called **artificial variables** since they have no meaning in the original problem. (They are not slack variables and they are not variables representing the real quantities in the problem.)

$$\left. \begin{array}{l} 3x_1 - 2x_2 - x_3 + x_4 + 0x_5 = 7 \\ 5x_2 + 2x_3 + 0x_4 + x_5 = 10 \end{array} \right\} \quad (9.7)$$

The matrix of coefficients of these two new variables is the unit matrix

$$\begin{bmatrix} 1 & 0 \\ 0 & 1 \end{bmatrix}.$$

This means that the system is immediately solved with x_4 and x_5 as the basis. Furthermore, the fact that the right-hand sides of all of the equations in the system are positive means that this immediately available solution is *feasible* (all positive components).

The contributions made by the artificial variables to the objective function present a problem. What should the coefficients of these variables be in the objective function? We want to assign coefficients that will *not* affect the outcome in the maximization or minimization of the objective function when evaluated for the actual variables in the system.

The coefficients needed to keep the artificial variables from interfering with the actual solution are determined as follows:

1. If the objective is to be maximized, make the coefficients of the artificial variables in the objective function so small (extremely negative) that there will be no chance that they will be selected in obtaining the maximum of the objective.

2. If the objective is to minimized, make the coefficients of the artificial variables in the objective function so large (extremely large positive) that there will be no chance that they will be in the final basis furnishing the minimum.

MAXIMIZING PROBLEM

We now continue with our example. Suppose we wish to maximize the objective function $P = 2x_1 + 4x_2 + 3x_3$ subject to the constraints given in Equations (9.6) above. Then the linear programming problem (with artificial variables introduced) becomes:

Maximize the objective

$$P = 2x_1 + 4x_2 + 3x_3 - Mx_4 - Mx_5$$

subject to the constraints

9.3 Obtaining an Initial Feasible Solution

$$3x_1 + 2x_2 - x_3 + x_4 + 0x_5 = 7$$
$$0x_1 + 5x_2 + 2x_3 + 0x_4 + x_5 = 10.$$

Where M is a large positive number ($-M$ is extremely negative) compared to the coefficients of x_1, x_2, and x_3 in the objective function. (Say, for example, that M is ten million; then the coefficients of the artificial variables, x_4 and x_5, are both $-M$, or minus ten million.)

Let us construct a table for this problem using x_4 and x_5 as the basis. (See Table 9.13.) Notice that since the system is already solved for x_4 and x_5 (the coefficient matrix for these variables is already the identity) and since the solution is feasible (this is the reason for making the right-hand side positive before introducing the artificial variables), the opportunity cost z_i and the net potential $C_i - z_i$ could be computed immediately. That is, there is no need to use trial and error to find the first feasible basic solution.

Table 9.13

C_i	2	4	3	$-M$	$-M$		

V	x_1	x_2	x_3	x_4	x_5	x_0	C_j
x_4	3	2	-1	1	0	7	$-M$
x_5	0	⑤	2	0	1	10	$-M$

						z	
z_i	$-3M$	$-7M$	$-M$	$-M$	$-M$	$-17M$	
$C_i - z_i$	$2 + 3M$	$4 + 7M$	$3 + M$	0	0		

Examining the table, we find that the largest positive net potential $C_i - z_i$ is $4 + 7M$ in the x_2 column; therefore, we should pivot the variable x_2 into the system. Now find the replacement quotients for the x_2 column. We find that for x_4 the quotient is $7/2 = 3\frac{1}{2}$, while for x_5 it is $10/5 = 2$. Replacing the variable with the smallest quotient, we find that the new basis should be x_2 and x_4. The pivot 5 is in the column of the incoming variable x_2 and the row of the outgoing variable x_5. Dividing the x_5 row by the pivot, we get the new x_2 row. Put this together with the x_4 row and get

V	x_1	x_2	x_3	x_4	x_5	x_0
x_2	0	1	2/5	0	1/5	2
x_4	3	2	−1	1	0	7

If we solve for the new basis x_2, x_4, then Table 9.14 is the complete table.

Table 9.14

C_i	2	4	3	−M	−M		
V	x_1	x_2	x_3	x_4	x_5	x_0	C_j
x_2	0	1	2/5	0	1/5	2	4
x_4	③	0	−9/5	1	−2/5	3	−M
						z	
z_i	−3M	4	(8 + 9M)/5	−M	(4 + 2M)/5	8 − 3M	
$C_i - z_i$	2 + 3M	0	(7 − 9M)/5	0	(−4 − 7M)/5		

The x_1 column is the pivotal column (it has the largest positive value of $C_i - z_i$), and the x_4 row is the pivotal row (smallest positive replacement quotient). The pivot element 3 is circled. The basis for the next system will be x_1, x_2. Dividing the x_4 row by the 3 in the pivot position converts this row into the new x_1 row as in the following tableau.

x_1	1	0	−3/5	1/3	−2/15	1
x_2	0	1	2/5	0	1/5	2

This is, fortunately, already solved for the new basis; so our complete table is Table 9.15. Notice that the system has been pivoted into one in which the regular

9.3 Obtaining an Initial Feasible Solution

Table 9.15

C_i	2	4	3	$-M$	$-M$		
V	x_1	x_2	x_3	x_4	x_5	x_0	C_j
x_1	1	0	$-3/5$	$1/3$	$-2/15$	1	2
x_2	0	1	$2/5$	0	$1/5$	2	4
						z	
z_i	2	4	$2/5$	$2/3$	$8/15$	10	
$C_i - z_i$	0	0	$13/5$	$-M - 2/3$	$-M - 8/15$		

variables provide a feasible basis; this means that we can virtually ignore the artificial variables from now on. (At least we can round off the net potential each time for the artificial variables. For example, $-M - 2/3$ may be written simply as $-M$.) This is because M is so large and $-M$ is such an extremely negative number that if $C_i - z_i$, for an artificial variable, is $-M +$ some small amount, then we could never select that variable as incoming. Thus, an artificial variable cannot be pivoted back into the system.

Proceeding from Table 9.15, we apply the rules for determining the incoming and outgoing variables and we get the new basis x_1, x_3. Solving for this basis, we get Table 9.16.

Table 9.16

C_i	2	4	3	$-M$	$-M$		
V	x_1	x_2	x_3	x_4	x_5	x_0	C_j
x_1	1	$3/2$	0	$1/3$	$1/6$	4	2
x_3	0	$5/2$	1	0	$1/2$	5	3
						z	
z_i	2	$21/2$	3	$2/3$	$11/6$	23	
$C_i - z_i$	0	$-13/2$	0	$-M - 2/3$	$-M - 11/6$		

There is no positive net potential. The maximum profit is 23 obtained from the basic solution $(x_1, x_3) = (4,5)$.

MINIMIZING PROBLEM

Consider the same constraints as in Equation (9.5) and now suppose we wish to *minimize* the objective function $P = 2x_1 + 4x_2 + 3x_3$ subject to those constraints. We introduce the artificial variables and give them large positive coefficients, M, in the objective function. Thus we have the linear programming problem: Minimize the objective

$$P = 2x_1 + 4x_2 + 3x_3 + Mx_4 + Mx_5$$

subject to the constraints

$$3x_1 + 2x_2 - x_3 + x_4 + 0x_5 = 7$$
$$0x_1 + 5x_2 + 2x_3 + 0x_4 + x_5 = 10.$$

Since this is a minimizing problem, we want to bring in a variable that will reduce the objective function. In Table 9.17, we look for the *most negative value of the*

Table 9.17

C_i	2	4	3	M	M		
V	x_1	x_2	x_3	x_4	x_5	x_0	C_j
x_4	3	2	−1	1	0	7	M
x_5	0	⑤	2	0	1	10	M
z_i	$3M$	$7M$	M	M	M	$17M$ (z)	
$C_i - z_i$	$2 - 3M$	$4 - 7M$	$3 - M$	0	0		

net potential $C_i - z_i$ rather than the largest positive value as we did in the maximizing problem. The number $4 - 7M$ is the most negative of the net potential numbers

9.3 Obtaining an Initial Feasible Solution

and it occurs in the x_2 column; therefore, x_2 is the incoming variable. Looking at the two replacement quotients for that column, we have

for x_4, $7/2 = 3.5$;

for x_5, $10/5 = 2$, smallest, replace this variable.

The new basis is x_2, x_4. Divide the x_5 row of Table 9.17 by the pivot element 5 (circled), relabel the row as x_2, and put it above x_4. The tableau is

x_2	0	1	$2/5$	0	$1/5$	2
x_4	3	2	-1	1	0	7

In Table 9.18, we have solved for the basis. The value of $C_1 - z_1$ is $2 - 3M$, *a negative number*; so the objective function *can* be decreased. The incoming variable that will cause the decrease is x_1 (the x_1 column is the one with the most negative net potential). The new basis is (x_1, x_2), which furnishes the minimum value of the objective function. (Check this for yourself. You should get $x_1 = 1$ and $x_2 = 2$, with $x_3 = x_4 = x_5 = 0$ as the minimizing point and $z = 10$ as the minimum value.)

Sometimes artificial variables are not needed in linear programming problems; for example, if the constraint inequalities are all *less than or equal to*, then the

Table 9.18

C_i	2	4	3	M	M		
V	x_1	x_2	x_3	x_4	x_5	x_0	C_j
x_2	0	1	$2/5$	0	$1/5$	2	4
x_4	3	0	$-9/5$	1	$-2/5$	3	M
						z	
z_i	$3M$	4	$(8 - 9M)/5$	M	$(4 - 2M)/5$	$8 + 3M$	
$C_i - z_i$	$2 - 3M$	0	$(9M + 7)/5$	0	$(7M - 4)/5$		

slack variables themselves can serve as the initial basis and the solution will be feasible (provided all the constants on the right-hand side are nonnegative). In most cases, however, the inequalities are *mixed* with some of the constraints *greater than* and others *less than* the right-hand sides. In such cases, the slack variables will be introduced in such a way that some of them will have negative coefficients in the constraint equations and they cannot serve as an initial feasible basis. When this happens we need artificial variables to provide an easy start to the pivoting process.

Example 9.10 Maximize

$$z = 3x_1 + 2x_2 - x_3$$

subject to

$$x_1 + x_2 + x_3 \leq 100$$
$$2x_1 - x_2 \geq 35.$$

Solution First introduce slack variables x_4 and x_5:

$$\left. \begin{array}{l} x_1 + x_2 + x_3 + x_4 = 100 \\ 2x_1 - x_2 - x_5 = 35. \end{array} \right\} \quad (9.8)$$

Here the coefficient of x_5 is negative, so the slack variables will not be a feasible basis. We want to introduce artificial variables to obtain an initial feasible solution, but first we check to make sure that the constant values on the right-hand sides are all nonnegative. They are. Let x_6 and x_7 denote the artificial variables; then Equations (9.8) become

$$\left. \begin{array}{l} x_1 + x_2 + x_3 + x_4 + x_6 = 100 \\ 2x_1 - x_2 - x_5 + x_7 = 35. \end{array} \right\} \quad (9.9)$$

In the objective function, the coefficients of the slack variables will be 0 and the coefficients of the artificial variables will be $-M$. Thus, the objective function becomes

$$z = 3x_1 + 2x_2 - x_3 + 0x_4 + 0x_5 - Mx_6 - Mx_7.$$

With the artificial variables as the basis, we get Table 9.19 solved for its feasible basis. The largest value of $C_i - z_i$ in Table 9.19 is $3 + 3M$ in the x_1 column. This means that x_1 will be the incoming variable. The replacement quotients are

for x_6, $100/1$;

for x_7, $35/2$, smallest, replace this variable.

9.3 Obtaining an Initial Feasible Solution

Table 9.19

C_i	3	2	−1	0	0	−M	−M		
V	x_1	x_2	x_3	x_4	x_5	x_6	x_7	$\mathbf{x}_0$	C_j
x_6	1	1	1	1	0	1	0	100	−M
x_7	2	−1	0	0	−1	0	1	35	−M
								z	
z_i	−3M	0	−M	−M	M	−M	−M	−135M	
$C_i − z_i$	3 + 3M	2	M − 1	M	−M	0	0		

The new basis will be x_1, x_6. We again compute z_i and $C_i - z_i$ and pivot to the next basis, which is x_1, x_2, whose solution is $x_1 = 45$ and $x_2 = 55$. (Check this for yourself.) Continue the pivoting process and you finally arrive at the maximizing solution: $x_1 = 100$ and $x_5 = 165$.

One final word of caution on the use of artificial variables—they may actually alter the equation when they are *not zero*. That is, the original system may not be satisfied when one of the artificial variables is in the basis. Therefore, we *must* pivot them all out of the basis before we can claim that we have the solution.

EXERCISE 9.3

Maximize or minimize the objective function by pivoting. Introduce artificial variables as necessary.

1. Maximize
$$z = 6x_1 + 5x_2 + 2x_3$$
subject to
$$x_1 + x_2 + x_3 \leq 10$$
$$x_1 - 6x_2 \geq -1.$$

2. Maximize
$$z = 2x_1 + x_2 + 5x_3$$
subject to
$$x_1 - 6x_2 + x_3 \leq 22$$
$$-x_1 + 3x_2 - x_3 \geq -15.$$

3. Maximize
$$z = x_1 + x_2 + 3x_3$$
subject to
$$x_1 + 3x_2 - x_3 \leq 17$$
$$x_1 - x_2 + 5x_3 \geq 7.$$

4. Minimize
$$z = 2x_1 - x_2 + 6x_3$$
subject to
$$x_1 + x_2 + x_3 \leq 72$$
$$x_1 \quad\quad - 14x_3 \geq 15.$$

5. Minimize
$$z = 8x_1 + x_2 + x_3$$
subject to
$$x_1 + 3x_2 + 3x_3 \geq 86$$
$$2x_1 + 5x_2 + x_3 \geq 90.$$

6. Maximize
$$z = 10x_1 + 5x_2 + 3x_3 + x_4$$
subject to
$$x_1 + 2x_2 + 6x_3 \quad\quad \leq 40$$
$$2x_1 \quad\quad\quad\quad + 3x_4 \leq 20.$$

7. Maximize
$$z = 2x_1 + 5x_2$$

9.3 Obtaining an Initial Feasible Solution

subject to

$$x_1 + 2x_2 \leq 6$$
$$x_1 - 6x_2 \leq 3$$
$$x_1 + 4x_2 \geq 8.$$

8. Minimize

$$z = 3x_1 + 4x_2 + x_3$$

subject to

$$x_1 + x_2 + x_3 \leq 100$$
$$2x_1 \qquad - x_3 \geq 25$$
$$x_2 + x_3 \geq 37.$$

▲ **9. Trim loss problem** A wholesaler has a large supply of 12-inch-wide rolls of calculator paper tape, which he wants to cut into smaller widths to fill the following orders:

> At least 15,000 rolls of $4\frac{1}{2}$-inch tape
> At least 1,000 rolls of 5-inch tape
> At least 8,000 rolls of $6\frac{1}{2}$-inch tape

After cutting a 12-inch roll into some combination of the required widths, any leftover piece less than $4\frac{1}{2}$ inches wide is waste, called the *trim loss.* The objective will be to minimize waste subject to filling the above orders. The figure below shows an example of one possible cutting arrangement.

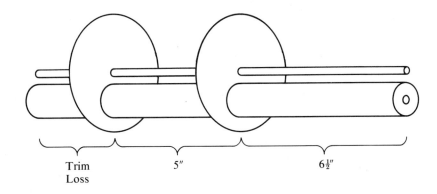

The only possible cutting arrangements are

1. Two $4\frac{1}{2}$-inch rolls   3-inch trim loss
2. One $4\frac{1}{2}$-inch roll and one 5-inch roll $2\frac{1}{2}$-inch trim loss
3. One $4\frac{1}{2}$-inch roll and one $6\frac{1}{2}$-inch roll   1-inch trim loss
4. Two 5-inch rolls   2-inch trim loss
5. One 5-inch roll and one $6\frac{1}{2}$-inch roll $\frac{1}{2}$-inch trim loss

Let x_1, x_2, x_3, x_4, and x_5 be the numbers of rolls cut by cutting arrangements 1, 2, 3, 4, and 5, respectively. The total possible waste is

$$z = 3x_1 + 2\tfrac{1}{2}x_2 + x_3 + 2x_4 + \tfrac{1}{2}x_5.$$

Write the three constraint inequalities (obtain them from the orders to be filled), and find the values of x_i that minimize the trim loss. (Hint: The number of $4\frac{1}{2}$-inch rolls from the x_1 rolls cut by arrangement 1 is $2x_1$, the number from the x_2 rolls cut by arrangement 2 is x_2, the number from the x_3 rolls cut by arrangement 3 is x_3, the number from the x_4 rolls cut by arrangement 4 is $0x_4$, and the number from the x_5 rolls cut by arrangement 5 is $0x_5$. But, by the given orders, the total number of the $4\frac{1}{2}$-inch rolls must exceed 15,000; therefore the first constraint inequality is $2x_1 + x_2 + x_3 + 0x_4 + 0x_5 \geqq 15{,}000$.)

10. Repeat Problem 9, this time for the following orders:

At least 4,000 rolls of $4\frac{1}{2}$-inch tape
At least 12,000 rolls of 5-inch tape
At least 8,000 rolls of $6\frac{1}{2}$-inch tape   ▲

9.4

THE SIMPLEX ALGORITHM— HISTORY AND SUMMARY

This section is primarily a general discussion of the linear programming techniques developed in this text.

The five topics that we have studied thus far: 1. slack variables, 2. basic and nonbasic variables, 3. Gauss-Jordan elimination, 4. pivoting (net potential and replacement quotients), and 5. artificial variables constitute a systematic method for solving linear programming problems. This method is called the *simplex method* or the **simplex algorithm.** The name comes from the geometric figures underlying the process.

9.4 The Simplex Algorithm—History and Summary

In a system of m equations in $m + k$ unknowns (m and k are positive integers), the simplex method finds the optimal values of a linear function whose domain is a convex geometric *solid* in n-dimensional space bounded by $(n - 1)$-dimensional *hyperplanes*. Such a solid is called an **n-simplex**. For example, a 2-simplex is a polygon, such as a triangle or rectangle, bounded by lines (one-dimensional *planes*). A 3-simplex is a convex polyhedron (such as a cube) bounded by two-dimensional planes.

Geometric properties of simplexes have been studied for a long time, but their use in linear programming has a very short history, less than forty years.

Applications of linear programming for the economical use of resources can be traced back to the 1940s. Much of the early work was done by a Russian, L. V. Kantorovich, and an American, George Dantzig. Dantzig's book; *Linear Programming and Extension* (Princeton, N.J.: Princeton University Press, 1963), gives a comprehensive history of linear programming and is the classical reference source for various methods of solution.

Many early uses of linear programming involved the distribution of military supplies in World War II. It is fair to claim that linear programming was one of the technical developments of that war. In the past thirty to forty years, it has grown to become one of the major fields of applied mathematics. It is quite a powerful tool in modern management and has good prospects for even broader applications through the use of computers.

We summarize below the steps of the simplex algorithm and accompany it with a flow chart (Figure 9.2). To read the chart, simply follow the arrows, performing the steps indicated, until the problem is solved. If you know something about computers, you could even use this flow chart to write a computer program for solving linear programming problems. Most computer centers already have such programs on file.

In the algorithm below, notice particularly Step 17. If all the replacement quotients are negative in a given pivotal column, then there is no feasible solution when the incoming variable replaces one of the basic ones. This may mean that the domain of the objective function is unbounded and the optimal solution cannot be attained as a corner point. The problem should be reexamined to see if anything can be done about the constraints. An alternative incoming variable might be selected, if possible, but it would not necessarily produce the optimal solution.

SIMPLEX ALGORITHM

1. Introduce slack variables.
2. Introduce artificial variables.

3. Is this a maximizing problem?
   a. If no, go to Step 8.
   b. If yes, continue.
4. Make the objective coefficients of the artificial variables $-M$.
5. Compute z, z_i, and $C_i - z_i$.
6. Is $C_i - z_i$ positive for some x_i?
   a. If no, stop.
   b. If yes, continue.
7. Select the most positive $C_i - z_i$. Let x_i enter the basis; go to Step 12.
8. Make the objective coefficients of the artificial variables $+M$.
9. Compute z, z_i, and $C_i - z_i$.
10. Is $C_i - z_i$ negative for some x_i?
    a. If no, stop.
    b. If yes, continue.
11. Select the most negative $C_i - z_i$. Let x_i enter the basis.
12. Does the incoming column have any positive component p?
    a. If no, go to Step 17.
    b. If yes, continue.
13. Compute the replacement quotients b/p for only the positive values p in the incoming column.
14. Denote by x_k the row with the smallest nonnegative replacement quotient b/p, with $p > 0$. This is the row of the outgoing variable. Replace x_k by x_i.

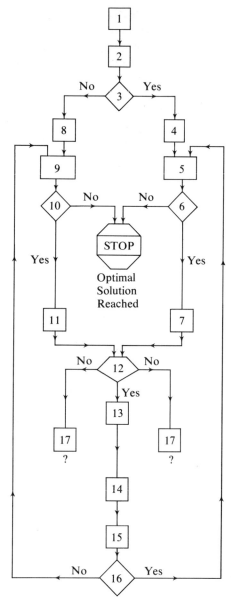

Figure 9.2

15. Apply Gauss-Jordan elimination to solve for the new basis.
16. Is this a maximizing problem?

    a. If no, loop back to Step 9.

    b. If yes, loop back to Step 5.

17. Analyze the problem. No feasible improvement is possible for this incoming variable. Some improvement may be achieved by returning to Step 7 (maximizing problems) or Step 11 (minimizing problems) and selecting a second incoming variable, if possible. The domain may be unbounded, or there may be no unique solution.

10
INTRODUCTION TO PROBABILITY

10.1

RANDOM EVENTS AND SIMULATION

Can anyone predict the future? No. How about predicting the weather? People try. In the past people have, by experience and intuition, developed certain folklore for forecasting natural phenomena and their severity. "Red sky in the morning, sailors take warning" or "A preponderance of ants in the fall means a harsh winter follows" are examples of such crude models or theories for predicting the weather.

Modern forecasters will make sophisticated predictions such as "Red sky in the morning—there is a 60 percent chance that sailors should take warning." They recognize that predictions are not one hundred percent accurate and therefore should be stated only as *probabilities*.

There are at least two reasons for using a probability model for predicting weather. First, there are so many varying conditions; such as, humidity, cloud cover, wind, atmospheric pressure, convection currents, and solar cycles, that no equation could possibly relate them all in any meaningful way; it may not even be possible to determine what all of the variables are. Second, nature is full of surprises and can produce changes in the variables *at random* (that is, in an unpredictable fashion).

This random nature of a large number of variables makes it impractical, if not impossible, to develop a *deterministic* mathematical model that will predict the weather with certainty. Weather prediction is not the only field in which a deterministic model is impractical.

People, in their roles as voters, gamblers, consumers, medical patients, air polluters, and parents, also behave in ways that have a large number of randomly changing variables. These phenomena can be effectively studied only with probability models. In other words, it is much more practical to develop a probability model than a deterministic model in these areas.

But how can researchers working in these fields conduct experiments? The answer is **simulation**. In simulation, experimenters imitate the possible conditions through some randomizing device that causes simulated events to occur in a random fashion.

Example 10.1 For testing the reaction of a student pilot, a flight simulator could have emergency situations occur at random times with a random degree of (simulated) danger, all governed by a *random number table*.

Example 10.2 A stock market game could have a random number table to determine price rises and drops and their amounts.

Example 10.3 An electronic poker machine could have its displayed poker hand generated by a built-in random circuit.

In general, most probability experiments rely on devices that produce numbers in a random fashion. Some randomizing devices that have been useful are:

1. Spinning a dial surrounded by a circle of numbers.
2. Rolling dice.
3. Drawing a card from a shuffled deck of cards.
4. Selecting numbered slips of paper from a container.
5. Electronically generating a random number table, usually done by a computer. See Table 10.1.

Table 10.1 A Short Random Number Table

05044	81229	41045	96884	07948	40318	10363	98926
10092	30502	77641	44521	69058	73608	76076	36460
81466	21454	92849	72758	42204	81122	19906	55749
27306	89222	04289	41559	40236	68103	67127	26238
37368	67907	55274	64887	23550	78577	33031	39633
A	B	C	D	E	F	G	H

10.1 Random Events and Simulation

The next example illustrates the use of a random number table in simulating a random process. The object will be to let the random number table deal a five-card poker hand.

Example 10.4 Let the cards in an ordinary deck of playing cards be coded as two-digit numbers as shown in Table 10.2 below. The random number table in Table 10.1 and the playing card code in Table 10.2 could be used to generate a five-card poker hand as follows: Go to one block of the random number table (Table 10.1) and start with the first number in that block. Then,

1. Pick the first two digits of that number.
2. If these two digits form a number between 01 and 52, then select the corresponding card from the coded deck in Table 10.2. Repeat Step 1 for the next random number.
3. If the two-digit number selected in Step 1 is not between 01 and 52, reject that number and go to the next random number, selecting its first two digits.
4. Repeat this process, each time repeating Step 2 for those numbers between 01 and 52 and rejecting those numbers that are not.
5. Continue until five different cards have been obtained.

Table 10.2 Two-Digit Code for a Deck of Playing Cards

Spades Card	Code	Hearts Card	Code	Diamonds Card	Code	Clubs Card	Code
ace	01	ace	14	ace	27	ace	40
2	02	2	15	2	28	2	41
3	03	3	16	3	29	3	42
4	04	4	17	4	30	4	43
5	05	5	18	5	31	5	44
6	06	6	19	6	32	6	45
7	07	7	20	7	33	7	46
8	08	8	21	8	34	8	47
9	09	9	22	9	35	9	48
10	10	10	23	10	36	10	49
Jack	11	Jack	24	Jack	37	Jack	50
Queen	12	Queen	25	Queen	38	Queen	51
King	13	King	26	King	39	King	52

Example 10.5 Use the simulation of the poker deal just described to generate a poker hand, starting with the random numbers in Block **C** of the random number table.

Solution The first number in Block **C** is 41045. Its first two digits, 41, form a number between 01 and 52; so, according to the coded deck, the corresponding card is the two of clubs. The next random number is 77641. The first two digits are 77, not between 01 and 52. Reject 77 and go to the next random number, 92849. Reject 92. Next is 04289, where the first two digits equal 04. Select the corresponding card from the coded deck, which is the four of spades. The next number is 55274; reject 55. Continuing into Block **D**, the next random number is 96884. Reject 96. Next is 44521, and 44 corresponds to the five of clubs. Next, reject 72. Next is the random number, 41559, but its first two digits equal 41, a number that has already been used to select a card, and *this same card cannot be picked again.* Continue to the next acceptable number, which is 07948 in Block **E**. The card is the seven of spades. The next acceptable random number is 42204, and 42 yields the three of clubs.

This completes the five-card poker hand: seven of spades, five of clubs, four of spades, three of clubs, and two of clubs. Not a bad hand! We could discard the seven and draw to a straight open at both ends. ▲

Randomness applies to unpredictable results encountered in any circumstance, natural or man-made, scientific or nonscientific, planned experiments or everyday experiences. The random results of a given set of circumstances are called **events**.

Example 10.6 An obstetrical nurse is on night duty; she goes to work five nights per week on a regular schedule. This schedule is a deterministic process, not random, but random events happen when she gets to work. She never knows how many babies are going to be born on her shift. One night she may find that she participates in three deliveries; the next night there may be two babies born on her shift; the next night there could be no births.

These random results of a deterministic process are called **random events of an experiment,** where the deliveries in which the nurse participates are the random events and the nurse's act of going to work for the night is the experiment.

If going on duty can be called an *experiment,* then apparently that word is being used in a very general way and requires a definition.

Definition **(Experiment)**
An experiment is defined as *any action* (whether deliberate, accidental, or natural) or *any set of conditions* that can produce observable results.

10.1 Random Events and Simulation

Note that this definition is very broad and does not conform to the usual notion of a *controlled experiment* or a *scientific experiment*.* Here, the word experiment is almost synonymous with *experience* and may include, for example, any natural disaster, such as a tornado. An observable result of a tornado might be the destruction of a given number of houses.

Returning to our definitions—if the results of an experiment are not predictable with certainty, then the experiment is a *random experiment*. A set of results of a random experiment is called a *random event*. More precisely, we have the following definitions:

Definition **(Sample point)**
Let n denote a positive integer. If $X_1, X_2, ..., X_n$ are n possible distinct results of an experiment, they are called *sample points* of the experiment. If an experiment has only a finite number of possible outcomes, the n sample points are described in such a way that they cover all the outcomes. A set of sample points for an experiment is not necessarily unique. One experimenter may wish to obtain finer subdivisions of the results than another, and he or she could describe the outcomes with sample points that give more details.

Definition **(Events)**
Any set that contains one or more sample points is called an *event*.

Definition **(Simple events)**
A *simple event* is a set that contains only one sample point. It is a singleton set whose element is a sample point. There are as many simple events as there are sample points.

Example 10.7 A tornado hits a town that has 1000 houses. The possible results (sample points) may be denoted as follows:

---

*In his book, *Introduction to Probability and Statistics*, 4th ed. (North Scituate, Mass.: Duxbury Press, 1975), William Mendenhall discusses this broad use of the word *experiment*. He states: "Data is obtained either by observation of uncontrolled events in nature or by controlled experimentation in the laboratory. To simplify our terminology, we seek a word which will apply to either method of data collection and hence define *experiment* to be the process by which an observation (or measurement) is obtained. Note that the observation need not be numerical."

X_0 = the result that no house is damaged.
X_1 = the result that exactly one house is damaged.
X_2 = the result that exactly two houses are damaged.
$\vdots$
X_k = the result that exactly k houses are damaged ($k < 1000$).
$\vdots$
X_{1000} = the result that exactly 1000 houses are damaged.

Each result X_i ($i = 0,1,2, ..., 1000$) is a sample point and each of the singleton sets, $\{X_0\}, \{X_1\}, \{X_2\}, ..., \{X_k\}, ..., \{X_{1000}\}$, is a simple event. The combination of simple events, for example, a set such as $A = \{X_1, X_3, X_5\}$, is an event. This is read "A is the event that exactly one or exactly three or exactly five houses were damaged."

These sets can be combined by unions and intersections just as for any other sets. Thus, for example, if B is the event that between three and five houses were damaged, then B is the event that either exactly three, exactly four, or exactly five houses were damaged, and the two sets A and B could be united into a set whose elements are X_1, X_3, X_4, and X_5.

Example 10.8 If B is the event $\{X_3, X_4, X_5\}$ and $A = \{X_1, X_3, X_5\}$, list the elements in the sets $A \cup B$ and $A \cap B$. Describe these sets in words.

Solution $A \cup B = \{X_1, X_3, X_4, X_5\}$ and $A \cap B = \{X_3, X_5\}$. The former is the event that either exactly one house was damaged or between three and five houses were damaged. The latter is the event that either exactly three or exactly five houses were damaged.

Example 10.9 Using this same tornado example, write a set statement for the events C = the event that at least 100 houses were damaged, and D = the event that no more than 30 houses were damaged. Describe $C \cup D$ and $C \cap D$.

Solution $C = \{X_{100}, X_{101}, ..., X_{1000}\}$ and $D = \{X_1, X_2, ..., X_{30}\}$. $C \cup D$ = the event that either 100 or more houses were damaged, or 30 or fewer were damaged. $C \cap D$ is the empty set; that is, it is an impossible event. It says that at least 100 houses were damaged and also that no more than 30 were damaged.

EXERCISE 10.1

1. Use the coded deck of cards given in Example 10.4 and (following the procedure outlined in that example) make up a five-card poker hand starting with Block **A** of the random number table.

2. Let the random number table deal another poker hand, this time use digits *three* and *four* of each five-digit random number instead of the first two digits. Start the process in Block **A**. (For example, in Block **A** the first five-digit random number is 05044; the third and fourth digits are 0 and 4. The first card selected is the one corresponding to 04 in the code.)

3. In manufacturing 100 items, the random numbers in Table 10.1 were used for selecting a sample of items to be tested. The table was used by taking the last two digits of each number *in reverse order*. For example, in the number 10092, the last two digits 92 reversed are 29: so the 29th item was tested. Using Blocks **C** and **D** in the above manner, make up a list of items selected for testing.

4. Suppose Fischer and Korchnoi play a game of chess. Let $\{K\}$ = the event that Korchnoi wins and let $\{F\}$ = the event that Fischer wins. What event is needed to *complete* the set of events? (A complete set of events is one that accounts for *all* possible outcomes.)

5. Suppose Spassky and Karpov play a game of chess. Let $\{S\}$ = the event that Spassky does not lose and $\{K'\}$ = the event that Karpov does not lose. What event is the intersection of these two events? (Under what conditions is it true that both Spassky and Karpov do not lose?)

6. A packager of shelled walnuts has a machine that will sometimes produce *whole* kernels and sometimes produce *broken* kernels; furthermore, each walnut kernel is rated as either good or poor quality. The shelling process, then, is an experiment that has four possible results, as summarized in the following table of sample points:

	Good	Poor
Whole	X_1	X_2
Broken	X_3	X_4

The first two simple events are $\{X_1\}$ = the event the kernel is good and whole, and $\{X_2\}$ = the event the kernel is poor and whole.

    a. Describe the remaining two simple events, $\{X_3\}$ and $\{X_4\}$, in words.

    b. Describe the event $\{X_1\} \cup \{X_2\}$ in words.

    c. Describe the event $\{X_1\} \cup \{X_3,X_4\}$ in words.

7. A sample of 1000 people, 500 men and 500 women, are selected at random. Some are color-blind and some have normal vision. The following is a table of the sample points:

	Normal	Color-Blind
Men	X_1	X_2
Women	X_3	X_4

    a. Describe in words the simple events $\{X_1\}$ and $\{X_2\}$.

    b. Describe in words the event $\{X_1,X_2\}$ and the event $\{X_1,X_3\}$.

8. Some ill effects of cigarette smoking are

    X_1 = lung cancer X_4 = weakened immune responses
    X_2 = heart disease X_5 = gum disease
    X_3 = emphysema X_6 = a complaining spouse

For each sample point X_i (i = 1,2,3,4,5,6) the singleton set $\{X_i\}$ = the event that the smoker has the ill effect X_i. For example, the set $\{X_3\}$ = the event that the smoker has emphysema.

    a. Describe in words the events $\{X_5\}$, $\{X_6\}$, and $\{X_5\} \cup \{X_6\}$.

    b. Describe in words the event $\{X_1,X_2,X_4\}$.

    c. Write the set equivalent to the description A = the event that the smoker has either heart disease or a complaining spouse.

9. Experiments with homing pigeons show that they basically use four independent methods for finding their way home: 1. the sun's position, 2. barometric pressure, 3. polarized light on cloudy days, or 4. the earth's magnetic field. Make up a complete set of sample points. Give an example of an event that is the union of two or more of the simple events. Describe that event in words.

10. A service department of an auto dealer has repair jobs whose sample points are given in the following table.

	Regularly Scheduled Maintenance	Drive-In Service	Tow-In Service
Keep Overnight	X_1	X_2	X_3
Finish Same Day	X_4	X_5	X_6

  a. List the simple events and describe two of them in words.

  b. Describe the event $\{X_1, X_4\}$ in words.

  c. Make up another set of random events for this service department that has three sample points, is complete, and does not have overlapping sample points. ▲

10.2
FINITE PROBABILITY MODEL

Assume that an experiment is repeated many times, all of its results are observed, and the frequency with which each result occurs is recorded. Let E be one particular result (E is a simple event), and suppose we are interested in computing how many times E occurs relative to the number of times the experiment is repeated. The need for such a computation would arise in an attempt to determine the likelihood that this particular result would occur at any time the given experiment is performed. Thus, we seek the ratio

$$\frac{\text{Number of times the event } E \text{ occurs}}{\text{Number of times the experiment is repeated}}.$$

This ratio is called the **relative frequency** of event E. Usually the relative frequency of an event varies according to the number of times the experiment is repeated. Sometimes, as the number of repetitions of the experiment is increased, the relative frequency of the given event may approach a constant value.

▲ **Example 10.10** Recall Example 10.6 of the previous section. For the obstetrical nurse, the repeated experiment would be going to work on her scheduled nights. Let the event E be having at least one delivery on her shift. Note that there are only two possible events; either E occurs or $-E$ occurs. (Recall from Chapter 1 that $-E$ denotes the complement of E.) In other words, the sample space has only two elements: either there is at least one baby born during her shift or there are none.

In four years she goes to work on 1000 nights; that is, the experiment has been repeated 1000 times. If, on 611 of these nights, she had at least one delivery during the night (event E), then the event E had the relative frequency

$$\frac{611}{1000} = 0.611.$$

After going to work on ten more nights, suppose there was at least one delivery on nine of them; so now the relative frequency (counting all 1010 nights) is

$$\frac{620}{1010} = 0.614.$$

In the next ten nights, the event E occurred four times:

$$\frac{624}{1020} = 0.612.$$

Then suppose she has 30 nights with at least one delivery in the next 50 nights. The relative frequency for the total 1070 repetitions is

$$\frac{654}{1070} = 0.611.$$

Assuming that this trend continues, the nurse could claim that there is a 61 percent chance that she will participate in at least one delivery when she goes to work on any given night. The probability is 0.611. ▲

In general, if an experiment is repeated over and over again and (as more and more repetitions are made) the relative frequency of an event E approaches some limiting value, then that limiting value, denoted by $P(E)$, is called the **probability of E**. This defines a probability as the limit of a set of relative frequencies; later we will give an axiomatic definition of a probability space.

From the above example we conclude that the probability of an event may be defined as a number paired with the given event. Here, the event E of having a delivery on the night shift is paired with the number 0.611. This means that the probability of an event is actually a *function* that associates a positive real number (not greater than one) with the event. (Recall the definition of a function in Chapter 3.)

How is this function to be applied to other sets, such as, for example, the complement of E? If the nurse has at least one delivery on a given night with probability 0.611—that is, $P(E) = 0.611$—then what is the probability that there will be no deliveries on a given night? The answer is 0.389, or $1 - 0.611$.

In general, if any event E has probability p (where p is a number such that $0 \leq p \leq 1$)—that is, if $P(E) = p$—then $P(-E) = 1 - p$.

In cases where there are more than two simple events, how is the probability function applied to combinations of them? We will answer this question and develop some ideas about probability in the rest of this chapter. First, we need the following definition of **sample space**.

Definition (Sample space)
A set of nonoverlapping simple events that cover all the possible outcomes of an experiment is called a *sample space* for that experiment. An event in a given sample space is the union of simple events.

Example 10.11 Suppose S is a sample space consisting of the sample points X_1, X_2, and X_3. Then S is the set $\{X_1, X_2, X_3\}$. The singleton subsets of S are the simple events $E_1 = \{X_1\}$, $E_2 = \{X_2\}$, and $E_3 = \{X_3\}$. Altogether, S has eight subsets. The above three simple events are three of them. The other five are $\{X_1, X_2\}$, $\{X_1, X_3\}$, $\{X_2, X_3\}$, S, and $\emptyset$ (the empty set). These are all the possible events that can occur in this sample space.

Example 10.12 Consider as an experiment a given set of meteorological conditions (temperature, atmospheric pressure, humidity, wind velocity, cloud formation, position of a weather front, cyclical considerations, etc.). Suppose that one weather forecaster lists as observable results the following four outcomes: X_1 = mist, X_2 = rain, X_3 = clouds without rain, and X_4 = mostly clear skies. The sample space S is the set $\{X_1, X_2, X_3, X_4\}$, and the simple events are $E_1 = \{X_1\}$ = the event it is misty, $E_2 = \{X_2\}$ = the event it is raining, $E_3 = \{X_3\}$ = the event it is cloudy, and $E_4 = \{X_4\}$ = the event it is clear. An example of a subset of this space would be a set $A = \{X_1, X_4\}$, which is the event that it is either misty or clear.

When these same meteorological conditions are repeated many times, then the relative frequency of each of the four simple events may be computed and reported in the form of a weather prediction. If, for example, a set of conditions has resulted in rain 85 percent of the time over the long run, then the weather report claims that there is an 85 percent chance of rain.

It should be noted that another weather forecaster may wish to compose a more detailed list of sample points. The sample space S' for these same conditions could be a set of simple events that would reveal the *amount* of rain, the *percent* of cloud cover, etc. Both S and S' would be *complete* sample spaces, in the sense that at least one event of S must occur and at least one event of S' must occur. Still another forecaster could compose a *less* detailed sample space, one in which there would be only two sample points: Y_1 = rain and Y_2 = no rain. All three sample spaces would be based on the same conditions and have the same accuracy, but they would give different information.

Return to Example 10.12 and suppose that the long-run relative frequencies for the given outcomes define the following probability values: $P(\{X_1\}) = 0.07$, $P(\{X_2\}) = 0.85$, $P(\{X_3\}) = 0.05$, and $P(\{X_4\}) = 0.03$. This sample space along with the probabilities for the four simple events is a probability model for the given meteorological conditions. If some possible outcome is the union of simple events, we can compute the probability of that outcome from the probabilities of the simple events. For example the event $A = \{X_1, X_4\}$ has the probability $P(A) = P(\{X_1, X_4\}) = 0.07 + 0.03 = 0.10$.

In general, any sample space, along with the probabilities of all the simple events, is called a probability model. The formal definition of this concept is stated below.

Definition (**Finite probability function on a sample space**)
Let $S = \{X_1, X_2, \ldots, X_n\}$ be a finite sample space, and $P =$ a function whose domain is S and whose range is a subset of the real numbers R. Then P is called *a finite probability function on S* if and only if it satisfies the following five properties:

Property 1 $P(\{X_i\}) \geq 0$ for all X_i in S.
The probability of each simple event is greater than or equal to zero.

Property 2 $P(\{X_i\}) \leq 1$ for all X_i in S.
The probability of each simple event is less than or equal to one.

Property 3 $P(\{X_1\}) + P(\{X_2\}) + \ldots + P(\{X_n\}) = 1$.
The sum of the probabilities of all the simple events is equal to one.

Property 4 $P(\emptyset) = 0$.
The probability of the empty set is zero. (The empty set is also called the impossible event.)

10.2 Finite Probability Model

Property 5 If E is any event, the union of simple events in S, then $P(E)$ is the sum of the probabilities of the simple events whose union is E.

To illustrate Property 5, consider the event A in Example 10.12. A is the event that it is either misty or clear, $A = \{X_1, X_4\}$; that is, $A = \{X_1\} \cup \{X_4\}$. Therefore, $P(A) = P(\{X_1\}) + P(\{X_4\})$, and from the probability value $P(\{X_1\}) = 0.07$ and $P(\{X_4\}) = 0.03$, we get $P(A) = 0.10$.

Definition **(Finite probability model)**
A finite sample space, along with one of its probability functions, is called a **finite probability model**.

Example 10.13 Let the experiment be the toss of a fair coin, and let the results be H, a head will appear, and T, a tail will appear. The sample space is $S = \{H, T\}$. The sample points are H and T, and the simple events are $\{H\}$ and $\{T\}$. Let the probability function P be defined as follows: P is the function $P: S \to R$ (R = the set of real numbers) such that $P(\{H\}) = \frac{1}{2}$ and $P(\{T\}) = \frac{1}{2}$. Does P satisfy the five properties of the definition? Yes, as shown here:

1. $P(\{H\}) = \frac{1}{2} \geq 0$ and $P(\{T\}) = \frac{1}{2} \geq 0$ for all simple events.
2. $P(\{H\}) = \frac{1}{2} \leq 1$ and $P(\{T\}) = \frac{1}{2} \leq 1$ for all simple events.
3. $P(\{H\}) + P(\{T\}) = \frac{1}{2} + \frac{1}{2} = 1$.
4. Here, the empty set is the event that neither H nor T occurs, which is impossible; so its probability is zero. $P(\emptyset) = 0$.
5. There are only four events because S has only four subsets. For example, consider the subset $B = \{T\} \cup \emptyset$, then $P(B) = P(\{T\}) + P(\emptyset)$. That is, $P(B) = \frac{1}{2} + 0 = \frac{1}{2}$.

In similar fashion, Property 5 can be verified for the other three subsets.

Definition **(Abuse of notation)**
Sometimes, for the sake of brevity, no distinction is made between a sample point X_i and the simple event $\{X_i\}$. An expression such as $P(\{X_i, X_j\})$ is sometimes written without the braces { }; that is, as $P(X_i, X_j)$. This minor sacrifice of accuracy is called an *abuse of notation* and may be used whenever the actual meanings are clear.

EXERCISE 10.2

1. A tornado hits a town with 1000 houses. Let the sample points be

   X_0 = no houses were damaged.
   X_1 = at most, 20 houses were damaged.
   X_2 = at least 21 houses but no more than 100 houses were damaged.
   X_3 = at least 101 but no more than 500 houses were damaged.
   X_4 = at least 501 houses were damaged.

   a. What sets are the simple events?

   b. What is the sample space S?

   c. Let $P: S \rightarrow R$ be the probability function $P(\{X_0\}) = 0.07$, $P(\{X_1\}) = 0.15$, $P(\{X_2\}) = 0.32$, and $P(\{X_3\}) = 0.38$. Find what value $P(\{X_4\})$ must have.

   d. Describe the event $A = \{X_3, X_4\}$ in words.

   e. What is the event that at least 21 houses were damaged?

   f. Find the probability that at least 101 houses were damaged.

   g. Find the probability that at most 100 houses were damaged.

   h. How many possible events (as subsets of this sample space) are there?

2. Certain meteorological conditions exist. Let the sample points be Y_1 = snow, Y_2 = sleet, Y_3 = hail, and Y_4 = none of these.

   a. What sets are the simple events?

   b. What is the sample space S?

   c. Let $P: S \rightarrow R$ be the probability function $P(\{Y_1\}) = 0.45$, $P(\{Y_2\}) = 0.10$, and $P(\{Y_3\}) = 0.08$. Find $P(\{Y_4\})$.

   d. Describe in words the event $\{Y_1, Y_2, Y_3\}$.

   e. What is the event that it will either snow or sleet?

   f. Find the probability that it will either hail or sleet.

   g. Find the probability that it will not snow.

   h. Since S has four elements, how many subsets does it have?

3. Hypnosis is tried as a treatment for warts. The simple events are

   $\{X\}$ = the event that the warts disappeared completely.
   $\{Y\}$ = the event that the warts faded gradually.
   $\{Z\}$ = the event that no improvement was noted.

10.2 Finite Probability Model

In a large experiment the relative frequencies suggest the probabilities $P(X) = 0.35$ and $P(Y) = 0.24$.

a. Find the probability that no improvement was noted.
b. Find the probability of the event $\{X,Y\}$.
c. Find the probability of $-\{X\}$, the complement of $\{X\}$.
d. Find the probability of $-\{Y\}$, the complement of $\{Y\}$.
e. How many subsets of $\{X,Y,Z\}$ are there?
f. Find the probabilities of each of the subsets of $\{X,Y,Z\}$.

Telephone switchboard Problems 4 through 19 refer to the following description. During any five-minute interval in a normal working day at a telephone switchboard, the following events may occur. (Here, by abuse of notation, we are letting sample points mean the same thing as simple events.)

E_0 = the event that no phone calls come into the switchboard.

E_1 = the event that exactly one phone call comes in.

$\vdots$

E_k = the event that exactly k phone calls come in ($k < 10$).

$\vdots$

E_{10} = the event that exactly ten phone calls come in.

If eleven *or more* calls come in, the event is denoted by E_{11}. Over the long run, the relative frequencies of these events suggest the probabilities:

$P(E_0) = 0.000 \quad P(E_3) = 0.105 \quad P(E_6) = 0.201 \quad P(E_9) = 0.003$

$P(E_1) = 0.001 \quad P(E_4) = 0.221 \quad P(E_7) = 0.108 \quad P(E_{10}) = 0.001$

$P(E_2) = 0.024 \quad P(E_5) = 0.311 \quad P(E_8) = 0.023 \quad P(E_{11}) = 0.002$

4. Express in words the event $A = \{E_1, E_2, E_3, E_4\}$ and the event $B = \{E_5, E_6, E_7, E_8\}$.

5. Express in words the event $C = \{E_3, E_4, E_5\}$ and the event $D = \{E_8, E_9, E_{10}, E_{11}\}$.

6. Find the probabilities $P(A)$ and $P(B)$. (A and B are given in Problem 4.)

7. Find the probabilities $P(C)$ and $P(D)$. (C and D are given in Problem 5.)

8. Explain why $P(E_0, E_1, E_2, \ldots, E_{10}, E_{11}) = 1$.

9. Show that $P(E_0) + P(E_1) + \ldots + P(E_{11}) = 1$.

10. If A and C are the sets defined in Problems 4 and 5, find $P(A \cap C)$.

11. If B and D are the sets defined in Problems 4 and 5, find $P(B \cap D)$.
12. Let Y_1 = the event that four or fewer calls come in, Y_2 = the event that five or six calls come in, and Y_3 = the event that seven or more calls come in. Show that $P(Y_1) = 0.351$.
13. Show that $P(Y_1, Y_2, Y_3) = 1$. (See Problem 12.)
14. Find $P(-Y_1)$. (See Problem 12.)
15. Find $P(-Y_2)$. (See Problem 12.)
16. Let S' = the event $\{Y_1, Y_2, Y_3\}$ (see Problem 12) and S = the event $\{E_0, E_1, \ldots, E_k, E_{10}, E_{11}\}$.

    a. Do S and S' cover the same random results?

    b. Are S and S' both sample spaces for the same experiment?

17. Make a list of all the probabilities for the simple events in S'. (See Problem 16.)
*18. Use the elements of S to construct a five-element sample space S'' for this same experiment. Make a list of the probabilities for this new sample space. (See Problems 16 and 17.)
*19. Use the elements of S to construct a seven-element sample space S'''. Make a list of the probabilities for this new sample space. (see Problems 16 and 17.)
▲

10.3

SAMPLE SPACES WITH EQUIPROBABLE EVENTS

In a bingo game, what is the probability that the first number drawn will be 28? To answer this question, we assume that the numbers are all equally likely to be drawn; that is, all 75 numbers have exactly the same probability of being selected. Therefore, the number 28 has only one chance in 75 of being drawn on the first draw. If E is the event of selecting 28 on the first draw, then $P(E) = 1/75$.

Given any sample space, if the probability function assigns the *same value* to every simple event in the space, then the space is called an **equiprobable space.** If n is the number of elements in the space, then the probability of each simple event is $1/n$.

Most of the examples discussed in the previous section were *not* equiprobable

10.3 Sample Spaces with Equiprobable Events

spaces. In order to solve those problems it was necessary to be given the probabilities for all (but one) of the simple events. By contrast, the problems in this section involve equiprobable spaces that can be solved with knowledge of only the *number* of sample points in the space.

Example 10.14 A single die (one-half of a pair of dice) has six sides, with each side containing a set of spots representing the numbers 1, 2, 3, 4, 5, and 6. One set of possible results for a toss of the die is

X_1 = the 1 shows X_4 = the 4 shows
X_2 = the 2 shows X_5 = the 5 shows
X_3 = the 3 shows X_6 = the 6 shows

If we assume it a fair die and it is randomly thrown, then each of the simple events, $\{X_1\}, \{X_2\}, \ldots, \{X_6\}$, has the same chance of occurring. Since there are six simple events, then the probability of each is $1/6$ (Property 3, page 302).

From the definition of a finite probability model, any other event is the union of simple events, and the probability of any given event is the sum of the probabilities of the simple events making up the given event. In this example, the probability for an event $A = \{X_1, X_4\}$ (the probability that either a 1 or a 4 is thrown) can be computed from the sum of the probabilities of the simple events, $\{X_1\}$ and $\{X_4\}$. Therefore, $P(\{X_1, X_4\}) = (1/6) + (1/6) = 1/3$.

Alternatively, the probability of A can be computed by counting the elements in A and dividing that number by the number of elements in S. Here, A has two elements and S has six; hence, the probability $P(A)$ is the ratio

$$\frac{\text{Number of elements in } A}{\text{Number of elements in } S} = \frac{2}{6} = \frac{1}{3}.$$

This method of computing the probability for a set works for *any* equiprobable space and is summarized in the following theorem and corollary.

Theorem 10.1 If S is a nonempty sample space of n simple events and each *simple* event has the same probability of occurring, then the probability of each simple event is $1/n$.

Corollary If S is a nonempty equiprobable space of n sample points, A is any event that is a subset of S, and A has m elements in it, then the probability that the event A occurs is $P(A) = m/n$.

Notational definitions $n(S)$ = the number of elements in a nonempty equiprobable space S. $n(A)$ = the number of elements in an event $A \subseteq S$. $P(A)$ or $P(x \in A)$ = the probability that x, chosen at random, is in A. Therefore, the corollary may be written as follows:

$$P(A) = \frac{n(A)}{n(S)}, \text{ given } A \subseteq S.$$

Example 10.15 If a card is selected at random from a standard deck of 52 playing cards, what is the probability that it will be a face card?

Solution Let S be the space whose sample points are the whole deck of cards and let F be the event that a face card is selected. $F \subseteq S$. The number of elements in these two sets are $n(S) = 52$ and $n(F) = 12$ (there are twelve face cards in the deck). The probability is $P(F) = {}^{12}\!/_{52} = 0.23$.

Since the probability of an event in an equiprobable space is completely determined by the number of sample points, then it is clear that counting procedures are important for computing such probabilities. The following counting principle is useful in equiprobable spaces but is not necessarily restricted to them.

MULTIPLICATIVE COUNTING PRINCIPLE

Suppose an experiment can be done in several stages. If the first stage has n_1 possible outcomes, the second stage has n_2 possible outcomes, the third stage has n_3 possible outcomes, and so on until the final stage with n_k possible outcomes, then the number of possible results (of the entire multiple-stage experiment) in the order indicated is the product of the numbers $n_1 n_2 n_3 \ldots n_k$, where k is the total number of stages.

Example 10.16 A television writer has in his files seven story plots that he uses for the middles of stories. He also has six beginnings and four endings. He chooses, at random, a beginning, a middle, and an ending. How many different stories can he write?

Solution Each story can start six different ways, and after each beginning there are seven ways the middle can be put in. Following this, there are four possible endings. Therefore, $7 \cdot 6 \cdot 4 = 168$ different stories can be written.

10.3 Sample Spaces with Equiprobable Events

Suppose five of the beginnings are weak, five of the middles are pointless, and two of the endings are insipid. How many ways can he write stories with weak beginnings, pointless middles, and insipid endings? The solution is $5 \cdot 5 \cdot 2 = 50$.

Let A be the event that this author writes a story with all three flaws: a weak beginning, a pointless middle, and an insipid ending. Assuming an equiprobable model, there are 50 ways that event A can happen, $n(A) = 50$. He can write a total of 168 stories, $n(S) = 168$; therefore, the probability of event A is

$$P(A) = \frac{n(A)}{n(S)} = \frac{50}{168} = 0.298.$$

Example 10.17 If a coin is tossed twice, the first toss has two possible outcomes, heads H or tails T. The second toss has two outcomes. The tosses have four possible outcomes by the multiplication principle. We may picture this entire two-stage experiment by the following diagram, called a tree diagram (Figure 10.1). In the two-coin toss, let the four sample points be $X_1 = (H,H)$; $X_2 = (H,T)$; $X_3 = (T,H)$; and $X_4 = (T,T)$.

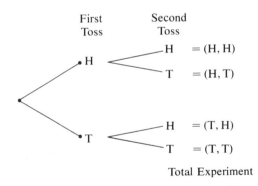

Total Experiment

Figure 10.1

a. What is the probability of each simple event $\{X_1\}, \ldots, \{X_4\}$?
b. Describe in words the event $\{X_1, X_4\}$.
c. What is the probability of $\{X_1, X_4\}$?
d. What is the probability of the event that at least one head appears?

Solution a. Each simple event has the probability $1/4$.

b. $\{X_1, X_4\}$ is the event that the coins match; that is, they are both heads or both tails.

c. $P(X_1, X_4) = P(X_1) + P(X_4) = (1/4) + (1/4) = 1/2$.

d. At least one head comes from the event $\{X_1, X_2, X_3\}$. Its probability is $P(X_1) + P(X_2) + P(X_3) = 3/4$.

EXERCISE 10.3

In Problems 1 and 2 use the census data given in Chapter 1, Exercise 1.2, Problem 8.

▲ 1. If a household in Alaska is selected at random, what is the probability that it has a flush toilet? (Let S = the set of all households in Alaska, and F = the set of households in Alaska with flush toilets. Compute $P(F)$.)

2. What is the probability that a household selected at random in Arkansas will be rural?

3. What is the probability that a card selected at random from a standard 52-card deck will be a red card with a number nine or less on it. (Count the ace as a one.)

4. What is the probability that a card selected at random from a standard 52-card deck is yellow? (Assume only red and black cards in the deck.)

5. What is the probability that the next California earthquake will be on a Saturday?

6. In the year 1998, there are 52 Fridays and three of them are Friday the thirteenth (February, March, and November). In 1999, there are 53 Fridays and only one of them (August) is on the thirteenth. What is the probability that a Friday selected at random in 1998 will be on the thirteenth? What is the probability that a Friday is on the thirteenth in 1999? ▲

7. The equation $P(A) = [n(A)]/[n(S)]$ can be solved for either of the quantities $n(A)$ or $n(S)$. Suppose $P(A) = 0.81$ and $n(S) = 3,653,000$, find $n(A)$.

8. Suppose $P(B) = 0.43$ and $n(B) = 724$, find $n(S)$.

▲ 9. A certain manufacturing process is 99.44 percent reliable; that is, the probability that it will produce a defective item is 0.0056. If it was found to produce 43 defective items, what is the expected number of items produced? ▲

10. If some event has the probability 0.102, what probability does its denial have?

10.3 Sample Spaces with Equiprobable Events

11. What is the probability that an element will be in the intersection of a set and its complement?

12. What is the probability that an element will be in the union of a set and its complement?

13. An event X can occur in five ways, after which another event Y can occur in three ways. Make up a tree diagram for the two events to occur in succession.

▲ 14. If a coin is tossed three times, each of the tosses has the possible outcomes heads H or tails T. Make up a tree diagram for the three tosses.

15. In the three-coin toss, let the eight sample points be $X_1 = $ (H,H,H); $X_2 = $ (H,H,T); $X_3 = $ (H,T,H); $X_4 = $ (H,T,T); $X_5 = $ (T,H,H); $X_6 = $ (T,H,T); $X_7 = $ (T,T,H); and $X_8 = $ (T,T,T).

   a. What is the probability of each of the simple events $\{X_i\}$, $i = 1, ..., 8$.

   b. Describe in words the event $\{X_2, X_3, X_5\}$.

   c. What is the probability of the set in part b.

   d. Write as a set of sample points the event that at least two tails occur.

   e. What is the probability that *exactly* two tails occur?

   f. What is the probability that *at least* two tails occur?

16. A magnetic tape manufacturer makes recording tapes in three different materials: polyester, acetate, and polyvinyl chloride. The tapes are made in two thicknesses: 1.0 mil and 1.5 mil. Finally, the tapes are sold in two different reel sizes: 5 inches and 7 inches.

   a. How many different types of tape packages are produced?

   b. Draw a tree diagram for these outcomes.

17. If in Problem 16 above, only the polyester can also be made in a third thickness, 0.5 mil. Draw a tree diagram for the entire process.

18. When a baseball player comes up to bat, there are eight possible ways that his team can have runners on the bases. (One of them is that there will be no runners on base.) Describe the other seven. Make a tree diagram.

Probabilities in dice In Problems 19 through 24 use the following description. Let $S = $ the set of all possible outcomes from throwing a pair of dice (see the table below). The first die can have six results and the second die can also have six. Both dice together have $6 \cdot 6 = 36$ outcomes. All 36 sample points are listed in the table below. The notation (3,5) means that three is the number showing on the top of the first die and five is the number on the second. Note that the designations of the dice as "first" and "second" have nothing to do with the order in which they are thrown but are merely for the convenience of identifying the two dice so that the sample space can be computed.

Set B — Four or five on second die.

	Second Die					
First Die	1	2	3	4	5	6
1	(1,1)	(1,2)	(1,3)	(1,4)	(1,5)	(1,6)
2	(2,1)	(2,2)	(2,3)	(2,4)	(2,5)	(2,6)
3	(3,1)	(3,2)	(3,3)	(3,4)	(3,5)	(3,6)
4	(4,1)	(4,2)	(4,3)	(4,4)	(4,5)	(4,6)
5	(5,1)	(5,2)	(5,3)	(5,4)	(5,5)	(5,6)
6	(6,1)	(6,2)	(6,3)	(6,4)	(6,5)	(6,6)

Set A — Total of six, seven, or eight.

19. How many ways can a total (on both dice) of seven be thrown?

20. How many ways can a total of ten be thrown?

For Problems 21 and 22, use the sets A and B defined by A = the event that a total of six, seven, or eight is thrown and B = the event that a four or five occurs on the second die.

21. Find $n(S)$, $n(A)$, $n(B)$, $P(A)$, and $P(B)$.

22. If $A \cap B$ = the event that a throw will total six, seven, or eight, *and* have a four or five on the second die, find $P(A \cap B)$.

23. What is the probability that the total on both dice is *not* nine?

24. What are the odds against throwing a total of three on the dice? ▲

10.4

ARRANGING AND COUNTING EVENTS—PERMUTATIONS AND COMBINATIONS

The relation between counting and probability was established in the last section. Two problems associated with counting are **permutations** and **combinations**.

10.4 Arranging and Counting Events—Permutations and Combinations

A *permutation* is an arrangement of a set of things in a specific order. The same set of elements may have a variety of permutations. For example, the letters X, Y, and Z can be arranged into the six permutations XYZ, XZY, YXZ, YZX, ZXY, and ZYX. Each such arrangement is counted as a different permutation.

A *combination* of elements is the collection of elements, without counting the different arrangements as different combinations. For example, there is only one combination of the three letters X, Y, and Z; it is just the set {X,Y,Z}.

Four elements can be arranged into 24 permutations and five elements can be arranged into 120 permutations. In general, k elements can be arranged into $k!$ permutations, where $k!$ (read "k factorial") is, loosely, the product of all the integers from k down to one. The precise definition of $k!$ includes the statement that $0! = 1$. (There is only one way to arrange no elements.)

Definition (Factorial)
If k is a nonnegative integer, then,

$$k! = \begin{cases} 1 \text{ if } k = 0. \\ k \cdot (k-1)! \text{ if } k \geq 1. \end{cases}$$

Example 10.18 $0! = 1.$

$1! = 1 \cdot 0! = 1 \cdot 1 = 1.$

$2! = 2(2-1)! = 2 \cdot 1! = 2 \cdot 1 = 2.$

$3! = 3 \cdot 2! = 3 \cdot 2 = 6.$

$4! = 4 \cdot 3! = 4 \cdot 6 = 24.$

$5! = 5 \cdot 4! = 120.$

$6! = 6 \cdot 5! = 720.$

Alternate ways to write factorials for $n \geq 2$ are:

$2! = 2 \cdot 1.$

$3! = 3 \cdot 2 \cdot 1.$

$4! = 4 \cdot 3 \cdot 2 \cdot 1.$

$5! = 5 \cdot 4 \cdot 3 \cdot 2 \cdot 1$, etc.

Example 10.19 A manufacturer uses a five-step process that can be done in any order. He wants to try out every possible permutation. He decides to try each one for a month. How many years will it take for him to complete this experiment?

Solution Since there are five steps, the number of permutations is $5! = 120$, so at one permutation per month, it will take him ten years.

It is important for the reader to understand why $k!$ is the formula for the number of permutations of k objects. If k distinct items are to be arranged into k positions, how many choices are available for the first position? The answer is k. Now, after one item has been placed, how many choices are left for the next position? The answer is $k - 1$. Next, there are only $k - 2$ of the items left for the third position, etc. Finally after $k - 1$ positions have been filled, there is only one choice left for the last position. Applying the multiplication principle, all k positions can be filled in $k \cdot (k - 1) \cdot (k - 2) \cdot \ldots \cdot 3 \cdot 2 \cdot 1$ different ways. This expression is equal to $k!$ for $k \geq 1$.

Example 10.20 A developer has six house plans. He wants to build them all on one block of model homes in a housing development. How many different arrangements of these six houses can he make?

Solution 720.

Usually permutation problems are complicated by having fewer positions to fill than there are items with which to fill them.

Example 10.21 A little league baseball manager has five good hitters and is required to make a list of three lead-off batters in a specific order. How many such batting lineups for the first three positions can he compose?

Solution Here there are three positions to be filled and five people to fill them. The first position can be filled in five ways. After that is done, the second position can be filled by any one of the four remaining hitters, and finally, the third position can be filled in three ways. Therefore, he has $5 \cdot 4 \cdot 3 = 60$ batting lineups possible for the permutations of five hitters, three at a time. ▲

Notice that the answer $5 \cdot 4 \cdot 3$ is *not* $5!$ (which is $5 \cdot 4 \cdot 3 \cdot 2 \cdot 1$). Only three positions were to be filled, although there were five choices at the beginning. This expression, $5 \cdot 4 \cdot 3$, is a truncated factorial cut off before the last two factors, $2 \cdot 1$. It may be written as $5!/2!$. In other words,

$$5 \cdot 4 \cdot 3 = \frac{5 \cdot 4 \cdot 3 \cdot (2 \cdot 1)}{2 \cdot 1} = \frac{5!}{2!}.$$

10.4 Arranging and Counting Events—Permutations and Combinations

Incidentally, there are several other symbols representing the choice of five things permuted three at a time. Some of them are $P(5,3)$, $_5P_3$, $P_{5,3}$, P_3^5, or most useful $5!/(5-3)!$. More examples of truncated factorials for permutations are given below.

Example 10.22 The permutation of eight things, two at a time is

$$8 \cdot 7 = \frac{8 \cdot 7 \cdot (6 \cdot 5 \cdot 4 \cdot 3 \cdot 2 \cdot 1)}{6 \cdot 5 \cdot 4 \cdot 3 \cdot 2 \cdot 1} = \frac{8!}{6!} \quad \left(\text{or} \quad \frac{8!}{(8-2)!}\right).$$

Permutation of ten things, six at a time:

$$10 \cdot 9 \cdot 8 \cdot 7 \cdot 6 \cdot 5 = \frac{10 \cdot 9 \cdot 8 \cdot 7 \cdot 6 \cdot 5 \cdot (4 \cdot 3 \cdot 2 \cdot 1)}{4 \cdot 3 \cdot 2 \cdot 1} = \frac{10!}{4!}.$$

Permutation of thirteen things, five at a time:

$$\frac{13!}{8!} \quad \text{or} \quad 13 \cdot 12 \cdot 11 \cdot 10 \cdot 9.$$

In general, we state the definition:

Definition (**Permutation of n things k at a time**)
The permutation of n things k at a time is $n!/(n-k)!$, where n and k are nonnegative integers and $k \leq n$.

Example 10.23
a. Find the number of permutations of seven things taken three at a time.
b. Find the number of permutations of fifteen things taken twelve at a time.
c. Find the number of permutations of twenty things taken eight at a time.

Solution
a. Since the formula for the number of permutations of n things taken k at a time is $n!/(n-k)!$, and here $n = 7$ and $k = 3$, then

$$\frac{7!}{(7-3)!} = \frac{7!}{4!} = \frac{7 \cdot 6 \cdot 5 \cdot 4 \cdot 3 \cdot 2 \cdot 1}{4 \cdot 3 \cdot 2 \cdot 1}.$$

Cancelling the last four factors, 4, 3, 2, and 1, in the numerator and denominator, we get $7!/4! = 7 \cdot 6 \cdot 5 = 210$. There are 210 permutations of seven things taken three at a time.

b. Here $n = 15$ and $k = 12$; so $n!/(n-k)! = 15!/3!$; that is,

$$\frac{15 \cdot 14 \cdot 13 \cdot \ldots \cdot 4 \cdot 3 \cdot 2 \cdot 1}{3 \cdot 2 \cdot 1}$$

which is, by pocket calculator, the number 217,945,728,000.

c. Here $n = 20$ and $k = 8$; so $n!/(n - k)! = 20!/12! = 5{,}079{,}110{,}400$.

Example 10.24 If a developer has eight house plans and wants to build five of them in a block of model homes, how many different permutations could he make?

Solution $8!/(8 - 5)! = 8!/3! = 8 \cdot 7 \cdot 6 \cdot 5 \cdot 4 = 6720$.

In the above example, each different arrangement was counted as important; this does not seem realistic. Suppose, instead, that the developer had wanted to know only how many ways he could build five of the eight house plans without counting the different orders in which the houses were to be arranged. The answer to this can be computed by noticing that the number 6720 is a count of the same five houses as many times as there are ways in which those five could be arranged. That is, the same five houses are counted 5! times in the 6720 permutations. Dividing by 5! will reduce the number of times the same combination of houses are counted so that each combination is counted only once: $6720/5! = 6720/120 = 56$. In other words, there are 56 combinations of eight houses taken five at a time.

In general, the number of combinations defined in terms of the number of permutations is as follows:

Definition **(Combination of n things taken k at a time)**
The number of ways n things can be chosen k at a time, not counting different orders of arrangement as different, is $n!/(n - k)! \cdot k!$.

In this definition, the $k!$ ways that the same k objects can be arranged is divided out, so that the same combination of k objects is counted only once.

Example 10.25
a. Find the number of combinations of seven things taken three at a time.

b. Find the number of combinations of fifteen things taken twelve at a time.

c. Find the number of combinations of twenty things taken eight at a time.

10.4 Arranging and Counting Events—Permutations and Combinations

Solution   **a.** The formula for the combination of n things taken k at a time is $n!/[(n-k)!k!]$. Here $n = 7$ and $k = 3$; so the number of combinations is

$$\frac{7!}{4!\,3!} = \frac{7\cdot 6\cdot 5\cdot 4\cdot 3\cdot 2\cdot 1}{(4\cdot 3\cdot 2\cdot 1)(3\cdot 2\cdot 1)} = 35.$$

There are 35 different combinations of seven things taken three at a time.

b. Here $n = 15$ and $k = 12$; so the number of combinations is

$$\frac{15!}{3!\,12!} = \frac{15\cdot 14\cdot 13\cdot 12\cdot 11\cdot 10\cdot 9\cdot 8\cdot 7\cdot 6\cdot 5\cdot 4\cdot 3\cdot 2\cdot 1}{(3\cdot 2\cdot 1)(12\cdot 11\cdot 10\cdot 9\cdot 8\cdot 7\cdot 6\cdot 5\cdot 4\cdot 3\cdot 2\cdot 1)} = 455.$$

c. Here $n = 20$ and $k = 8$; so the number of combinations is

$$\frac{20!}{12!\,8!} = 125{,}970.$$

Example 10.26   The 26 letters of the alphabet correspond to 26 flags in the international flag code. Messages consist of two flags displayed together. For example, some messages are

   RY = The crew has mutinied.

   IX = I have been seriously damaged in a collision.

   AD = I must abandon ship (see Figure 10.2).

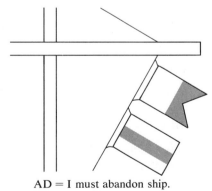

AD = I must abandon ship.

Figure 10.2

How many such two-letter messages are possible under the two assumptions given below?

Assumption 1 Assume that the *order* (one above the other) of the flags as they appear on the halyards makes a difference in the message. (The same two flags in different order are a different message.)

Assumption 2 Assume that the order of the flags on the halyards is not important and that the same message is given by the two flags regardless of which one is above the other.

Solution In Assumption 1, the number of permutations of 26 things taken two at a time is the number of messages. There are 650 permutations from the formula $26!/(26 - 2)!$, which is $26!/24! = 26 \cdot 25 = 650$ messages. In Assumption 2, the same two flags can be permuted $2!$ ways; therefore, any one combination is counted as two permutations. To count each combination only once, divide by $2!$; hence, the number of combinations possible is $650/2 = 325$ messages. ▲

Sometimes a collection of n elements consists of several different types of elements, a sampling is made and questions are asked concerning the number of each type in the sample. Usually the answers require applications of both the combination formula and the multiplication principle.

▲ **Example 10.27** A four-member community council is to be made up of eleven citizens who have volunteered. The volunteers consist of seven businessmen and four college professors.

    a. How many four-member councils can be named from the eleven volunteers?

    b. How many councils could there be with no college professors on them?

    c. How many could there be with just one college professor on them?

    d. How many could have just two professors on them?

Solution a. $11!/[(11 - 4)! \cdot 4!] = 330 =$ the number of combinations (eleven choose four).

    b. This question is asking: How many ways can the seven businessmen be chosen for the four-member council? Answer: seven choose four, or $7!/[(7 - 4)! \cdot 4!] = 35$ councils, each consisting of only businessmen.

    c. This question asks: How many ways can three of the council members be chosen from the seven businessmen, and then how many ways can one of

10.4 Arranging and Counting Events—Permutations and Combinations

the members be chosen from the four college professors. It uses the multiplication principle. *Seven choose three* multiplied by *four choose one*, in symbols:

$$\frac{7!}{(7-3)! \cdot 3!} \cdot \frac{4!}{3! \cdot 1!} = 140.$$

d. This question asks: How many ways can two of the council members be chosen from the seven businessmen, and then how many ways can two members be chosen from the four college professors? Answer:

$$\frac{7!}{5! \cdot 2!} \cdot \frac{4!}{2! \cdot 2!} = 126.$$ ▲

EXERCISE 10.4

1. In how many ways can six elements be arranged?

2. In how many ways can seven elements be arranged?

3. Find the number of ways ten things can be taken four at a time:

   a. As permutations (counting the various orders as different).

   b. As combinations (not counting the various orders as different).

4. Find the number of ways eleven things can be taken eight at a time:

   a. As permutations.

   b. As combinations.

▲ 5. The order in which candidates' names appear on a ballot influences the outcome of the election. Suppose that election officials try to eliminate this influence by printing ballots with every possible permutation (one ballot for each permutation). In a five-candidate race:

   a. How many different types of ballots must be printed in order to cover all possible permutations?

   b. If 30,000 ballots are to be printed, how many of each type described in part **a** will there be?

6. A college faculty of 5000 wishes to rank candidates in an election in which five vacancies are to be filled by choosing from a ballot of ten candidates

(that is, the voters will rank five of the ten candidates from 1 through 5). Show that there are more ways to rank the candidates than there are voters.

7. A ballot has twelve candidates and the voters are to vote for three of them to fill vacancies on the city council. In how many ways can voters select combinations of three candidates?

8. A stock exchange lists 2000 companies. If a group of 100 companies is chosen as an index (stocks whose averages are quoted everyday as a price indicator for the entire stock market), how many such indices are there? Write your answer in terms of factorials.

9. In linear programming problems, there are n equations with $n + k$ unknowns. Suppose the unknowns are made nonbasic k at a time, while the remaining n are basic. How many such systems of basic and nonbasic variables are there?

10. A town has a volunteer fire department. The firehouse dispatcher can sound three types of signals on his horn to call the volunteers. The three signals are long blast (L), short blast (S), and warbling blast (W).

    a. How many permutations of the three different blasts are possible? (Assume the same type of blast is not repeated.)

    b. How many permutations of two different blasts are there?

11. If a person tosses a penny, a nickel, a dime, a quarter, a half-dollar, and a silver dollar, how many ways can each of the following combinations occur?

    a. six heads
    b. five heads and one tail
    c. four heads and two tails
    d. three heads and three tails
    e. two heads and four tails
    f. one head and five tails
    g. six tails

12. A sociologist wishes to conduct an experiment using a model with four variables. (He wants to test his data in a theory that has four variables or *parameters*.) He has collected data that measures seven variables. Assume that the order in which the variables are put into the model makes a difference in the results. How many different four-variable models could the sociologist construct from his seven-variable data?

*13. An artist painted 28 different 1-foot-by-1-foot abstract paintings, which he continually rearranges in a large 4-feet-by-7-feet rectangle. He calls each different arrangement a different painting. How many different possible paintings does he have? (Write the answer in factorial notation.) Show that the number of different paintings is larger than the number of tons that

10.4 Arranging and Counting Events—Permutations and Combinations

the earth weighs. (The earth weighs approximately 6.6 times 10^{21} tons.) *Use a calculator.*

*14. a. See Problem 13. Show that if the artist continually rearranges his pictures every twenty seconds, he will not go through every possible arrangement before the universe doubles its present age (approximately 4.5 times 10^9 years).

   b. See Problem 13. If each small square painting can be rotated to four different positions with a different result in each position, how many different 4-feet-by-7-feet paintings does the artist have?

*15. A set of twelve objects has eight in category I and four in category II.

   a. Suppose samples of six objects are to be selected at random, how many of these six-item samples are there?

   b. How many six-item samples contain no members from category II?

   c. How many six-item samples contain exactly one member from category II?

*16. A stimulus to a sense receptor with 400 neurons can activate 100 of them.

   a. How many ways can the 400 neurons be activated 100 at a time?

   b. If 150 of the neurons are in category A and 250 are in category B (such as *rods* and *cones* in the eye), how many ways can the 100 neurons activated be in category A?

   c. How many ways can 50 of the neurons activated be in category A and 50 be in category B?

ESP problems In Problems 17 through 24 use the following description. An experiment to test extrasensory perception (ESP) uses a deck of sixteen cards, which have printed on their faces either a star, a circle, or wavy lines. A deck with the distribution ten star cards, two circle cards, and four wavy line cards is shuffled and dealt out face down. The person being tested for ESP studies the backs of the cards, trying to pick out only the star cards. Assuming random selection, answer the following questions concerning possible combinations.

   17. How many ways can seven of the sixteen cards be picked? (As a combination.)

   18. How many combinations of eight cards are there?

   19. If seven cards are picked, how many ways can all seven of them be star cards?

   20. If eight cards are picked, how many ways can all of them be star cards?

21. Use the data in Problems 17 and 19 and find the relative frequency of picking all star cards in a seven-card experiment.

22. Use the data in Problems 18 and 20 and find the relative frequency of picking all star cards in an eight-card experiment.

23. a. In a four-card experiment, how many ways can three of them be star cards?

    b. What is the chance of this occurring?

24. a. In a five-card experiment, how many ways can four of them be star cards?

    b. What is the chance of this occurring? ▲

11
CONDITIONAL PROBABILITIES

11.1

CONDITIONAL PROBABILITIES AND INDEPENDENT EVENTS

Certain conditions are more favorable for the occurrence of a given event than others are. For example, a drunk driver has a greater chance of getting into an auto accident than does a sober one. In bowling, with two pins left standing after the first ball, the bowler has a better chance to make a spare if the two pins are the 1 and 2 than he does if the pins are the 7 and 10. (See Figure 11.1.)

In the above examples, it is clear when a given event favors a subsequent event, but in some other cases, it may not be so obvious that one event is favorable to another. For example, is it true that a doctor's son has a better chance to get into medical school than does a person who is not a doctor's son? Does a given advertising campaign increase the probability of a larger profit? There are many other instances in which conditional probabilities are sought or the independence of events are questioned. The more complex questions concerning conditional probabilities and independent events require the use of hypotheses-testing techniques, which will be discussed in later chapters. Here we consider only the simplest cases.

In order to compute the probability of one event, given another event, it is important to know the number of ways in which both events can occur simultaneously, as well as the number of ways each individual event can occur.

In general, *the probability that two events E_1 and E_2 both occur, or are simultaneous, is the probability of their intersection $P(E_1 \cap E_2)$.* In an equiprobable space, the probability that E_1 and E_2 are simultaneous equals the number of elements in the intersection of E_1 and E_2 divided by the total number of elements

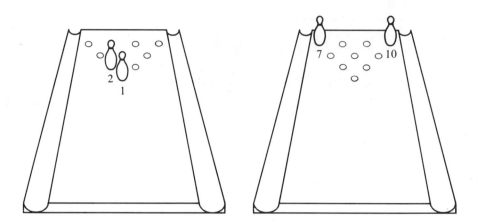

Figure 11.1
The probability of a spare, given that the 1 and 2 pins are standing, is greater than the probability of a spare, given that the 7 and 10 pins are standing.

in S. In symbols,

$$P(E_1 \cap E_2) = \frac{n(E_1 \cap E_2)}{n(S)}.$$

Example 11.1 What is the probability that a household in Delaware will both be rural and have a telephone available? (See the data in Chapter 1, Exercise 1.2, Problem 8.)

Solution Let S = the set of households in Delaware and $R \cap T$ = the intersection of the set R (rural households) and the set T (households with telephones available). Then

$$P(R \cap T) = \frac{n(R \cap T)}{n(S)} = \frac{40{,}300}{175{,}000} = 0.23,$$

that is, only 23 percent of the Delaware households both are rural and have a telephone available. The probability of the simultaneous event is 0.23.

From the probability of simultaneous events, we obtain the probability that an event A will occur, given that the event B has already occurred. This is defined as the *conditional probability of A given B*.

11.1 Conditional Probabilities and Independent Events

Definition **(Conditional probability)**
The probability of the event A, given the event B, is equal to the *number of elements in the intersection of A and B divided by the number of elements in the set B* (for equiprobable spaces), $n(A \cap B)/n(B)$.

The conditional probability is denoted by $P(A|B)$ and is read as follows: "the probability that event A occurs given that event B occurred."
We write this definition as Formula (I) for a conditional probability:

(I) $$P(A|B) = \frac{n(A \cap B)}{n(B)}.$$

Example 11.2 The conditional probability that a household in Delaware has a telephone available, given it is rural, is

$$P(T|R) = \frac{n(R \cap T)}{n(R)} = \frac{40,300}{46,400} = 0.87.$$

An important alternative to Formula (I) is obtained by dividing both the numerator and the denominator by $n(S)$, where S is the sample space:

(II) $$P(A|B) = \frac{n(A \cap B)}{n(B)} = \frac{\frac{n(A \cap B)}{n(S)}}{\frac{n(B)}{n(S)}} = \frac{P(A \cap B)}{P(B)}.$$

Note that in either formula, the second set (the *given* set) is the one that appears in the denominator.
Continuing with Example 11.2, find the conditional probability $P(T|R)$ by Formula (II).

Solution $P(R) = {}^{46,400}\!/_{175,000} = 0.264$, and $P(R \cap T) = 0.23$ as before. Therefore,

$$P(T|R) = \frac{P(R \cap T)}{P(R)} = \frac{0.23}{0.264} = 0.87.$$

Additional interesting information can be obtained by *reversing the condition* and asking what is the probability of B, given A? For example, what is the probability that a household is rural, given that it has a telephone? The answer to such a question would be

$$P(R|T) = \frac{P(R \cap T)}{P(T)},$$

and since $P(T) = 0.855$, then $P(R|T) = 0.23/0.855 = 0.27$.

Example 11.3 Let E = the event that an automobile trip of ten miles is made at excessive speeds, and let A = the event that an accident occurs. Suppose the probability that a car both speeds (for ten miles) and has an accident is 0.0006. In symbols, $P(A \cap E) = 0.0006$. Suppose the probability that a driver will exceed the speed limit on such a trip is 0.11; that is, $P(E) = 0.11$.

a. Find the probability that a car will have an accident, given that the driver speeds on a ten-mile trip.

b. Suppose, in addition, it is known that the probability that a driver is not speeding when he has an accident is 0.002. In other words, $P(A \cap -E) = 0.002$. Find the probability of an accident, given that he was not speeding.

Solution The conditional probability sought is

a. $P(A|E) = P(A \cap E)/P(E) = 0.0006/0.11 = 0.0055$.

b. $P(A|-E) = P(A \cap -E)/P(-E)$.

To get $P(A|-E)$, we need the value for $P(-E)$, which we find from knowing $P(E)$. $P(E) = 0.11$; therefore, $P(-E) = 1 - P(E) = 1 - 0.11 = 0.89$. We now have $P(A|-E) = 0.002/0.89 = 0.0022$; that is, the probability of having an accident while speeding is more than twice as great as the probability of having an accident while not speeding. ▲

Sometimes a given event will not affect the probability of a second event. In those cases, the events are said to be **independent.**

Example 11.4 What is the probability that a card selected at random is a face card, given that it is not a heart?

Solution Let F = the event that a face card is selected, and $-H$ = the event that a heart is not selected. Then by Formula (I),

$$P(F|-H) = \frac{n(F \cap -H)}{n(-H)} = \frac{9}{39} = 0.23.$$

11.1 Conditional Probabilities and Independent Events

But 0.23 is also the probability of selecting a face card without regard to suit. (Since 3 of 13 cards in any suit are face cards and $3/13 = 0.23$.) Reversing the condition, what is the probability that a card selected at random is not a heart, given that it is a face card? The solution is

$$P(-H|F) = \frac{n(-H \cap F)}{n(F)} = \frac{9}{12} = 0.75.$$

But three-fourths of the suits are not hearts; therefore, the condition that the card selected is a face card did not affect the outcome.

These results occur because the same proportion of each suit is face cards ($3/13 = 0.23$) and because the face cards have the same proportion of nonheart suits as the entire deck does ($3/4 = 0.75$). This means that the two events, F and $-H$, are independent of each other. A general definition of independence is given below, but first let us introduce some simplifying notation.

Notational definition A convenient substitute notation for the intersection symbol $\cap$ is the *dot notation*. Thus, for any two sets, A and B, we write

$$A \cap B = A \cdot B.$$

The dot notation was, historically, the first notation for intersection. (Also, $+$ was used for unions.) With the use of the dot notation, the Formulas (I) and (II) can be written

(I) $$P(A|B) = \frac{n(A \cdot B)}{n(B)}$$

(II) $$P(A|B) = \frac{P(A \cdot B)}{P(B)}.$$

We now state the general definition of independence.

Definition **(Independent events)**
If A and B are events for which the probability of A, given B, is the same as the probability of A itself, then A is *independent* of B. In symbols, if $P(A|B) = P(A)$, then A is independent of B.

Example 11.5 In a sample of 175 people, it was found that 105 of them had a certain personality characteristic. (Call having this characteristic the event X.) Also 70 of them had a certain disease. (Call having the disease the event Y.) Furthermore, 42 of them had both the personality characteristic and the disease. Is having the disease independent of having the personality characteristic?

Solution To see if Y is independent of X, we need to see if $P(Y|X) = P(Y)$. Here $n(S) = 175$, $n(X) = 105$, $n(Y) = 70$, and $n(X \cdot Y) = 42$. Now,

$$P(Y|X) = \frac{n(X \cdot Y)}{n(X)} = \frac{42}{105} = 0.40, \text{ and } P(Y) = \frac{n(Y)}{n(S)} = \frac{70}{175} = 0.40.$$

Therefore, Y is independent of X; that is, a person had the same probability of having the disease whether or not he had the personality characteristic.

Figure 11.2 illustrates the concept that a set A is independent of a set B. Notice that this occurs only when the number of elements in $A \cdot B$ has the same ratio to $n(B)$ as $n(A)$ has to $n(S)$.

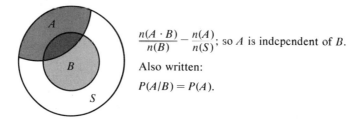

$\frac{n(A \cdot B)}{n(B)} = \frac{n(A)}{n(S)}$; so A is independent of B.

Also written:

$P(A/B) = P(A)$.

Figure 11.2

In Example 11.5, having the disease (event Y) was found to be independent of having the personality trait (event X), but is having the personality trait independent of having the disease? Yes, because

$$P(X|Y) = \frac{n(X \cdot Y)}{n(Y)} = \frac{42}{70} = 0.60, \text{ and } P(X) = \frac{n(X)}{n(S)} = \frac{105}{175} = 0.60.$$

$P(X|Y) = P(X)$; so X is independent of Y.

In general, if any event is independent of a second event, then the second event is independent of the first, as demonstrated below. Suppose A is independent of B. Then $P(A|B) = P(A)$, or

$$\frac{n(A \cdot B)}{n(B)} = \frac{n(A)}{n(S)}. \tag{11.1}$$

11.1 Conditional Probabilities and Independent Events

Now multiply both sides of Equation (11.1) by $n(B)$ and divide by $n(A)$:

$$\frac{n(A \cdot B)}{n(A)} = \frac{n(B)}{n(S)}. \tag{11.2}$$

Equation (11.2) says the same thing as $P(B|A) = P(B)$, which proves that B is independent of A.

Furthermore, when A and B are independent events (that is, independent of each other), then $P(A \cdot B) = P(A) \cdot P(B)$. This is called the **product rule** for independent events. A proof may be derived from Equation (11.1) above. Multiply both sides by $n(B)$, getting

$$n(A \cdot B) = \frac{n(A)}{n(S)} \cdot n(B). \tag{11.3}$$

Now divide both sides of Equation (11.3) by $n(S)$; this yields

$$\frac{n(A \cdot B)}{n(S)} = \frac{n(A)}{n(S)} \cdot \frac{n(B)}{n(S)}. \tag{11.4}$$

Equation (11.4) says the same thing as $P(A \cdot B) = P(A) \cdot P(B)$.

From Example 11.5, having the disease and having the personality trait are independent; so it should be true that $P(X \cdot Y) = P(X) \cdot P(Y)$, and it is:

$$P(X \cdot Y) = \frac{42}{175} = 0.24, \quad \text{and} \quad P(X) \cdot P(Y) = (0.60) \cdot (0.40) = 0.24.$$

The product rule is also true for more than two independent events. For example, if A, B, and C are independent events, then $P(A \cdot B \cdot C) = P(A) \cdot P(B) \cdot P(C)$.

EXERCISE 11.1

1. a. Compute the conditional probability $P(X|Y)$, if $P(X \cap Y) = 0.03$ and $P(Y) = 0.035$.
   b. If, furthermore, $P(X) = 0.24$, find $P(Y|X)$.
2. Given $P(M) = 1/4$, $P(N) = 1/3$, and $P(M \cap N) = 1/4$, find $P(M|N)$ and $P(N|M)$.
3. If $P(A|B) = 0.056$ and $P(B) = 0.08$, find $P(A \cap B)$.
4. If $P(X|Y) = 0.37$ and $P(Y) = 0.68$, find $P(X \cap Y)$.

5. If $P(A \cap B) = 0.15$ and $P(-B) = 0.10$, find $P(A|B)$.
6. If $P(X \cap Y) = 0.60$ and $P(-Y) = 0.35$, find $P(X|Y)$.
7. Compute, by either formula, the conditional probabilities $P(A|B)$ and $P(B|A)$, if A and B are subsets of S, and $n(S) = 800$, $n(A) = 413$, $n(B) = 265$, and $n(A \cap B) = 135$. Hint: Use the Venn diagram in the figure on the right.

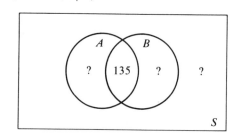

8. Compute the conditional probabilities $P(X|Y)$ and $P(Y|X)$ if X and Y are subsets of S, and $n(S) = 500$, $n(X) = 240$, $n(Y) = 270$, and $n(X \cap Y) = 210$. See the accompanying figure.

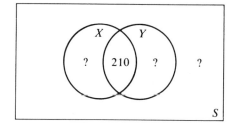

9. From the information in Problem 7, compute the conditional probabilities $P(A|-B)$, $P(-A|B)$, $P[(A \cap B)|B]$, and $P[(A \cap B)|-B]$.

10. From the information in Problem 8, compute the probabilities $P(X|-Y)$, $P(-X|Y)$, $P[(X \cap Y)|Y]$, and $P[(X \cap Y)|-Y]$.

11. From Problem 7, how many elements are in $A \cup B$? Is $n[A \cap (A \cup B)] = n(A)$? Compute $P[A|(A \cup B)]$.

12. From Problem 8, compute $P[X|(X \cup Y)]$.

*13. If $P(M|L) = 0.60$, $P(L) = 0.40$, and $P(M) = 0.80$, find $P(M \cdot L)$, $P(L|M)$, $P(L \cup M)$, and $P[(L \cup M)|-M]$. Hint: Compute $P(M \cdot L)$, then introduce Venn diagrams with M and L as subsets of S. Let $P(M \cdot L) = n(M \cap L)/n(S)$ and choose $n(S)$ to be 100.

*14. If $P(A) = 0.37$, $P(B) = 0.50$, $P(C) = 0.70$, $P(A \cdot B) = 0.25$, $P(A \cdot C) = 0.20$, $P(B \cdot C) = 0.40$, and $P(A \cdot B \cdot C) = 0.15$, then find

    a. $P[A|(B \cdot C)]$ c. $P[(A \cdot C)|B]$ e. $P[(A \cup B)|C]$

    b. $P[C|(B \cdot C)]$ d. $P[(A \cdot B)|C]$ f. $P[(A \cup B)|-C]$

▲ 15. The probability that an accident will occur in ten miles of night-time driving (for one vehicle) is 0.015. The probability that a night-time driver is drunk is 0.012. The probability that a person driving ten miles at night will both

11.1 Conditional Probabilities and Independent Events

be drunk and have an accident is 0.008. Suppose a person goes on a ten-mile drive at night.

  a. Find the probability that he will have an accident, given that he is drunk.

  b. Find the probability that he is drunk, given that he has an accident.

16. Let A = the event that a stockholder consults his horoscope before doing any trading on the stock market, and B = the event that he makes a successful transaction. Suppose $P(A \cap B) = 0.49$, $P(A) = 0.88$, $P(B) = 0.56$, and $P(A \cap -B) = 0.39$.

  a. Compute the probability that he will make a successful transaction, given that he consulted his horoscope.

  b. Compute the probability that he consulted his horoscope, given that he did not make a successful transaction.

  c. Compute the probability that he consulted his horoscope, given that he had a successful transaction.

  d. Do the events A and B seem to be independent, at least for probabilities accurate to two decimal places?

17. If three events, A, B, and C, are independent, then $P(A \cdot B \cdot C) = P(A) \cdot P(B) \cdot P(C)$. Let three dice be thrown and let

   A = the event that a 1 appears on the first die.

   B = the event that a 1 appears on the second die.

   C = the event that a 1 appears on the third die.

   All three events are independent, and if the dice are fair $P(A) = P(B) = P(C) = 1/6$.

  a. Compute the probability of the simultaneous event $A \cdot B \cdot C$.

  b. Describe in words the event $-A$.

  c. The events, $-A$, $-B$, and $-C$, are independent. Compute the probability of each and compute the probability $P[(-A) \cdot (-B) \cdot (-C)]$.

  d. $-A$, B, and $-C$ are independent events. Compute $P[(-A) \cdot B \cdot (-C)]$.

18. If three coins are tossed, let

   X_1 = the event that heads appears on the first coin.

   X_2 = the event that heads appears on the second coin.

   X_3 = the event that heads appears on the third coin.

All three events are independent and $P(X_i) = \frac{1}{2}$, for $i = 1, 2,$ and 3.

a. Compute $P(X_1 \cdot X_2 \cdot X_3)$.

b. Describe $-X_2$ in words.

c. $-X_1, -X_2,$ and $-X_3$ are independent. Compute $P[(-X_1) \cdot (-X_2) \cdot (-X_3)]$.

d. Compute $P[(-X_1) \cdot X_2 \cdot (-X_3)]$. ▲

11.2

TOTAL PROBABILITY

The probability that two events occur simultaneously was discussed in the preceding section. We now wish to examine the probability that one or another of two events occurs, either singly or together.

First, let us consider the case in which the two events cannot occur simultaneously. In other words, what is the probability that one of two logically incompatible events will occur? Such events may be represented by two disjoint subsets of a sample space, and the probability that one or the other of them occurs is the probability of the union of the two sets.

In symbols, this question becomes: If A and B are events and $A \cap B = \emptyset$, what is $P(A \cup B)$? In this situation, A and B have no points in common, so there is no way that both events can occur simultaneously. We know from properties of sets that the number of elements in the union of two *disjoint* sets is equal to the sum of the number of elements in each of the two sets; that is,

$$n(A \cup B) = n(A) + n(B), \quad \text{when} \quad A \cap B = \emptyset.$$

If S is the sample space of which A and B are subsets, then dividing both sides of the above equation by $n(S)$ yields

$$\frac{n(A \cup B)}{n(S)} = \frac{n(A) + n(B)}{n(S)} = \frac{n(A)}{n(S)} + \frac{n(B)}{n(S)}, \quad \text{when} \quad A \cap B = \emptyset$$

or

$$P(A \cup B) = P(A) + P(B), \quad \text{when} \quad A \cap B = \emptyset.$$

▲ **Example 11.6** Most chess players have a favorite first move, but they sometimes try a variety of opening moves. It is impossible, however, for them to make two different first moves in the same game; that is, the event that a player makes any one first move is logically incompatible with the event that he makes a different first move.

Suppose Anatoly Karpov opens 1. P-K4 with a probability of 0.75 and opens 1. P-QB4 with a probability of 0.10. What is the probability that he will play either 1. P-K4 or 1. P-QB4?

Solution Since the two events are disjoint, the probability of their union is the sum of the two probabilities, so $0.75 + 0.10 = 0.85$ is the probability that the world chess champion, Karpov, will move either 1. P-K4 or 1. P-QB4 on his first move.
▲

Example 11.7 What is the probability that a pair of dice will show a seven or a ten?

Solution Let T_7 = the set of throws in which a seven appears; that is, T_7 = the subset $\{(1,6), (2,5), (3,4), (4,3), (5,2), (6,1)\}$ of the set S of all 36 possible dice throws. (See Exercise 10.3.) Since $n(S) = 36$ and $n(T_7) = 6$, then $P(T_7) = 6/36 = 1/6$. Let T_{10} be the set of throws showing a ten on the pair of dice. $T_{10} = \{(4,6), (5,5), (6,4)\}$. Therefore, $P(T_{10}) = 3/36 = 1/12$.

These two events are incompatible since they cannot occur simultaneously (the same throw cannot be both a seven and a ten); that is $T_7 \cap T_{10} = \emptyset$. By the above equation for the probability of the union of two disjoint sets, $P(T_7 \cup T_{10}) = 1/6 + 1/12 = 1/4$.

As we have just seen, it is not difficult to compute the probability for the union of two disjoint sets. Unfortunately, *when the sets overlap, we cannot use this same method. If $A \cap B$ is not empty, then the probability of the union $P(A \cup B)$ is not the sum, $P(A) + P(B)$.*

Example 11.8 If there is an 80 percent chance of rain tomorrow and a 60 percent chance that the temperature will drop, what is the probability that it will either rain or the temperature will drop? Obviously, the answer *is not* $0.80 + 0.60$, since this sum is greater than one, and the probability must be less than or equal to one. The difficulty in this problem arises from the fact that *there is some chance that both of these events may occur at the same time.* The problem can not even be solved until more information is known.

In general, when two sets, A and B, are not disjoint ($A \cap B \neq \emptyset$), we need to know the probability of the intersection before we can compute the probability of the union $A \cup B$.

Example 11.9 In Exercise 10.3, Problem 21 (page 312), set $A =$ all dice throws showing a total of six, seven, or eight; and $B =$ throws having a four or five on the second die. In this case, $n(A) = 16$ and $n(B) = 12$, but $n(A \cup B)$ is not 28. To obtain the number of elements in $A \cup B$ by adding the number of elements in A to the number in B would be a mistake (counting the elements in $A \cap B$ twice). By looking at the table of sample points in Exercise 10.3 and counting, we find that $n(A \cup B) = 22$; so $P(A \cup B) = {}^{22}/_{36} = {}^{11}/_{18}$.

As an alternative to counting the number of elements in $A \cup B$, we could add the number of elements in A to the number of elements in B, *and subtract the number of elements* (counted twice) in $A \cap B$. So,

$$n(A \cup B) = n(A) + n(B) - n(A \cap B).$$

Dividing by $n(S)$ gives

$$\frac{n(A \cup B)}{n(S)} = \frac{n(A)}{n(S)} + \frac{n(B)}{n(S)} - \frac{n(A \cap B)}{n(S)}.$$

This gives us the general formula for the probability that an element is in the union of two given sets:

$$P(A \cup B) = P(A) + P(B) - P(A \cap B).$$

This formula is valid, even in the previous case where the intersection of A and B is empty. In logical terms, it is a formula for the probability of the disjunction of two events (the probability that one, or another, or both of two events occurs).

Example 11.10 Find the probability that a toss of a pair of dice will result in a number greater than seven *or* will have at least one of the dice come up a four.

Solution Let $M =$ the set of throws in which the total is greater than seven. Let $K =$ the set of throws in which either the first or the second die comes up a four. By counting the elements in the subsets, we get $n(M) = 15$, $n(K) = 11$, and $n(M \cap K) = 5$. So, $P(M) = {}^{15}/_{36}$, $P(K) = {}^{11}/_{36}$, and $P(M \cap K) = {}^{5}/_{36}$. Therefore, from the general formula,

$$P(M \cup K) = \frac{15}{36} + \frac{11}{36} - \frac{5}{36} = \frac{21}{36} = \frac{7}{12}.$$

A surprising amount of information can be derived from knowing just a few probabilities. For example, suppose that the three probabilities $P(A)$, $P(B)$, and

11.2 Total Probability

$P(A \cap B)$ are known; then the probabilities $P(-A)$, $P(-B)$, $P[-(A \cap B)]$, and $P[-(A \cup B)]$ can be computed. Also, we can compute $P(-A|-B)$ by using the two general set relationships called *De Morgan's laws*, stated below.

DE MORGAN'S LAWS

Given any two sets

1. The complement of their union is the intersection of their complements.
$$-(A \cup B) = -A \cap -B, \quad \text{for any two sets} \quad A \text{ and } B.$$

2. The complement of their intersection is the union of their complements.
$$-(A \cap B) = -A \cup -B, \quad \text{for any two sets} \quad A \text{ and } B.$$

Thus, by De Morgan's law 1,

$$P(-A|-B) = \frac{P(-A \cap -B)}{P(-B)} = \frac{P[-(A \cup B)]}{P(-B)}.$$

Example 11.11 Let R = the event it will rain tomorrow and S = the event it will snow tomorrow. Suppose $P(R) = 0.60$, $P(S) = 0.30$, and $P(R \cap S) = 0.10$.

a. What is the probability that it will either not snow or not rain?

b. What is the probability it will either snow or rain?

c. What is the probability that it will not rain, given that it does not snow?

Solution **a.** $P(-S \cup -R) = P[-(S \cap R)]$ by De Morgan's law 2.
$= 1 - P(S \cap R)$ from the probability of complements.
$= 0.90$ from the given data that $P(S \cap R) = 0.10$.

b. $P(S \cup R) = P(S) + P(R) - P(S \cap R) = 0.60 + 0.30 - 0.10 = 0.80$.

c. $P(-R|-S) = P(-R \cap -S)/P(-S)$.
But from De Morgan's law 1, $P(-R \cap -S) = P[-(R \cup S)]$, and this is equal to $1 - P(R \cup S) = 1 - 0.80 = 0.20$ (part b). Also, we know $P(-S) = 1 - P(S) = 1 - 0.30 = 0.70$. Now

$$P(-R|-S) = \frac{1 - 0.80}{1 - 0.30} = \frac{0.20}{0.70} = \frac{2}{7} = 0.29.$$

Another interesting (and valuable) formula that can be derived from knowing $P(A)$, $P(B)$, $P(A \cap B)$ is

$$P(A \cap -B) = P(A) - P(A \cap B).$$

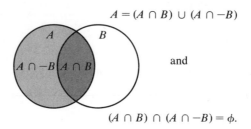

$A = (A \cap B) \cup (A \cap -B)$

and

$(A \cap B) \cap (A \cap -B) = \phi.$

Figure 11.3

Proof

1. By the Venn diagram in Figure 11.3, the set $A = (A \cap B) \cup (A \cap -B)$, for any sets A and B.

2. $P(A) = P[(A \cap B) \cup (A \cap -B)]$, by the equality in Step 1. Also, $A \cap B$ is disjoint from $A \cap -B$, as seen in Figure 11.3.

3. Now from the formula for the probability of the union of two disjoint sets, we get

$$P(A) = P(A \cap B) + P(A \cap -B).$$

4. Solve this equation for $P(A \cap -B)$ to get $P(A \cap -B) = P(A) - P(A \cap B)$, which is what we wanted to prove. Similarly, we can derive the following formulas

$$P(B \cap -A) = P(B) - P(A \cap B),$$

$$P(A|-B) = \frac{P(A) - P(A \cap B)}{P(-B)} = \frac{P(A) - P(A \cap B)}{1 - P(B)}.$$

Example 11.12 A man goes fishing; the simple events are A = the event he catches a perch and B = the event he catches a bass. Suppose the probabilities are $P(A) = 0.83$, $P(B) = 0.22$, and $P(A \cap B) = 0.16$.

11.2 Total Probability

a. What is the probability that he will catch a perch and no bass?
b. What is the probability that he will catch a bass and no perch?
c. What is the probability that he catches a bass, given he catches no perch?
d. What is the probability that he catches a perch, given he catches no bass?
e. What is the probability that he either catches a perch or catches no bass?

Solution
a. $P(A \cap -B) = 0.83 - 0.16 = 0.67$.
b. $P(B \cap -A) = 0.22 - 0.16 = 0.06$.
c. $P(B|-A) = [P(B) - P(A \cap B)]/[1 - P(B)] = {}^{0.06}\!/\!_{0.17} = 0.35$.
d. $P(A|-B) = [P(A) - P(B \cap A)]/[1 - P(A)] = {}^{0.67}\!/\!_{0.78} = 0.86$.

Table 11.1 Table of Probability Formulas Combining Conditional and Total Probabilities for Two Sets, A and B.
Given $P(A) = a$, $P(B) = b$, $P(A \cap B) = c$.

Number	Expression	Any A and B (Assume $a^2 \neq a$ and $b^2 \neq b$ where necessary.)	A and B independent	$A \cap B = \emptyset$
1	$P(A)$	a	a	a
2	$P(B)$	b	b	b
3	$P(A \cap B)$	c	ab	0
4	$P(A \mid B)$	c/b	a	0
5	$P(B \mid A)$	c/a	b	0
6	$P(-A)$	$1 - a$	$1 - a$	$1 - a$
7	$P(-B)$	$1 - b$	$1 - b$	$1 - b$
8	$P(A \cup B)$	$a + b - c$	$a + b - ab$	$a + b$
9	$P[-(A \cap B)]$	$1 - c$	$1 - ab$	1
10	$P(-A \cup -B)$	$1 - c$	$1 - ab$	1
11	$P[-(A \cup B)]$	$1 + c - a - b$	$1 + ab - a - b$	$1 - a - b$
12	$P(-A \cap -B)$	$1 + c - a - b$	$1 + ab - a - b$	$1 - a - b$
13	$P(A \cap -B)$	$a - c$	$a(1 - b)$	a
14	$P(B \cap -A)$	$b - c$	$b(1 - a)$	b
15	$P(A \cup -B)$	$1 + c - b$	$1 + ab - b$	$1 - b$
16	$P(B \cup -A)$	$1 + c - a$	$1 + ab - a$	$1 - a$
17	$P(-A \mid -B)$	$(1 + c - a - b)/(1 - b)$	$1 - a$	$(1 - a - b)/(1 - b)$
18	$P(-B \mid -A)$	$(1 + c - a - b)/(1 - a)$	$1 - b$	$(1 - a - b)/(1 - a)$
19	$P(-A \mid B)$	$(b - c)/b$	$1 - a$	1
20	$P(A \mid -B)$	$(a - c)/(1 - b)$	a	$a/(1 - b)$
21	$P(B \mid -A)$	$(b - c)/(1 - a)$	b	$b/(1 - a)$
22	$P(-B \mid A)$	$(a - c)/a$	$1 - b$	1

e. $P(A \cup -B) = P(A) + P(-B) - P(A \cap -B) = 0.83 + 0.78 - 0.67 = 0.94$. ▲

There is really an amazing amount of information that can be derived from knowing just $P(A)$, $P(B)$, and $P(A \cap B)$ for two sets A and B. Table 11.1 summarizes some of it. Notice that the solutions to parts **a, b, c, d,** and **e** in Example 11.12 above correspond to table entries 13, 14, 21, 20, and 15, respectively.

With more than two sets, the formulas become more complicated. As an illustration and for reference, we give the following two statements.

1. If A, B, and C are subsets of S, then

   a. $P[(B \cup C)|A] = P(B|A) + P(C|A) - P(B \cdot C|A)$.

   b. If, in addition, B and C are disjoint, then $P[(B \cup C)|A] = P(B|A) + P(C|A)$.

   c. Furthermore, if $C = -B$, then $P[(B \cup C)|A] = P(S|A) = 1$.

2. If A, B, and C are events, then $P(A \cdot B \cdot C) = P(A) \cdot P(B|A) \cdot P(C|A \cdot B)$. In general, for a collection $A_1, A_2, ..., A_n$ of sets,

$$P(A_1 \cdot A_2 \cdot A_3 \cdot ... \cdot A_n) = P(A_1) \cdot P(A_2|A_1) \cdot P(A_3|A_1 \cdot A_2) \cdot ... \cdot P(A_n|A_1 \cdot A_2 \cdot ... \cdot A_{n-1}).$$

EXERCISE 11.2

1. If $P(A) = 1/6$, $P(B) = 1/9$, and $P(A \cap B) = 1/10$, find

   a. $P(A \cup B)$
   b. $P(A|B)$
   c. $P[A|(A \cup B)]$
   d. $P[(A \cup B)|B]$

2. If $P(X) = 0.47$, $P(Y) = 0.14$, and $P(X \cap Y) = 0.10$, find

   a. $P(-X \cap -Y)$
   b. $P(X \cup -Y)$
   c. $P[X|(X \cup -Y)]$
   d. $P[(-X \cap -Y)|-X]$

▲ 3. Refer to Exercise 1.1, Problem 11. In the sample of 1000 people, what is the probability that a person selected at random will have blood type AB positive? Type A negative? Type O?

11.2 Total Probability

4. Refer to Exercise 1.1, Problem 9. In the survey of tape recorder buyers, what is the probability that a person selected at random will own either TV or FM? What is the probability that a person will own both TV and FM? ▲

5. If $P(A) = 2/5$, $P(B) = 1/3$, and $P(A \cdot B) = 1/4$, find
   a. $P(A \cup B)$
   b. $P[(A \cup B) \cdot B]$
   c. $P[(A \cup B)|B]$
   d. $P[B|(A \cup B)]$
   e. $P[(-A) \cdot (-B)]$
   f. $P(-A \cup -B)$

6. If $P(X|Y) = 0.82$, $P(X|-Y) = 0.65$, and $P(Y) = 0.48$, find
   a. $P(X \cdot Y)$
   b. $P[X \cdot (-Y)]$
   c. $P(X)$
   d. $P(Y|X)$

7. Show by Venn diagrams the De Morgan's law: $-(A \cup B) = (-A) \cdot (-B)$.

8. Show by Venn diagrams the De Morgan's law: $-(A \cdot B) = (-A) \cup (-B)$.

▲ 9. On third down, Gene, the quarterback, will pass with probability 0.70. If he passes, he will complete the pass for a first down with probability 0.60. If he does not pass, he will make a first down with probability 0.50. What is the probability that he will make a first down?

10. Use the information in Problem 9. Suppose you missed the last third-down play, but you heard that Gene had made the first down, what was the probability that he passed? ▲

11. a. Show by Venn diagrams the distributive law: $A \cdot (B \cup C) = A \cdot B \cup A \cdot C$.
    b. Show by Venn diagrams the distributive law: $A \cup (B \cdot C) = (A \cup B) \cdot (A \cup C)$.
    c. Show by Venn diagrams that $(A \cdot B) \cdot (B \cdot C) = A \cdot B \cdot C$.
    d. Is it true that $P[A \cdot (B \cup C)] = P(A \cdot B \cup A \cdot C)$?
    e. Show that $P[A \cdot (B \cup C)] = P(A \cdot B) + P(A \cdot C) - P(A \cdot B \cdot C)$.

12. Observe that $P(A \cup B \cup C) = P[A \cup (B \cup C)]$.
    a. Show that $P(A \cup B \cup C) = P(A) + P(B \cup C) - P[A \cdot (B \cup C)]$.
    b. Use parts d and e of Problem 11 and part a of this problem to derive the following result:
    $$P(A \cup B \cup C) = P(A) + P(B) + P(C) - P(B \cdot C) - P(A \cdot B) - P(A \cdot C) + P(A \cdot B \cdot C).$$

13. Show that $P[-(A \cup B)] + P[-(A \cap B)] = P(-A) + P(-B)$ for any sets A and B.

14. Show that $P(A \cap -B) + P(B) = P(B \cap -A) + P(A)$.

▲ **Meristic variability** Problems 15 through 18 refer to the following description. Sometimes plants and animals vary from generation to generation in the number or location of certain of their parts. Guinea pigs, for example, can have varying numbers of toes, and even toeless paddle-footed guinea pigs can have a variable number of toenails. In some flowers, the number of petals varies widely between generations, as does the number of vertebrae in certain fish and the number of scales on the toes of a particular lizard. This phenomenon is called *meristic variability*. Enough data on the frequency of meristic variability have been collected to establish the probabilities indicated in the following problems.*

15. Certain strains of guinea pigs have meristic variability in the number of toes on their feet. For the right front foot of a randomly selected animal, let

    X_1 = the event it has only four toes, all in good shape.

    X_2 = the event it has one imperfect toe and four good ones.

    X_3 = the event it has exactly five good toes.

    The probabilities are $P(X_1) = 0.77$, $P(X_2) = 0.13$, and $P(X_3) = 0.10$.

    a. Explain how we know that $X_1 \cap X_2 = \emptyset$ for any one given animal.
    b. Compute $P(X_1 \cup X_2)$. Is $X_1 \cap X_3 = \emptyset$? Compute $P(X_1 \cup X_3)$.
    c. Compute $P(X_1 \cup -X_2)$. Hint: Use Table 11.1.
    d. Compute $P(-X_1 \cap -X_3)$.
    e. What is the probability that a guinea pig will not have exactly five good toes?
    f. What is the probability that it will not have exactly four good toes?

16. Meristic variability is exhibited in the number of petals on the flower, *ranunculus bulbosus*. Let the various petal numbers be grouped as follows:

    X_1 = event there are exactly five or six petals.

    X_2 = event there are exactly seven or eight petals.

---

*The data for Problems 15 and 17 were derived from Wright, Sewall. 1934. "An Analysis of Variability in the Number of Digits in an Inbred Strain of Guinea Pigs," *Genetics* 19: 506-36. The data for Problems 16 and 18 were derived from De Vries, H. 1910. The Mutation Theory. Chicago: Open Court.

11.3 Partitions and Cross Partitions

X_3 = event there are exactly nine or ten petals.

X_4 = event there are exactly eleven or twelve petals.

X_5 = event there are more than twelve petals.

Studies show that $P(X_1) = 0.20$, $P(X_2) = 0.27$, $P(X_3) = 0.30$, $P(X_4) = 0.12$, and $P(X_5) = 0.11$.

   a. Explain how we know that $X_3 \cap X_4 = \emptyset$ for any one given flower.

   b. Compute $P(X_1 \cup X_2)$, $P(X_3 \cup X_4)$, and $P(X_2 \cup X_5)$.

   c. Compute $P(X_4 \cup -X_3)$ and $P(-X_1 \cup -X_2)$.

   d. What is the probability that a flower will not have just seven or eight petals?

17. Use the data in Problem 15 to compute the probability that a guinea pig will not have exactly five good toes, given that he does not have exactly four good toes. [Find $P(-X_3 | -X_1)$.]

18. Refer to Problem 16. Use Table 11.1 to answer the following questions:

   a. What is the probability that a given flower will have exactly nine or ten petals, given that it does not have exactly eleven or twelve?

   b. What is the probability that a flower will not have eleven or twelve petals, given that it does not have seven or eight? ▲

11.3
PARTITIONS AND CROSS PARTITIONS

For any given experiment, there is a given sample space of outcomes, *but it is not the only possible sample space*. There are many possible alternative sets of simple events for describing the *same* experiment.

Example 11.13   Suppose three coins are tossed and the following eight sample points are given as the outcomes:

$X_1 = (H,H,H)$ $X_3 = (H,T,H)$ $X_5 = (T,H,H)$ $X_7 = (T,T,H)$

$X_2 = (H,H,T)$ $X_4 = (H,T,T)$ $X_6 = (T,H,T)$ $X_8 = (T,T,T)$

The space S is $\{X_1, X_2, X_3, X_4, X_5, X_6, X_7, X_8\}$. Assume the coins are fair; then the space is equiprobable and the probability function is $P(X_i) = 1/8$, for all $i = 1, 2, 3, \ldots, 8$.

This sample space is extremely detailed; it tells which way each coin falls on each toss. But suppose we want to know only *how many* of the coins come up heads and how many come up tails on each toss without knowing exactly which are which? Then, we are interested in a sample space whose simple events are

Y_1 = event there are no heads.

Y_2 = event there is exactly one head.

Y_3 = event there are exactly two heads.

Y_4 = event there are exactly three heads.

Denote the sample space $\{Y_1, Y_2, Y_3, Y_4\}$ by S'.

These sample points represent an alternative sample space for the same experiment, but *this sample space is not equiprobable*. Fortunately, however, the probability function for the sample space S' can easily be determined by the relationship between S and S'. Table 11.2 shows the relationship between S and S' and the probability functions for each sample space.

Table 11.2 The Two Spaces S and S′ and Their Probability Functions

New Sample Space S'	Old Sample Space S	Probability Function P
Y_1	$\{X_8\}$	$P(Y_1) = P(X_8) = 1/8$
Y_2	$\{X_7, X_6, X_4\}$	$P(Y_2) = P(X_7) + P(X_6) + P(X_4) = 3/8$
Y_3	$\{X_5, X_3, X_2\}$	$P(Y_3) = P(X_5) + P(X_3) + P(X_2) = 3/8$
Y_4	$\{X_1\}$	$P(Y_4) = P(X_1) = 1/8$

Notice that $P(Y_1) + P(Y_2) + P(Y_3) + P(Y_4) = 1$. Also, two important features of the new space S' are

11.3 Partitions and Cross Partitions

1. The points of S' cover every possible outcome in the experiment.
2. No two sets in S' overlap; that is, $Y_1 \cap Y_2 = \emptyset$, $Y_1 \cap Y_3 = \emptyset$, $Y_1 \cap Y_4 = \emptyset$, $Y_2 \cap Y_3 = \emptyset$, $Y_2 \cap Y_4 = \emptyset$, and $Y_3 \cap Y_4 = \emptyset$.

In the language of formal logic, the elements Y_i of S' satisfy the following two properties:

Property 1 The Y_i's are **conjointly exhaustive** subsets of S because they jointly "use up" all of S. In symbols, $Y_1 \cup Y_2 \cup Y_3 \cup Y_4 = S$.

Property 2 The Y_i's are **mutually exclusive** subsets of S because they do not overlap. In symbols, $Y_i \cap Y_j = \emptyset$ for i and $j = 1,2,3,4$ and $i \neq j$.

Any collection of subsets of S with Properties 1 and 2 is called a **partition** of S. Here S' is a partition of S, and it may be written

$$S' = [\{X_8\}, \{X_7, X_6, X_4\}, \{X_5, X_3, X_2\}, \{X_1\}].$$

The general definition of partition is now stated.

Definition (Partition of a set)
Given a finite set $S = \{X_1, X_2, ..., X_n\}$, then a partition of S is a collection S' of subsets of S: $S' = [Y_1, Y_2, ..., Y_k]$, such that $Y_1 \cup Y_2 \cup ... \cup Y_k = \{X_1, X_2, ..., X_n\} = S$ and such that no two sets in S' have any point in common.

Furthermore, if S is a sample space with probability function P, then the elements of S' also constitute a sample space for the same experiment, and the probability function P for S' is inherited from S through the formula $P(Y_i)$ = the sum of the probabilities for all points X_j that are in Y_i. For example, if $Y_i = \{X_j, X_m, X_n\}$, then

$$P(Y_i) = P(X_j) + P(X_m) + P(X_n).$$

In particular,

$$P(Y_1) + P(Y_2) + ... + P(Y_k) = 1.$$

A partition of a given set S also creates a partition of *every subset* of S; that is, any "chopping up" of S also chops up the subsets of S as in the following example.

Example 11.14 Let $S = \{1,2,3,4,5,6,7,8,9,10\}$ and let S' be the partition

$$S' = [\{2, 3, 5\}, \{4\}, \{1\}, \{6, 7, 8, 9\}, \{10\}].$$

Denote the elements of S' by Y_1, Y_2, Y_3, Y_4, and Y_5, respectively. Now, if A is any subset of S, then the collection A' defined by

$$A' = [A \cap Y_1, A \cap Y_2, A \cap Y_3, A \cap Y_4, A \cap Y_5]$$

is a partition of A. In particular, suppose $A = \{2,3,4,5,6,8,10\}$, then A' consists of the intersections $A \cap Y_i$ (denoted by A_i) listed here:

$$A \cap Y_1 = \{2,3,5\} = A_1$$
$$A \cap Y_2 = \{4\} \quad\ = A_2$$
$$A \cap Y_3 = \emptyset \quad\ \ = A_3$$
$$A \cap Y_4 = \{6,8\} \ = A_4$$
$$A \cap Y_5 = \{10\} \ \ = A_5$$

Notice that the A_i's are mutually exclusive and conjointly exhaustive subsets of A. This means A' is a partition of A. When this happens A is said to be *partitioned by S'*.

It is useful to write the partition of A by S' in tabular form as given in Table 11.3. (Always omit the empty sets that occur in partitioning.)

Table 11.3 The Partition A' of A Created by the Partition S' of S

S' \ A	Y_1	Y_2	Y_3	Y_4	Y_5
	$\{2,3,5\}$	$\{4\}$	$\{1\}$	$\{6,7,8,9\}$	$\{10\}$
$\{2,3,4,5,6,8,10\}$	$A \cap Y_1$ $\{2,3,5\}$	$A \cap Y_2$ $\{4\}$	$A \cap Y_3$	$A \cap Y_4$ $\{6,8\}$	$A \cap Y_5$ $\{10\}$

Next, we look at the partition of a partition. Let S' and S'' denote two given partitions of the same space S, then all the sets in S' can be partitioned by S'', and all the sets in S'' can be partitioned by S'. The result is a partition of S called the *cross partition* of S by S' and S''.

11.3 Partitions and Cross Partitions

Definition (Cross partition)

Suppose S is a finite set $S = \{X_1, X_2, ..., X_n\}$. If S_1 and S_2 are the two partitions of S defined by $S_1 = [Y_1, Y_2, ..., Y_r]$ and $S_2 = [Z_1, Z_2, ..., Z_s]$ then the collection $C = [Y_i \cap Z_j]$ for all i and j, $i = 1, 2, ..., r$, and $j = 1, 2, ..., s$, is called the *cross partition* $S_1 \times S_2$ of S.

The cross partition $S_1 \times S_2$ of S can usually be written in the form of a table. See Table 11.4.

Table 11.4 The Cross Partition $S_1 \times S_2$, Using the Dot Notation for Intersections

$S_1 \backslash S_2$	Z_1	Z_2	...	Z_s
Y_1	$Y_1 \cdot Z_1$	$Y_1 \cdot Z_2$	...	$Y_1 \cdot Z_s$
Y_2	$Y_2 \cdot Z_1$	$Y_2 \cdot Z_2$	...	$Y_2 \cdot Z_s$
⋮	⋮	⋮		⋮
Y_r	$Y_r \cdot Z_1$	$Y_r \cdot Z_2$	...	$Y_r \cdot Z_s$

Example 11.15 Let $S = \{1,2,3,4,5,6,7,8,9,10,11\}$ and let S_1 and S_2 be the two partitions

$$S_1 = [\{1,3,5,7,9\}, \{2,4,6,8\}, \{10,11\}]$$
$$S_2 = [\{1,2\}, \{3,5,6\}, \{4,7,8,9,10,11\}].$$

The cross partition $S_1 \times S_2$ is $[\{1\}, \{2\}, \{3,5\}, \{6\}, \{7,9\}, \{4,8\}, \{10,11\}]$, as the reader can easily verify by constructing for these two partitions, a table similar to Table 11.4. Cross partitions are useful in extracting detailed information from a set of data.

Example 11.16 Suppose that a set of six college students $\{X_1, X_2, X_3, X_4, X_5, X_6\}$ are separated into two partitions, one according to their major subject and the other according to their favorite form of recreation, as follows:

Partition S_1 Partition S_2

$Y_1 = \{X_1, X_3, X_6\}$ Economics major $Z_1 = \{X_2, X_3\}$ Tennis player

$Y_2 = \{X_4, X_5\}$ English major $Z_2 = \{X_1, X_4\}$ Chess player

$Y_3 = \{X_2\}$ P. E. major $Z_3 = \{X_5, X_6\}$ Frisbee player

a. Compute the cross partition $S_1 \times S_2$.

b. Describe the results in words.

Solution
a. $S_1 \times S_2 = [\{X_1\}, \{X_2\}, \{X_3\}, \{X_4\}, \{X_5\}, \{X_6\}]$.

b. X_1 = a chess-playing economics major.
X_2 = a tennis-playing P. E. major.
X_3 = a tennis-playing economics major.
X_4 = a chess-playing English major.
X_5 = a frisbee-playing English major.
X_6 = a frisbee-playing economics major.   ▲

Finally, we introduce the following important theorem, which is useful in determining the probability of an event when knowing only certain conditional probabilities involving that event.

Theorem 11.1 If A is any subset of a given sample space S, and $S' = [Y_1, Y_2, ..., Y_k]$ is a partition of S, then,

$$P(A) = P(Y_1) \cdot P(A|Y_1) + P(Y_2) \cdot P(A|Y_2) + ... + P(Y_k) \cdot P(A|Y_k).$$

Proof $A' = [A \cdot Y_1, A \cdot Y_2, ..., A \cdot Y_k]$ is a partition of A; therefore, A is the union of the disjoint subsets $A \cdot Y_i$, and

$$P(A) = P(A \cdot Y_1) + P(A \cdot Y_2) + ... + P(A \cdot Y_k). \tag{11.5}$$

But from the definition of conditional probabilities,

$$P(A|Y_i) = \frac{P(A \cdot Y_i)}{P(Y_i)}$$

or

$$P(A \cdot Y_i) = P(Y_i) \cdot P(A|Y_i). \tag{11.6}$$

Replacing each term on the right in Equation (11.5) by the corresponding value from Equation (11.6), we get

$$P(A) = P(Y_1) P(A|Y_1) + P(Y_2) P(A|Y_2) + ... + P(Y_k) P(A|Y_k).$$

▲ **Example 11.17** Let $S' = [Y_1, Y_2, Y_3]$ be a partition of Congress into the subsets

Y_1 = the event the Congress member is conservative.

Y_2 = the event the Congress member is liberal.

Y_3 = the event the Congress member is a middle-of-the-roader.

11.3 Partitions and Cross Partitions

The same Congress can be partitioned into two sets: those who favor a certain measure and those who are against it. Let A = the event that a Congress member will vote to support a massive experiment in global atmospheric research. Suppose the following probabilities hold: $P(Y_1) = 0.30$, $P(Y_2) = 0.50$, and $P(Y_3) = 0.20$. Also, the probability that a Congress person supports the global experiment, given that he or she is conservative is 0.40. (In symbols, $P(A|Y_1) = 0.40$.) The other known probabilities are $P(A|Y_2) = 0.60$ and $P(A|Y_3) = 0.45$. Will the measure to support the global experiment pass?

Solution The question is: Will $P(A)$ be greater than 50 percent? From the above theorem, we get

$$P(A) = P(Y_1) \cdot P(A|Y_1) + P(Y_2) \cdot P(A|Y_2) + P(Y_3) \cdot P(A|Y_3)$$

and with the given data, this becomes

$$P(A) = (0.30)(0.40) + (0.50)(0.60) + (0.20)(0.45)$$

$$= 0.12 + 0.30 + 0.09 = 0.51.$$

The measure passed. ▲

EXERCISE 11.3

1. Suppose S is a sample space $\{E_1, E_2, E_3, E_4, \ldots, E_{10}\}$ with ten elements and a uniform probability function (S is an equiprobable space). Let $S' = [A_1, A_2, A_3]$ be the partition of S:

   $$[\{E_1, E_7, E_8\}, \{E_2, E_6\}, \{E_3, E_4, E_5, E_9, E_{10}\}].$$

   a. Write the probabilities for all three sample points in S'.

   b. Show that the sum of the probabilities for all three points equals one.

2. Suppose S is the sample space $\{X_1, X_2, X_3, X_4, X_5\}$ with five elements and the probability function $P(X_1) = 0.03$, $P(X_2) = 0.15$, $P(X_3) = 0.25$, $P(X_4) = 0.27$, and $P(X_5) = 0.30$. Let S be partitioned into the sample space S':

   $$S' = [A_1, A_2] = [\{X_1, X_2, X_4\}, \{X_3, X_5\}].$$

   a. Write the probability function value for both points in S'.

   b. Show that their sum, $P(A_1) + P(A_2)$, is equal to one.

3. In Problem 1, make up a different three-element partition S'' of S and repeat parts **a** and **b** for S''.

4. In Problem 2, make up a different two-element partition S'' of S and repeat parts **a** and **b** for S''.

5. In Problem 1, if $B = \{E_3, E_6, E_7\}$ and A_1, A_2, A_3 are the elements of S', let $B' = [A_1 \cdot B, A_2 \cdot B, A_3 \cdot B]$, and show that B' is a partition of B.

6. In Problem 2, let $C = \{X_1, X_4, X_5\}$ and for A_1 and A_2 in S', let $C' = [A_1 \cdot C, A_2 \cdot C]$. Show that C' is a partition of C.

▲ 7. In the sample space S of dice throws (see Exercise 10.3, page 312), there are 36 points.

   a. Partition this space into an eleven-point space S',
   $$S' = [Y_2, Y_3, Y_4, \ldots, Y_{11}, Y_{12}],$$
   as follows. (Notice that the subscripts start at 2, not 1, and go to 12, not 11.) Y_i represents all the sample points of S whose total value on both dice is equal to i. For example, $Y_5 =$ the sample points whose total from both dice is five. They are (1,4), (2,3), (3,2), and (4,1).

   b. Make up a list of the probabilities for this eleven-point space.

   c. Let $S'' = [Z_1, Z_2]$ be another partition of S, with $Z_1 =$ the set of points of S with an odd number on the first die, and $Z_2 =$ the set of points of S with an even number on the first die. Show that S'' is a partition of S.

   d. Compute the cross partition $S' \times S''$ of S.

8. In the toss of four coins, let the 16-point space S be

$X_1 =$ HHHH	$X_5 =$ HTHH	$X_9 =$ THHH	$X_{13} =$ TTHH
$X_2 =$ HHHT	$X_6 =$ HTHT	$X_{10} =$ THHT	$X_{14} =$ TTHT
$X_3 =$ HHTH	$X_7 =$ HTTH	$X_{11} =$ THTH	$X_{15} =$ TTTH
$X_4 =$ HHTT	$X_8 =$ HTTT	$X_{12} =$ THTT	$X_{16} =$ TTTT

   a. Partition S into a space $S' = [Y_0, Y_1, Y_2, Y_3, Y_4]$, where the events Y_i are defined as $Y_i =$ the event that exactly i of the coins will be tails.

   b. Make up a list of the probabilities for the five points in S'.

   c. Let S'' be the partition $[Z_1, Z_2, Z_3]$, where $Z_1 =$ the event there are more heads than tails, $Z_2 =$ event there are the same number of heads and tails, and $Z_3 =$ event of more tails than heads. Show that S'' is a partition of S.

   d. Compute $S' \times S''$ and write the probabilities for all of the elements in the cross partition $S' \times S''$.

11.3 Partitions and Cross Partitions

9. Suppose the growing conditions, rainfall, temperature, length of growing season, etc., can be classified as the events

Y_1 = perfect growing conditions.

Y_2 = excellent growing conditions.

Y_3 = good growing conditions.

Y_4 = poor growing conditions.

Suppose the probability function P is distributed over the sample space as follows: $P(Y_1) = 0.15$, $P(Y_2) = 0.20$, $P(Y_3) = 0.50$, and $P(Y_4) = 0.15$. The event A, that a farmer will grow a superior crop, satisfies the following conditional probabilities: $P(A|Y_1) = 0.90$, $P(A|Y_2) = 0.65$, $P(A|Y_3) = 0.25$, and $P(A|Y_4) = 0.01$. Find the probability that the farmer will grow a superior crop; that is, find $P(A)$. Hint: Use Theorem 11.1.

10. Suppose a Senate committee is partitioned into geographical regions:

Y_1 = event the Senator is from the South.

Y_2 = event the Senator is from the East.

Y_3 = event the Senator is from the Midwest.

Y_4 = event the Senator is from the far West.

Say that the committee probability function is $P(Y_1) = 0.10$, $P(Y_2) = 0.07$, $P(Y_3) = 0.63$, and $P(Y_4) = 0.20$. Let A be the event that a certain bill will get a favorable vote. The committee members' attitudes toward the bill are: Given a southern Senator, he is against the bill; that is, the probability of event A, given a southern Senator, is zero. $P(A|Y_1) = 0$. The other conditional probabilities are $P(A|Y_2) = 0.60$, $P(A|Y_3) = 0.51$, and $P(A|Y_4) = 0.50$. What is the probability of A? ▲

12
PROBABILITY ANALYSIS

12.1

DISCRETE RANDOM VARIABLE

The number of mammae (feeding stations) on swine can vary from 8 to 18, which is fortunate since the number of baby pigs in a litter varies from 1 to 16. The events that a female hog has either 8, 10, 12, 14, 16, or 18 mammae (which usually occur in pairs, so only *even numbers* are considered) are listed below:

X_8 = the event that there are exactly 8 mammae on a female hog.
X_{10} = the event that there are exactly 10 mammae on a female hog.
$\vdots$
X_{18} = the event that there are exactly 18 mammae on a female hog.

Data collected show that a probability function can be distributed over these events as follows: $P(X_8) = 0.006$, $P(X_{10}) = 0.118$, $P(X_{12}) = 0.581$, $P(X_{14}) = 0.229$, $P(X_{16}) = 0.065$, and $P(X_{18}) = 0.001$ (Wright, Sewall. *Evolution and the Genetics of Populations.* Vol. I. Chicago: University of Chicago Press, 1968).

At this point, we could introduce some problems similar to those stated on meristic variability in Exercise 11.2, but the reason for presenting the example here is to examine the numerical properties inherent in a space of random events and to discuss the concept of **random variables.**

What does the equation $P(X_{10}) = 0.118$ say? It reads: "The probability of *the event that there are 10 mammae on a female hog is 0.118.*" This reference to the occurrence of an event involving a certain number is awkward. Why couldn't we just refer to the number itself without dragging along all that verbiage about the "event that a number of whatchamacallits are involved"? We can. One way is to state the result of the experiment and then simply label all the outcomes so that the probabilities are given in terms of numbers. (The events would not be eliminated by this plan; they would only be relegated to a lesser role. They would serve as background for the problem rather than appearing in a prominent place, right in the thick of things.)

Example 12.1 Let X be a variable that ranges over the possible number of mammae that a female hog can have, and let each sample point X_n discussed above be referred to by the equation $X = n$. The possible values of X are $X = 8$, $X = 10$, $X = 12$, $X = 14$, $X = 16$, and $X = 18$. These equations represent the events X_8, X_{10}, etc. Now, the sample space S is the set of *numbers* $\{8,10,12,14,16,18\}$ and the probabilities P can be written $P(X = 8) = 0.006$; $P(X = 10) = 0.118$; $P(X = 12) = 0.581$, and so on.

Aside from its value in streamlining the notation, the use of a variable to stand for the random sample points allows us to use the coordinate plane to draw a graph of the probability function. (See Figure 12.1.) Other mathematical concepts, such as vectors, matrices, and calculus, are available when probabilities are expressed as functions of variables rather than events.

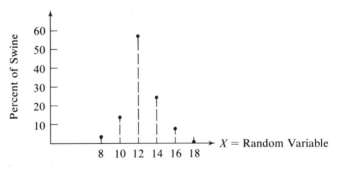

Figure 12.1
Distribution graph for swine mammae.

In the above example it was obvious what number was to be assigned to each simple event, but in some other cases, it will not be so easy to determine the numerical values to be assigned to given random events. Sometimes the process of "quantifying" a set of events is arbitrary.

Example 12.2 The quantifying of a deck of cards as a set of two-digit numbers (as in Table 10.2) was used along with a random number table to simulate the random events

12.1 Discrete Random Variable

of dealing a poker hand. In that case, each sample point was the deal of a card, and the variables were the two-digit codes (numbers) from 01 to 52 that were associated with the cards. The *pairing off* of numbers with the cards depended on an arbitrary arrangement of the cards into some order.

In these two examples, we speak of pairing off sample points with real numbers or variables. This means we are speaking of a *function from the sample points to the real numbers*. (Such pairings make the random variable a *point function*, as we see in the next definition.)

Definition (Random variable)
Let S be a sample space whose simple events are results of a random experiment, and let X be a function from S into the real numbers R; that is, $X:S \to R$. Then X is called a *random variable* on the sample space S. (It should be called a function, but it is called a variable because *its range will be used as a variable in the domain of a probability function*. See the definitions in Chapter 3 for the domain and range of a function.)

This definition is sometimes abbreviated: A random variable X is a *real-valued point function* from a space of random sample points.

Note 1 A random variable X can be a many-to-one function; that is, several different sample points can have the same value of the random variable assigned to them. *Several points can map into the same value of the random variable.*

Note 2 There are two kinds of random variables: *discrete random variables* and *continuous random variables*. We will work with the discrete random variables since they are the ones applied to finite sets.

Note 3 If $X:S \to R$ is a random variable on S, E is a point of S, and x is the real number paired off with E, then $X(E) = x$, and x is usually represented as a number on the x-axis (rather than on the y-axis). In other words, the *range* of the function X poses as a subset of the x-axis so that a probability function can use it for a domain. One of the main reasons random variables exist is to serve in the domain of a probability function. (See Figure 12.2.)

Example 12.3 In the toss of two coins, the space S has sample points $E_1 = (H,H)$; $E_2 = (H,T)$; $E_3 = (T,H)$; and $E_4 = (T,T)$. Assign a random variable X as follows. For any

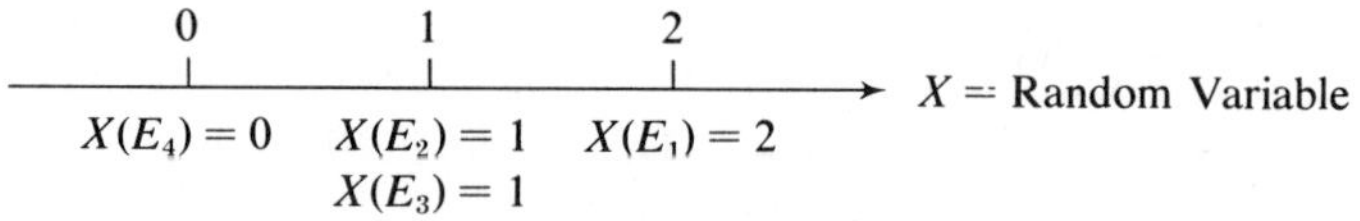

```
    0           1           2
    |           |           |                → X = Random Variable
 X(E₄)=0    X(E₂)=1     X(E₁)=2
            X(E₃)=1
```

Figure 12.2
Random variables on the x-axis.

point E in S, $X(E)$ = the number of heads that appear in E. The range of values for this random variable is $\{0,1,2\}$ because $X(E_1) = 2$, $X(E_2) = 1$, $X(E_3) = 1$, and $X(E_4) = 0$. In other words, the event E_1 maps into 2, the events E_2 and E_3 both map into 1, and E_4 maps into 0. $P(X = x)$ = the probability that the random variable will be x (or take on x), and it is the sum of the probabilities of the events that map into x. Therefore, $P(X = 0) = P(E_4) = ¼$, $P(X = 2) = P(E_1) = ¼$, and since both E_2 and E_3 map into 1, $P(X = 1) = P(E_2) + P(E_3) = ½$. The probabilities assigned to the values of the random variable are called the **distribution of the random variable** (by the given probability function of the underlying sample space). We can present the distribution in the form of a table, as shown below.

Values of the Random Variable x	0	1	2
Probabilities $P(X = x)$	¼	½	¼

In this example, the probability function of the sample space exposes the relative frequency of each random variable.

The procedure is similar in probability spaces that are not equiprobable. The idea is that the probabilities for the random variable are obtained from the probabilities for the simple events in the sample space. If any one value a of the random variable X comes from more than one simple event, then the probability for the random variable $X = a$ is the sum of the probabilities for all the simple events mapping into a. This is how the probability of the underlying sample space distributes the random variable, and the distribution is called the **distribution induced by the probability function.**

Example 12.4 A sample space S consists of the events $\{E_1, E_2, E_3, E_4\}$ with the probability function $P(E_1) = 0.50$, $P(E_2) = 0.25$, $P(E_3) = 0.20$, and $P(E_4) = 0.05$. Let $X: S \to R$ be the random variable defined by $X(E_1) = 2$, $X(E_2) = 3$, $X(E_3) = 5$, and $X(E_4) = 2$.

12.1 Discrete Random Variable

a. Represent the real number line by the x-axis and mark the points in the range of the random variable on it.

b. Find the probability that the random variable takes on the value 3.

c. What are the sample points that map into the value 2 for the random variable? Find $P(X = 2)$.

d. Find the probability that $X = 3$ or $X = 5$.

e. Find $P(X = 2 \text{ or } X = 5)$.

f. In the x-y coordinate plane, plot the *distribution graph* (the points defined by the equation $y = P(X = x)$ for all x in the range of the random variable).

Solution a. See Figure 12.3 below.

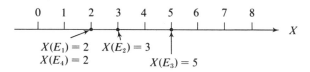

Figure 12.3

b. $P(X = 3) = P(E_2) = 0.25$.

c. The random variable has value $X = 2$ for the two sample points E_1 and E_4. $P(X = 2) = P(E_1, E_4) = P(E_1) + P(E_4) = 0.55$.

d. Since $X = 3$ and $X = 5$ represent disjoint sets (the sample points E_2 and E_3), then the probability of the union of the two random variables $P(X = 3$ or $X = 5) = P(X = 3) + P(X = 5)$. From the given data, $P(X = 5) = 0.20$ and $P(X = 3) = 0.25$; therefore $P(X = 3 \text{ or } X = 5) = 0.45$.

e. $P(X = 2 \text{ or } X = 5) = 0.75$.

f. See Figure 12.4 below.

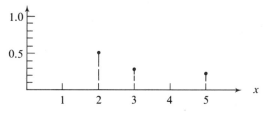

Figure 12.4
Distribution graph.

Example 12.5   In the toss of two dice (see Exercise 10.3, page 312), let a random variable $T =$ the total shown on both dice.

a. What are the values in the range of the random variable?

b. How many sample points in the underlying sample space correspond to the value $T = 8$ of the random variable?

c. What is $P(T = 8)$?

d. What is the relative frequency with which the random variable takes on the value 5? [That is, what is $P(T = 5)$?]

e. What is the probability of $T = 6$?

f. What is the probability of the union of the two values, $T = 6$ and $T = 5$?

g. Find $P(T = 3, 8, 10, \text{ or } 11)$.

h. What is meant by the random variable inequality $T \leq 4$?

i. What is meant by the random variable inequality $4 \leq T \leq 8$?

j. Find $P(T \leq 4)$ and $P(4 \leq T \leq 8)$.

Solution

a. The range of values for the random variable is $\{2,3,4,5,6,7,8,9,10,11,12\}$.

b. The five sample points (2,6); (3,5); (4,4); (5,3); and (6,2) correspond to $T = 8$.

c. $P(T = 8) = 5/36$.

d. The relative frequency with which the random variable takes on the value 5 is 4 out of 36. The four points, (1,4); (2,3); (3,2); and (4,1) correspond to $T = 5$. This means $P(T = 5) = 4/36$ or $1/9$.

e. $P(T = 6) = 5/36$.

f. The probability of the union of $T = 6$ and $T = 5$ is the sum of the probabilities $P(T = 6)$ and $P(T = 5)$ since these two values of the random variable represent nonoverlapping subsets of the sample space. Therefore, $P(T = 6 \text{ or } T = 5) = 5/36 + 4/36 = 1/4$.

g. $P(T = 3,8,10,11) = P(T = 3) + P(T = 8) + P(T = 10) + P(T = 11)$
$= 2/36 + 5/36 + 3/36 + 2/36 = 12/36 = 1/3$.

h. $T \leq 4$ means $T = 2$, $T = 3$, or $T = 4$.

i. $4 \leq T \leq 8$ means $T = 4, 5, 6, 7,$ or 8.

j. $P(T \leq 4) = P(T = 2) + P(T = 3) + P(T = 4) = 1/36 + 2/36 + 3/36 = 1/6$.
$P(4 \leq T \leq 8) = P(T = 4) + P(T = 5) + P(T = 6) + P(T = 7) + P(T = 8) = 23/36$.

12.1 Discrete Random Variable

It is often very useful to construct a table for the distribution induced by the probability function.

Example 12.6 Using the information from Example 12.5:

a. Construct the table of the distribution of the random variable T.

b. Use this table to draw the graph of the equation $y = P(T = t)$ for all t (called the distribution).

Solution a. See the table below.

t	2	3	4	5	6	7	8	9	10	11	12
$P(T=t)$	1/36	2/36	3/36	4/36	5/36	6/36	5/36	4/36	3/36	2/36	1/36

b. See Figure 12.5.

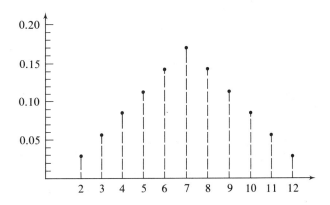

Figure 12.5
Distribution graph of the random variable for the totals on a pair of dice.

Example 12.7 In a lot of twelve items, two are defective. Suppose a sample of four items is randomly selected. Let a random variable X stand for the number of defective items in the sample.

a. What is the range of values for the random variable?

b. Assume an equiprobable model and compute the distribution of the random variable. (That is, what is the probability for each value of the random variable?)

Solution a. There can be either zero, one, or two defective items in the sample of four. (There cannot be as many as three or four defective items in the sample since there are only two defective items in the entire lot.)

b. Use the method developed in Section 10.4. The number of ways that twelve things can be combined four at a time is $12!/(8!)(4!)$, which is 495. For the values of the random variable, if $X = 0$, then the number of ways zero defective items can be chosen from two is $2!/(2!)(0!) = 1$. When four items in the sample are good, the number of ways they can be chosen from the 10 good items is $10!/(6!)(4!) = 210$. This means that the number of samples with no defective items = (number of ways to choose zero defective items) × (number of ways to choose four good ones) = $1 \cdot 210 = 210$, and the probability $P(X = 0)$ is $210/495 = 0.424$. If $X = 1$, then the number of ways one defective item can be chosen from two is $2!/(1!)(1!) = 2$. The remaining three items in the sample are chosen from the ten good items in $10!/[(7!)(3!)] = 120$ ways. So a sample containing one defective item can be chosen $2 \cdot 120 = 240$ ways, and the probability $P(X = 1) = 240/495 = 0.485$. Finally, for $X = 2$, the number of ways two defective items can be chosen from two is $2!/[(0!)(2!)] = 1$. The number of ways two remaining sample items can be chosen from the ten good lot items is $10!/[(8!)(2!)] = 45$; hence $P(X = 2) = 45/495 = 0.091$. ▲

EXERCISE 12.1

1. A sample space S consists of the events E_1, E_2, E_3, E_4, E_5 with probability function $P(E_1) = 0.07$, $P(E_2) = 0.33$, $P(E_3) = 0.35$, $P(E_4) = 0.15$, and $P(E_5) = 0.10$. Let $X: S \to R$ be the random variable assigned to these points as follows: $X(E_1) = -2$, $X(E_2) = 0$, $X(E_3) = 1$, $X(E_4) = 10$, and $X(E_5) = 0$.

   a. Represent the real number line by the x-axis and mark points in the range of the random variable on it.

   b. Find the probability that the random variable is -2.

   c. What are all the sample points for which the random variable has the value zero? Find $P(X = 0)$.

12.1 Discrete Random Variable

d. Find the probability that $X = 1$ or $X = 10$.

e. Find $P(X = -2$ or $X = 10)$.

f. In the x-y coordinate plane, plot the points defined by the equation $y = P(X = x)$ for all x in the range of the random variable.

2. A sample space S consists of the events $A_1, A_2, A_3, A_4, A_5, A_6, A_7$ with probability functions $P(A_1) = 0.05$, $P(A_2) = 0.12$, $P(A_3) = 0.14$, $P(A_4) = 0.30$, $P(A_5) = 0.26$, $P(A_6) = 0.10$, and $P(A_7) = 0.03$. Let $X:S \to R$ be the random variable defined by $X(A_1) = 1$, $X(A_2) = 2$, $X(A_3) = 4$, $X(A_4) = 2$, $X(A_5) = 4$, $X(A_6) = 3$, and $X(A_7) = 5$.

   a. Represent the real number line by the x-axis and mark the points in the range of the random variable on it.

   b. Find the probability that the random variable takes on the value 3.

   c. What are all the sample points corresponding to $X = 4$?

   d. Find $P(X = 1)$, $P(X = 2)$, $P(X = 3)$, $P(X = 4)$, and $P(X = 5)$.

   e. Draw the graph of the distribution for this random variable.

▲ 3. In the toss of two dice, let a random variable X be defined as the *product* of the two numbers showing. For example, if the throw is the event (3,5), then the random variable associated with that event is $X = 3 \cdot 5 = 15$. Note that the underlying sample space is still the set of 36 sample points for two dice.

   a. Make up a multiplication table to determine all values of the random variable and list them all as a set S.

   b. Assume fair dice; compute the probabilities for all values of the random variable.

   c. Find $P(X \geq 18)$.

4. In the toss of two dice, let a random variable Y be defined as the *difference* of the two numbers showing on the dice. For example, if the throw is the event (3,5), then the random variable associated with that event is $Y = 3 - 5 = -2$.

   a. Make up a subtraction table to determine all values of the random variable and list them all as a set S.

   b. Assume fair dice; compute the probabilities for all values of the random variable.

c. Find $P(Y \leq 0)$. ▲

5. Assume the set S of random variables and its distribution from Problem 3. Let $A = \{1,3,5,9,15\}$ and $B = \{3,4,5,6,8,9,10\}$ be subsets of S. Define $P(A)$ to mean $P(X = x)$ for all x in A, with similar meanings for $P(B)$ and for P(combinations of A and B). Solve the following:

   a. Find $P(A)$, $P(B)$, and $P(A \cap B)$.

   b. Find $P[A|(-B)]$ and $P(B \cap -A)$.

   c. Find $P(-A \cup B)$.

6. Assume the set S of random variables and its distribution of Problem 4. Let $A = \{-4,-3,-2,0,1,2,5\}$ and $B = \{-3,-1,0,3,4,5\}$.

   a. Find $P(A)$, $P(B)$, and $P(A \cap B)$.

   b. Find $P(A \cap -B)$ and $P(-A|-B)$.

   c. Find $P[A|(A \cup -B)]$.

7. Suppose a random variable X takes on the range of values $\{-4,-2,0,3\}$ with probabilities $(t + 2)/15$, $(2t - 7)/15$, $(t + 4)/15$, and $(t - 4)/15$, respectively.

   a. Solve for t (by setting the sum of the probabilities equal to one).

   b. List the distribution of the random variable.

   c. Graph the distribution for the random variable.

8. If the range of a random variable X is $\{0,1,2,3,4,5\}$ with probabilities

   $$P(X = 0) = \frac{3t + 10}{100} \quad P(X = 2) = \frac{t + 6}{100} \quad P(X = 4) = \frac{t + 1}{100}$$

   $$P(X = 1) = \frac{2t - 4}{100} \quad P(X = 3) = \frac{t + 7}{100} \quad P(X = 5) = \frac{5t + 15}{100}$$

   a. Solve for t.

   b. List the distribution of the random variable.

   c. Plot the distribution graph.

9. In a lot of ten items, three are defective. Suppose samples of four items are randomly selected. Let the random variable X stand for the number of defective items in the samples.

   a. What is the range of values of the random variable?

   b. Assuming an equiprobable model, compute the distribution of X.

12.2 Expected Values

10. In a lot of eleven items, six are defective. Suppose samples of seven items are selected at random. Let the random variable X stand for the number of defective items in the samples.

    a. What is the range of values of the random variable?

    b. Compute the distribution of X. ▲

12.2
EXPECTED VALUES

The following six problems all have the same solution; that is, the same formula can be used to solve all of them.

Example 12.8 A gambler contemplates entering a game in which he has a 0.12 chance to win $50, a 0.14 chance to win $10, a 0.25 chance to win $1, and a 0.49 chance to win nothing.

a. Should he pay $10 to enter the game?

b. What is the game worth to a player?

Example 12.9 A king wants to have a son for his heir. He would like to keep having children forever, but due to inflation and overpopulation, he must quit when a son is born or after five daughters are born. He (and his queen, of course) continue to have children until a son is born or until five daughters in a row are born. How many children is the royal family expected to have?

Example 12.10 A mathematics student is computing his final course grade. He knows that homework counts $1/4$ of the grade, semester exams count $5/12$ of the grade, and the final exam counts $1/3$ of the grade. His homework average was 72, his semester exam average was 86, and his final exam grade was 64. What is his course grade?

Example 12.11 An insurance company collects data over a long time and with a large number of policies in force. They find that in one year they can operate a hospital insurance policy as if each policyholder has the probabilities shown in the table below of filing a payable claim for benefits. Also shown in the table is the average amount

(1) Probability of Claim	(2) Benefit Claimed	(3) Average Amount the Company Pays on Such a Claim per Year
0.42	Hospitalization benefit	$540
0.02	Outpatient benefit	22
0.04	Home care benefit	100
0.18	Extended special care	410
0.34	No benefits claimed	—

paid on such claims. What should the yearly premium be on this policy, just to cover the probable claims?

Example 12.12 For a large number N of cell divisions, let $M =$ the number of mutations after N cell divisions. Suppose the distribution of this random variable M is according to the probability equation

$$P(M = m) = \frac{(1-p)^{N-m}(N+1)}{m+m^2}, \ 0 < p < \frac{1}{N} \text{ and } 0 < m \leq N.$$

Find the expected number of cell mutants after N cell divisions for a given value of p. (See Murphy et al., pp. 207-22 in the *American Journal of Human Genetics*, March 1974.) ▲

Example 12.13 Let X denote a random variable with range $\{x_1, x_2, ..., x_k\}$, whose distribution is induced by the probabilities $P(X = x_1) = p_1$, $P(X = x_2) = p_2$, ..., $P(X = x_k) = p_k$. What is the expected value of X?

Although it doesn't seem obvious, every one of the above problems asked for the same thing—the *expected value* of a random variable. We now state the definition.

Definition **(Expected value)**
Let X be a discrete variable with range $R_X = \{x_1, x_2, ... x_i, ..., x_n\}$ and let the distribution of X be given by the probabilities $P(X = x_1), P(X = x_2), ..., P(X = x_i),$

12.2 Expected Values

..., $P(X = x_n)$. Then the expected value of X is the number $E(X)$ defined by the summation

$$E(X) = \sum_{i=1}^{n} x_i \cdot P(X = x_i).$$

This equation is the solution to each of the above six problems, as shown below.

▲ Solution **12.8** In the gambling problem, the possible payoffs are the values of the random variable; they are 50, 10, 1, and 0. Their probabilities are $P(X = 50) = 0.12$, $P(X = 10) = 0.14$, $P(X = 1) = 0.25$, and $P(X = 0) = 0.49$. By the definition, the expected value is

$$E(X) = (50)(0.12) + (10)(0.14) + (1)(0.25) + (0)(0.49)$$
$$= \$6.00 + \$1.40 + \$0.25 + 0 = \$7.65.$$

a. He should not pay $10 to enter the game.

b. The game is worth only $7.65 to a player. This means that if 100 people enter the game, they will pay $1000. The total take for all 100 people (winners and losers) would be only $765—not a very good return on their investment. In the long run, it is not possible to come out even.

12.9 The royal children ending in either a male or five females can be represented by the set of the repeated events

$$E_1 = M \qquad E_3 = FFM \qquad E_5 = FFFFM$$
$$E_2 = FM \qquad E_4 = FFFM \qquad E_6 = FFFFF$$

Let the random variable X be the number of children that the king and queen could have under the conditions given in the problem. Then the range of the random variable is $\{1,2,3,4,5\}$. These values are assigned to the events as follows: $X(E_1) = 1$, $X(E_2) = 2$, $X(E_3) = 3$, $X(E_4) = 4$, $X(E_5) = 5$, and $X(E_6) = 5$. Notice that the two events E_5 and E_6 both map into the same value 5 of the random variable. Assume $P(F) = P(M) = 1/2$. A repeated event, such as FM, is actually a two-stage experiment in which the first stage can happen in two ways and the second can happen in two ways, so the entire event FM is one of four possible outcomes. These events are independent, and their probability can be obtained as the product of the probabilities of the individual events; that is, $P(FM) = P(F) \cdot P(M) = (1/2)(1/2) = 1/4$. The other probabilities are $P(FFM) = 1/8$, $P(FFFM)$

$= \frac{1}{16}$, $P(FFFFM) = \frac{1}{32}$, and $P(FFFFF) = \frac{1}{32}$. This means that the probabilities for the random variable are $P(X = 1) = \frac{1}{2}$, $P(X = 2) = \frac{1}{4}$, $P(X = 3) = \frac{1}{8}$, $P(X = 4) = \frac{1}{16}$, and $P(X = 5) = (\frac{1}{32}) + (\frac{1}{32}) = \frac{1}{16}$. Now, by the expected value formula, the number of children expected is

$$E(X) = (1)(\tfrac{1}{2}) + (2)(\tfrac{1}{4}) + (3)(\tfrac{1}{8}) + (4)(\tfrac{1}{16}) + (5)(\tfrac{1}{16})$$
$$= \tfrac{31}{16} = 1\,\tfrac{15}{16}.$$

They are expected to have around $1\,\frac{15}{16}$ children to get a male heir. (It was whispered throughout the kingdom that the second child was "not quite all there.") This number actually means that for large numbers of similar situations, the average number of children will be $\frac{31}{16}$. For example, in 1600 kingdoms there will be 3100 royal children. Some of the families would have five girls, some would have four girls and a boy, etc. (This model ignores multiple births.)

12.10 In the mathematics student's problem, the random variable X ranges over the three scores, 72, 86, and 64. Their distribution is $P(X = 72) = \frac{1}{4}$ (since 72 is the homework average and it is worth $\frac{1}{4}$ the course grade), $P(X = 86) = \frac{5}{12}$, and $P(X = 64) = \frac{1}{3}$. The course grade expected is

$$E(X) = (72)(\tfrac{1}{4}) + (86)(\tfrac{5}{12}) + (64)(\tfrac{1}{3}) = 75.16.$$

12.11 Here the random variable is the amount paid on a claim; column (3) gives the range of X. The distribution is the set of probabilities of filing a claim, column (1). The expected value is

$$E(X) = (540)(0.42) + (22)(0.02) + (100)(0.04) + (410)(0.18) + (0)(0.34)$$
$$= \$226.80 + \$0.44 + \$4.00 + \$73.80 = \$305.04.$$

The company can expect to pay out $305.04 per year per policyholder, so the premium should be at least this amount.

12.12 The expected number of mutations after N cell divisions is

$$E(M) = \sum_{m=1}^{N} \frac{m \cdot (1-p)^{N-m}(N+1)}{m + m^2}.$$

To solve for a particular case, such as $N = 10^5$ and $p = 3 \cdot 10^{-6}$, we would need a computer. ▲

12.13 This is the general problem and it is exactly like the definition of expected value; its solution is

$$E(X) = \sum_{i=1}^{k} x_i \cdot p_i.$$

12.2 Expected Values

In the gambling problem, Example 12.8, the value was $7.65. What changes in the payoffs would make the game worth the $10 entry fee? Surprisingly, the addition of $2.35 to *every outcome* (even the zero outcomes) will raise the value of the game to $10.

Example 12.14 Suppose the random variables representing the payoffs in the gambling game of Example 12.8 are $52.35, $12.35, $3.35, and $2.35, with the same probabilities. Now $P(X = 52.35) = 0.12$, $P(X = 12.35) = 0.14$, $P(X = 3.35) = 0.25$, and $P(X = 2.35) = 0.49$.

   a. Now what is the expected value?
   b. Should a person pay $10 to enter the game?

Solution a. $E(X) = \sum_{i=1}^{4} x_i P(X = x_i)$
   $= (52.35)(0.12) + (12.35)(0.14) + (3.35)(0.25) + (2.35)(0.49)$
   $= \$6.28 + \$1.73 + \$0.84 + \$1.15 = \$10.00$.

   b. This is really up to the individual, but it is a fair game since it is possible to come out even in the long run.

Example 12.14 illustrates the concept of increasing each value of the random variable by a constant to make the game's value equal to the cost. If the cost is subtracted from each payoff, the value of the game could include both the cost and the payoff and should have a zero expected value to be a *fair game*. In general, the values of the random variable can have some constant added to or subtracted from each of them, and the resulting expected value will be increased or decreased by that amount. (Similarly, multiplication of each value of the random variable by a fixed constant will result in the expected value being multiplied by that constant.) In Example 12.10, if the mathematics student could have added six points to each of the grades, his course grade would have been increased by six points from 75.16 to 81.16. In general terms, this concept may be expressed as one of the **laws of expectation**.

If X is a random variable with range $\{x_1, x_2, \ldots x_k\}$ and b is any constant, then let the random variable $Y = X + b$ be such that the range of Y is $\{x_1 + b, x_2 + b, \ldots x_k + b\}$, and let Y have the same probability distribution as X; that is, $P(X = x_i) = P(Y = x_i + b)$. Then the expected value of Y is given by

$$E(Y) = E(X) + b.$$

This may also be written as

$$E(X + b) = E(X) + b.$$

The number b could be positive or negative, thus allowing an increase or decrease in the expected value. Another law is

$$E(bX) = b \cdot E(X).$$

A loss is indicated by a negative value of the random variable; gain is represented by a positive value of the random variable. The expected value formula may contain both gains and losses.

Example 12.15 If the range of a random variable X is $\{-2,-1,0,1,2\}$ and the distribution is $P(X = -2) = 1/4$, $P(X = -1) = 3/16$, $P(X = 0) = 3/8$, $P(X = 1) = 1/8$, and $P(X = 2) = 1/16$,

a. Find the expected value $E(X)$.

b. Find $E(X + 7/16)$. Find $E(2X)$.

Solution a. $E(X) = (-2)(1/4) + (-1)(3/16) + (0)(3/8) + (1)(1/8) + (2)(1/16)$
$= -1/2 - 3/16 + 0 + 1/8 + 1/8 = -7/16$.

b. $E(X + 7/16) = E(X) + 7/16 = -7/16 + 7/16 = 0.$
$E(2X) = 2E(X) = -14/16.$

EXERCISE 12.2

1. In parts a and b, find the expected value $E(X)$ of the random variable X:

   a. If the range of values of X is $\{0,1,2,3\}$ and the distribution is $P(X = 0) = 0.09$, $P(X = 1) = 0.61$, $P(X = 2) = 0.15$, and $P(X = 3) = 0.15$.

   b. If the range of values of X is $\{-10,-2,9\}$ and the distribution is $P(X = -10) = 0.16$, $P(X = -2) = 0.24$, and $P(X = 9) = 0.60$.

   c. Find $E(X + 5)$ for parts a and b.

   d. Find $E(3X - 1)$ for parts a and b.

2. In parts a and b below, find the expected value $E(Y)$ of the random variable Y:

   a. If the range of Y is $\{-3,-2,-1,1,2,3\}$ and the distribution is $P(Y = -3) = 1/32$, $P(Y = -2) = 5/32$, $P(Y = -1) = 10/32$, $P(Y = 1) = 10/32$, $P(Y = 2) = 5/32$, and $P(Y = 3) = 1/32$.

12.2 Expected Values

b. If the following table gives the range of values of the random variable Y and its distribution.

$Y = y_i$	-3	-2	-1	1	2	3
$P(Y = y_i)$	$(1/3)^5$	$5(1/3)^4(2/3)$	$10(1/3)^3(2/3)^2$	$10(1/3)^2(2/3)^3$	$5(1/3)(2/3)^4$	$(2/3)^5$

c. Find $E(Y + 3)$ for parts **a** and **b**.

▲ **3.** The three top prizes in a chess tournament are $100, $75, and $50. The next three prizes are chess clocks worth $30 each, and the next 14 prizes are chess sets worth $9 each. There are 55 players. Assume the probability of winning each cash prize is $1/55$, the probability of winning a chess clock is $3/55$, and the probability of winning a chess set is $14/55$.

a. What is the expected value of this tournament to a player?

b. It costs $7 to enter; is this to the player's advantage?

c. How much money do the sponsors of the tournament lose?

d. If the sponsors can change the variables representing the chess clocks and the chess sets by buying them wholesale for $25 and $7, respectively, instead of $30 and $9, has the expected value changed for the players? Has the expected cost changed for the sponsors?

4. A slot machine has a payoff of a jackpot of $25 with a probability of $1/1000$, a payoff of $12.50 with a probability of $2/1000$, a payoff of $5 with a probability of $20/1000$, a payoff of $3 with a probability of $50/1000$, a payoff of $1 with a probability of $150/1000$, and it pays nothing with a probability of $777/1000$.

a. What is the expected value of these payoffs?

b. If it costs 50 cents to play, can a player break even in the long run?

5. A man has a "system" to keep his losses under control at the roulette table. He will continue to bet $10 on a single number (say, 5) until it wins or until he loses five times in a row; then he will quit. The events are W = the event the wheel stops on 5 and L = the event the wheel stops on the wrong number (anything but 5). Since there are 36 numbers plus 0 and 00, the probability of stopping on 5 is $1/38$. Thus, $P(W) = 1/38$ and $P(L) = 37/38$. The different spins of the wheel are independent. Let the random variable X = the number of times he bets.

a. What are the values of the random variable?
b. What is $P(X = 1)$?
c. What value of the random variable does the event LW map into?
d. What value does LLLLW map into? What about LLLLL?
e. Show that $P(\text{LLW}) = P(X = 3) = (^{37}/_{38})^2(^{1}/_{38}) = 0.0249$.
f. Compute the expected value for X.

6. Another better at the roulette wheel bets \$5 on the set of numbers from 1 through 12 until he wins once or loses five times in a row. Let A = the event it stops on one of those first 12 numbers and B = the event it does not. $P(A) = {}^{12}/_{38}$ and $P(B) = {}^{26}/_{38}$. Let the random variable X = the number of bets placed.

   a. What are the values in the range of the random variable?
   b. What is $P(X = 2)$?
   c. What value does BBA map into?
   d. Which two events map into the same value of the random variable?
   e. Show that $P(X = 5) = (^{26}/_{38})^4(^{12}/_{38}) + (^{26}/_{38})^5$.
   f. Compute the expected value for X. ▲

12.3

BAYES' THEOREM

Suppose a sample space is partitioned into events $[E_1, E_2, E_3, E_4]$ and the probability for each is known. If A is any *known* nonempty event, then from the partition A', $[A \cap E_1, A \cap E_2, A \cap E_3, A \cap E_4]$, we can compute the probabilities $P(A|E_1)$, $P(A|E_2)$, $P(A|E_3)$, and $P(A|E_4)$. If A is an *unknown* nonempty set, but one for which the conditional probabilities $P(A|E_i)$ are *known*, then $P(A)$ can be computed. In both cases, once the probabilities $P(E_i)$ are known and the conditional probabilities $P(A|E_i)$ are known, we can find the probability that one of the partition events occurs, given that the event A occurs. For example, we can find $P(E_1|A)$.

This process of computing the reverse conditional probability is important in practical applications. The method is called *Bayes' theorem*, (Thomas Bayes,

12.3 Bayes' Theorem

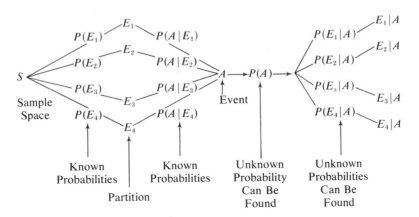

Figure 12.6
Diagram for a special case of Bayes' theorem.

1702-1761). In this special case (see Figure 12.6), Bayes' theorem has the form

$$P(E_i|A) = \frac{P(E_i) \cdot P(A|E_i)}{P(A)} = \frac{P(E_i)\, P(A|E_i)}{\sum_{i=1}^{4} P(E_i) \cdot P(A|E_i)}.$$

Example 12.16 A company requires all of its employees to take a certain test. A score in the 0 to 25th percentile is the event E_1, a score in the 26th to 50th percentile is called event E_2, the 51st to 75th percentile is event E_3, and the 76th to 100th percentile is event E_4. Suppose the long-run company records show that $P(E_1) = 0.20$, $P(E_2) = 0.40$, $P(E_3) = 0.25$, and $P(E_4) = 0.15$. Let A = the event an employee is rated above standard in his (or her) work. Also the company records show the probabilities $P(A|E_1) = 0.55$, $P(A|E_2) = 0.65$, $P(A|E_3) = 0.72$, and $P(A|E_4) = 0.80$.

a. What is the probability of the event A?

b. A supervisor rates a certain new employee as above standard. What is the probability that this employee's test score is E_2?

c. Find the probability that an employee scored between the 76th and 100th percentile, given that he or she is rated above standard.

d. Find the probabilities $P(E_2|-A)$ and $P(E_4|-A)$, and interpret the results.

Solution a. Recall Theorem 11.1, page 346. If A is any subset of a given sample space S, and $S' = [Y_1, Y_2, ..., Y_k]$ is a partition of S, then,

$$P(A) = P(Y_1) \cdot P(A|Y_1) + P(Y_2) \cdot P(A|Y_2) + ... + P(Y_k) \cdot P(A|Y_k).$$

Here, we have

$$P(A) = P(E_1) \cdot P(A|E_1) + P(E_2) \cdot P(A|E_2) + P(E_3) \cdot P(A|E_3)$$
$$+ P(E_4) \cdot P(A|E_4)$$
$$= (0.20)(0.55) + (0.40)(0.65) + (0.25)(0.72) + (0.15)(0.80)$$
$$= 0.11 + 0.26 + 0.18 + 0.12 = 0.67.$$

b. This question is asking for the probability of E_2 given A. By Bayes' theorem,

$$P(E_2|A) = \frac{P(E_2) \cdot P(A|E_2)}{P(A)} = \frac{(0.40)(0.65)}{0.67} = \frac{0.26}{0.67} = 0.39.$$

c. $P(E_4|A) = P(E_4) \cdot P(A|E_4)/P(A) = 0.12/0.67 = 0.18.$

d. From part a, $P(A) = 0.67$; therefore, $P(-A) = 0.33$. Now, by definition of conditional probability,

$$P(E_2|-A) = \frac{P(E_2 \cap -A)}{P(-A)}.$$

To get $P(E_2 \cap -A)$, we can use Table 11.1, page 337:

$$P(E_2 \cap -A) = P(E_2) - P(E_2 \cap A).$$

$P(E_2) = 0.40$ as given above, but what is $P(E_2 \cap A)$? We can answer that question by using the definition of conditional probability:

$$\frac{P(A \cap E_2)}{P(E_2)} = P(A|E_2), \text{ or } P(A \cap E_2) = P(E_2)P(A|E_2).$$

From the above data $P(A|E_2) = 0.65$; therefore, $P(A \cap E_2) = (0.40)(0.65) = 0.26$. Finally, we can find $P(E_2|-A)$; it is

$$P(E_2|-A) = \frac{P(E_2) - P(E_2 \cap A)}{1 - P(A)} = \frac{0.40 - 0.26}{0.33} = 0.42.$$

Similarly,

$$P(E_4|-A) = \frac{P(E_4) - P(E_4 \cap A)}{1 - P(A)} = \frac{P(E_4) - P(E_4)P(A|E_4)}{1 - P(A)}$$
$$= \frac{0.15 - 0.12}{0.33} = 0.091. \quad \blacktriangle$$

12.3 Bayes' Theorem

In parts **b** and **c**, the probability that an above-standard performer scored E_2 is large compared to those above-standard performers scoring E_4 because there are so few E_4 scorers. The people who are *not above standard* have an even smaller chance of coming from the E_4 scorers because $P(A|E_4) = 0.80$.

Theorem 12.1 (Bayes' theorem)
Let S be a finite sample space, $S' = [E_1, E_2, ..., E_k]$ is any partition of S, and A is any nonempty subset of S. If the conditional probability $P(A|E_i)$ for each $i = 1,2,3, ...,k$ is known or can be found, then the inverse conditional probability $P(E_i|A)$ can be found by the equation

$$P(E_i|A) = \frac{P(E_i)P(A|E_i)}{P(A)} = \frac{P(E_i)P(A|E_i)}{\sum_{i=1}^{k} P(E_i)P(A|E_i)}.$$

Proof From Theorem 11.1,

$$P(A) = \sum_{i=1}^{k} P(E_i)P(A|E_i). \tag{12.1}$$

From the definition of conditional probability,

$$P(A|E_i) = \frac{P(A \cap E_i)}{P(E_i)} \tag{12.2}$$

and

$$P(E_i|A) = \frac{P(A \cap E_i)}{P(A)}. \tag{12.3}$$

From Equation (12.2), we get

$$P(A \cap E_i) = P(E_i)P(A|E_i). \tag{12.4}$$

Now, substituting from Equation (12.4) into Equation (12.3), we get

$$P(E_i|A) = \frac{P(E_i)P(A|E_i)}{P(A)}. \tag{12.5}$$

The demoninator of Equation (12.4) can be replaced by the right-hand side of Equation (12.1) to give the results of the theorem.

In both the example and the theorem above, the partition of the sample space plays a part in the computation of the reverse conditional probabilities. Partitioning of a sample space is one of the things that a random variable does best because it does it automatically.

Example 12.17 In a three-child family, a sample space S can be composed of eight sample points listed here (M = male, F = female):

$$E_1 = \text{MMM} \qquad E_5 = \text{FMM}$$
$$E_2 = \text{MMF} \qquad E_6 = \text{FMF}$$
$$E_3 = \text{MFM} \qquad E_7 = \text{FFM}$$
$$E_4 = \text{MFF} \qquad E_8 = \text{FFF}$$

Let the random variable X = the exact number of girls in the family. The range of values of X is the set $S' = \{0,1,2,3\}$, and X maps S into S' by the following functional relationships: $X(E_1) = 0$, $X(E_2) = 1$, $X(E_3) = 1$, $X(E_4) = 2$, $X(E_5) = 1$, $X(E_6) = 2$, $X(E_7) = 2$, and $X(E_8) = 3$. The events in S' are $[X = 0, X = 1, X = 2, X = 3.]$ $X = 0$ could be read: "the event that the random variable *takes on* the value zero," or simply "the event that there are no girls in the family."

S' is a partition of S because no two events in S' overlap (the events in S' are mutually exclusive), and the events in S' cover all of the events in S (the events in S' are conjointly exhaustive).

An example illustrating that the events are mutually exclusive is: The event $X = 1$ does not overlap the event $X = 2$ since the same family cannot have exactly one girl *and* exactly two girls.

In general, the range of values of a random variable is *always* a partition of the sample space domain of that random variable. We can apply Bayes' theorem to sample spaces whose points are values of a random variable.

▲ **Example 12.18** This is a problem concerning a jury trial. It is somewhat artificial since every case is unique; but in a large number of court cases, enough data are collected to allow the relative frequencies to be regarded as probabilities in similar cases. (See Gelfand and Solomon's "Modeling Jury Verdicts in the American Legal System." *Journal of the American Statistical Association* 69 (March 1974): 32-37.) Let the random variable X stand for the number of jurors voting guilty on the *first* ballot. The range of values of X is the set $\{0,1,2,3,4,5,6,7,8,9,10,11,12\}$. Data from court cases show that the distribution is $P(X = 0) = 0.1160$, $P(1 \leq X \leq 4) = 0.1550$, $P(5 \leq X \leq 7) = 0.2050$, $P(8 \leq X \leq 11) = 0.3300$, and $P(X = 12) = 0.1940$. Let G = the event that the final, unanimous verdict is guilty. Suppose that the following conditional probabilities are known: $P(G|X = 0) =$

0, $P(G|1 \leq X \leq 4) = 0.0286$, $P(G|5 \leq X \leq 7) = 0.6956$, $P(G|8 \leq X \leq 11) = 0.8400$, and $P(G|X = 12) = 1.000$.

a. Compute the probability of a guilty verdict.

b. Given a guilty verdict, find the probability that the number of guilty votes on the first ballot was between one and four.

c. Given a guilty verdict, find the probability that the number of guilty votes on the first ballot was between eight and eleven.

d. Find $P(X = 12|G)$ and $P(5 \leq X \leq 7|G)$.

Solution **a.** $P(G) = P(X = 0)P(G|X = 0) + P(1 \leq X \leq 4)P(G|1 \leq X \leq 4) + P(5 \leq X \leq 7)P(G|5 \leq X \leq 7) + P(8 \leq X \leq 11)P(G|8 \leq X \leq 11) + P(X = 12)P(G|X = 12)$. Substituting the given numerical data and multiplying, we get

$$P(G) = 0 + 0.0044 + 0.1426 + 0.2772 + 0.1940$$
$$= 0.6182 \text{ or } 0.62 \text{ approximately.}$$

b. By Bayes' theorem,

$$P(1 \leq X \leq 4|G) = \frac{P(1 \leq X \leq 4)P(G|1 \leq X \leq 4)}{P(G)}$$

$$= \frac{(0.1550)(0.0286)}{0.62} = 0.007 \text{ approximately.}$$

c. $P(8 \leq X \leq 11|G) = (0.3300)(0.8400)/0.62 = 0.447$ approximately.

d. $P(X = 12|G) = 0.313$, $P(5 \leq X \leq 7|G) = 0.229$. ▲

EXERCISE 12.3

1. Let $[E_1, E_2, E_3, E_4, E_5]$ be a partition of a sample space S, and suppose $P(E_1) = 0.20$, $P(E_2) = 0.30$, $P(E_3) = 0.30$, $P(E_4) = 0.15$, and $P(E_5) = 0.05$. If A is an event in S such that $P(A|E_1) = 0.75$, $P(A|E_2) = 0.90$, $P(A|E_3) = 0.20$, $P(A|E_4) = 0.80$, and $P(A|E_5) = 1.00$,

   a. Compute $P(A)$.
   b. Compute $P(E_i|A)$ for $i = 1,2,3,4,5$.
   c. Compute $P(A \cap E_1)$.
   d. Compute $P(A \cap -E_2)$.
   e. Compute $P(E_3 \cap -A)$.
   f. Compute $P(E_3|-A)$.

2. Let $[Y_1, Y_2, Y_3, Y_4]$ be a partition of a sample space S and suppose $P(Y_1) = 0.13$, $P(Y_2) = 0.17$, $P(Y_3) = 0.25$, and $P(Y_4) = 0.45$. For an event B, let $P(B|Y_1) = 1.00$, $P(B|Y_2) = 1.00$, $P(B|Y_3) = 0.44$, and $P(B|Y_4) = 0.80$.

   a. Compute $P(B)$.

   b. Compute $P(Y_i|B)$ for $i = 1,2,3,4$.

   c. Compute $P(B \cap Y_1)$.

   d. Compute $P(B \cap -Y_2)$.

   e. Compute $P(Y_3 \cap -B)$.

   f. Compute $P(Y_3|-B)$.

3. Find the missing probabilities in the diagram below of Bayes' theorem.

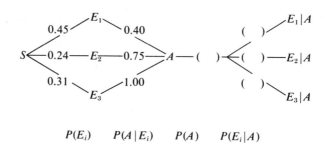

$P(E_i)$ $P(A|E_i)$ $P(A)$ $P(E_i|A)$

4. Repeat Problem 3 for the diagram below.

$P(Y_i)$ $P(B|Y_i)$ $P(B)$ $P(Y_i|B)$

5. A random variable X, its probability distribution, and the conditional probabilities for an event G are given in the following table.

12.3 Bayes' Theorem

$X = x$	0	1	2
$P(X = x)$	1/3	1/3	1/3
$P(G \mid X = x)$	1	1/2	0

a. Compute $P(G)$.

b. Compute $P[(X = x) \mid G]$ for $x = 0, 1, 2$.

6. A random variable X, its probability distribution, and the conditional probabilities for an event A are given in the following table.

$X = x$	-2	-1	1	2
$P(X = x)$	0.125	0.375	0.375	0.125
$P(A \mid X = x)$	0.500	1.000	1.000	0.500

a. Compute $P(A)$.

b. Compute $P[(X = x) \mid A]$ for $x = -2, -1, 1, 2$.

▲ **7.** A geneticist, working with experimental animals, observes that a certain physical trait T can occur with probability 0.70, given a congenital condition E_1; or T can occur with probability 0.50, given the recessive hereditary feature E_2; or T can occur with probability 0.02, given a mutation E_3. In other words, $P(T \mid E_1) = 0.70$, $P(T \mid E_2) = 0.50$, and $P(T \mid E_3) = 0.02$. The congenital condition is known to occur with a probability of 0.30, the hereditary feature with a probability of 0.40, and the mutation with a probability of 0.01. Assume these three conditions are mutually exclusive and conjointly exhaustive.

a. Find the probability that the trait T occurs.

b. Given T has occurred, what is the probability that it was hereditary?

c. Given T has occurred, what is the probability that a mutation has occurred?

8. A radio announcer is describing a boxing match. Suppose radio listeners know: The probability is 0.60 that one of the boxers will hit the other in a single round. The probability is 0.80 that the announcer will announce a hit, given there is one, during the round. The probability is 0.20 that the announcer will announce a hit in that round when one of the boxers misses with a wild swing, and there is no hit. Given that the reporter announces a hit, what is the probability that there actually was one? ▲

9. Suppose that $[E_1, E_2, E_3]$ is a partition of a sample space S, and $P(E_1)$, $P(E_2)$, $P(E_3)$, $P(A|E_1)$, $P(A|E_2)$, and $P(A|E_3)$ are known.

   a. Show that $P(A \cap E_1)$, $P(A \cap E_2)$, and $P(A \cap E_3)$ can be found.

   b. $P(A)$ can be found by Bayes' theorem; show that $P(A \cap -E_1)$ can be found. (Hint: Use Table 11.1, page 337.)

   c. Find $P(A \cap -E_2)$ and $P(A \cap -E_3)$ in terms of $P(A)$, $P(A \cap E_2)$, and $P(A \cap E_3)$.

   d. Can $P(A|-E_1)$, $P(A|-E_2)$, and $P(A|-E_3)$ be found?

   e. Can $P(-E_1|A)$, $P(-E_2|A)$, and $P(-E_3|A)$ be found?

10. If $[E_1, E_2, E_3]$ is a partition of S, and the probabilities are given as indicated in the following table,

$P(E_i)$	0.20	0.50	0.30
$P(A|E_i)$	0.75	0.40	0.60

    a. Find $P(A \cap E_1)$, $P(A \cap E_2)$, and $P(A \cap E_3)$.
    b. Find $P(A \cap -E_1)$.
    c. Find $P(A \cap -E_2)$ and $P(A \cap -E_3)$.
    d. Find $P(A|-E_1)$, $P(A|-E_2)$, and $P(A|-E_3)$.
    e. Find $P(-E_1|A)$, $P(-E_2|A)$, and $P(-E_3|A)$.

▲ **The false alarm problem** Problems 11 through 21 refer to the following description. Normally, the relative frequency of false alarms to the total of all fire alarms

12.3 Bayes' Theorem

ranges from 5 to 10 percent. On certain holiday nights, such as New Year's Eve and Halloween, the incidence is much higher. A high incidence also occurs at abnormal times, such as periods of riots, civil unrest, and natural disasters. Let F = the event that a fire occurs in one normal day, and A = the event that a fire alarm is turned in during the normal day. Essentially, there are three combinations of A and $-F$ that could be called *false alarms*. First is the conjunction $A \cap -F$, which is simply the event that an alarm is sounded, but there is no fire. This can happen either accidentally or deliberately. Second is the $A|-F$ event, which is the malicious false alarm—the person turning it in knows there is no fire. Third is the false alarm from the fireman's viewpoint; it is the $-F|A$ event, the fireman receives the alarm, but he doesn't know yet whether it is false. The relative frequencies of $A|-F$ and $-F|A$ are not necessarily the same.

Suppose the following probabilities are known: $P(F) = 0.60$, $P(F \cap -A) = 0.07$, and $P(A \cap -F) = 0.04$.

11. Find $P(-F)$.
12. Find the probability of the malicious fire alarm, $P(A|-F)$.
13. Find $P(A)$. (Hint: Use #13 and #14 in Table 11.1.)
14. Use $P(A)$ from Problem 13 to find the probability that there is no fire, given there is an alarm, $P(-F|A)$.
15. Find the probability of the unannounced fire, $P(F|-A)$.
16. Compute the probability $P(A \cap F)$.
17. Compute $P(-A \cap -F)$.
18. Find the probability of a failure to report a fire, $P(-A|F)$.
19. Find the probability of no fire, given no alarm, $P(-F|-A)$.
20. Find the probability of a fire, given an alarm, $P(F|A)$.
21. Find $P(A|F)$.

13
THE MARKOV PROCESS

13.1
PROBABILITY VECTORS AND MATRICES

The random variable is a versatile tool in probability analysis. It permits us to use counting techniques and coordinate geometry in solving applied problems. It is essential for solving expected value problems and it plays a major role in the partition of sample spaces, the composition of conditional probabilities, and the application of Bayes' theorem.

In this section, random variables and their distributions allow us to apply the concepts of vectors and matrices to probabilities. The presentation here assumes the reader has some background in the multiplication of vectors and matrices. This material is covered in Chapter 7.

Example 13.1 A precinct, V_1, has 8000 voters. Let $X =$ the number of votes a candidate receives from the precinct. The range of values of X is $S' = \{0,1,2,...,8000\}$. Suppose that this is partitioned into three groups of the random variable $[E_1, E_2, E_3]$, where E_1 is the event that $X \leq 2000$, E_2 is the event that $2000 < X \leq 5000$, and E_3 is the event that $X > 5000$ (and ≤ 8000). Let the distribution on this "grouped" random variable be $P(E_1) = 0.50$, $P(E_2) = 0.15$, and $P(E_3) = 0.35$. These probabilities can be expressed all at once as the vector $\mathbf{p}_1$:

$$\mathbf{p}_1 = (0.50, 0.15, 0.35).$$

In this case, $\mathbf{p}_1$ is called a **probability vector.** We can simplify the discussion if we let the random variable inequality $X \leq 2000$ be replaced by a single value between 0 and 2000, such as 1000. Now $X = 1000$ will stand for the entire group defined by $X \leq 2000$. Similarly, let the inequality $2000 < X \leq 5000$ be replaced

379

by $X = 3500$, and let $X > 5000$ be replaced by $X = 6500$. Now the partition $[E_1, E_2, E_3]$, representing the number of votes the candidate receives, can be replaced by the single numbers 1000, 3500, and 6500. This permits us to write the number of votes received as the vector **N**:

$$\mathbf{N} = (1000, 3500, 6500).$$

Actually, this vector could be considered as the original range of values of the random variable X.

The number of votes the candidate can expect to receive from this precinct is the scalar product of the two vectors $\mathbf{p}_1 \cdot \mathbf{N}$, which happens to be the expected value, $E(X)$:

$$E(X) = \mathbf{p}_1 \cdot \mathbf{N} = (0.50, 0.15, 0.35) \cdot (1000, 3500, 6500)$$
$$= 500 + 525 + 2275 = 3300.$$

Suppose another precinct, V_2, has 8000 voters, but in that precinct the random variable X, with a range of values as defined by vector **N**, has a different probability distribution vector, $\mathbf{p}_2$:

$$\mathbf{p}_2 = (0.48, 0.24, 0.28).$$

The expected value of X is the scalar product

$$E(X) = \mathbf{p}_2 \cdot \mathbf{N} = (0.48, 0.24, 0.28) \cdot (1000, 3500, 6500)$$
$$= 480 + 840 + 1820 = 3140.$$

In a third precinct, V_3, also with 8000 voters and the same range of the random variable, the distribution is the probability vector

$$\mathbf{p}_3 = (0.60, 0.30, 0.10).$$

The candidate's expected vote in precinct V_3 is

$$E(X) = \mathbf{p}_3 \cdot \mathbf{N} = (0.60, 0.30, 0.10) \cdot (1000, 3500, 6500)$$
$$= 600 + 1050 + 650 = 2300.$$

All three probability distributions of the random variable could be represented by one matrix **M**:

$$\mathbf{M} = \begin{bmatrix} 0.50 & 0.15 & 0.35 \\ 0.48 & 0.24 & 0.28 \\ 0.60 & 0.30 & 0.10 \end{bmatrix} \begin{array}{l} \text{Precinct } V_1 \\ \text{Precinct } V_2. \\ \text{Precinct } V_3 \end{array}$$

The range of values of the random variable X is the vector **N**, which we write as a column vector in order to be able to multiply it by **M**. The result is a vector of expected values, $E(X)$, whose components are the expected values in the precincts:

13.1 Probability Vectors and Matrices

$$E(X) = M \cdot N = \begin{bmatrix} 0.50 & 0.15 & 0.35 \\ 0.48 & 0.24 & 0.28 \\ 0.60 & 0.30 & 0.10 \end{bmatrix} \cdot \begin{bmatrix} 1000 \\ 3500 \\ 6500 \end{bmatrix} = (3300, 3140, 2300).*$$

▲

A square matrix, in which each row is a probability vector, is called a **stochastic matrix**. The matrix **M** above is an example of a stochastic matrix. We now present a formal definition of a probability vector.

Definition (Probability vector)
The statement that "the vector $\mathbf{v} = (v_1, v_2, \ldots, v_n)$ is a *probability vector*" means that $v_i \geq 0$ for all i, and

$$\sum_{i=1}^{n} v_i = 1;$$

that is, all the components of **v** are nonnegative and their sum is one.

Note that if only $n - 1$ of the components of a probability vector are known, then the missing term can be deduced from knowing that the sum of all the components must be one.

Example 13.2
a. If the vector $\mathbf{v} = (1/10, v_2, 2/11, 17/330)$ is a probability vector, what value must v_2 have?

b. If $x > 0$ and $y > 0$, is the vector $\mathbf{w} = (x, y, 1 - x - y)$ a probability vector? If, in addition, $1 - x - y \geq 0$, is **w** a probability vector?

c. Let $\mathbf{u} = (u_1, u_2, u_3, u_4)$ be a vector in which all the components are nonnegative, and $k = u_1 + u_2 + u_3 + u_4$. If $k > 0$, show that $(1/k)\mathbf{u}$ is a probability vector.

Solution
a. If **v** is a probability vector, then the sum of the components equals one. Therefore, $(1/10) + v_2 + (2/11) + (17/330) = 1$. Solving for v_2, we get $v_2 = 2/3$.

b. Not necessarily because, for example, if $x = 2$ and $y = 3$, then $1 - x - y = -4$, negative. If, in addition, $1 - x - y$ is given as nonnegative, then **w** *is* a probability vector because all components are nonnegative and their sum is one.

---

*We are writing all the answers as row vectors.

c. $(1/k)\mathbf{u} = [(1/k)u_1, (1/k)u_2, (1/k)u_3, (1/k)u_4]$ by the definition of multiplication of a scalar by a vector. If $k > 0$ and each component $u_i \geq 0$ for all $i = 1,2,3,4$, then every component of $(1/k)\mathbf{u}$ is nonnegative. Now, does the sum of all of the components of $(1/k)\mathbf{u}$ equal one? Yes, since

$$\sum_{i=1}^{4} \frac{1}{k} u_i = \frac{1}{k}(u_1 + u_2 + u_3 + u_4) = \frac{1}{k} k = 1.$$

Definition (Stochastic matrix)
The statement that "the matrix $\mathbf{M} = (m_{ij})$, i and $j = 1,2,3,...,n$, is a *stochastic matrix*" means that $\mathbf{M}$ is a square matrix such that for each fixed row index r, the row $(m_{r1}, m_{r2}, m_{r3}, ..., m_{rn})$ is a probability vector. In other words, a matrix is stochastic if and only if all of its rows are probability vectors.

Example 13.3   **a.** If

$$\mathbf{B} = \begin{bmatrix} 0 & 1 & 0 \\ 3/5 & 1/10 & b_{23} \\ b_{31} & 1/7 & 2/7 \end{bmatrix}$$

is a stochastic matrix, find b_{23} and b_{31}.

b. Compute the product

$$(0.20, 0.80) \cdot \begin{bmatrix} 0.33 & 0.67 \\ 0.40 & 0.60 \end{bmatrix}.$$

c. Show that the product of a one-by-two probability vector $\mathbf{v} = (v_1, v_2)$ and a two-by-two stochastic matrix $\mathbf{A} = (a_{ij})$ is a probability vector.

d. If $\mathbf{A}$ and $\mathbf{B}$ are the stochastic matrices

$$\mathbf{A} = \begin{bmatrix} 0.10 & 0.90 \\ 0 & 1 \end{bmatrix} \text{ and } \mathbf{B} = \begin{bmatrix} 0.50 & 0.50 \\ 0.25 & 0.75 \end{bmatrix},$$

find the product $\mathbf{A} \cdot \mathbf{B}$. Is it a stochastic matrix?

Solution   **a.** Each row is a probability vector and must have a sum of one. Hence, $3/5 + 1/10 + b_{23} = 1$ and $b_{31} + 1/7 + 2/7 = 1$. That is $b_{23} = 3/10$, $b_{31} = 4/7$.

b. $(0.386, 0.614)$.

c. $(v_1, v_2) \begin{bmatrix} a_{11} & a_{12} \\ a_{21} & a_{22} \end{bmatrix} = (v_1 a_{11} + v_2 a_{21}, v_1 a_{12} + v_2 a_{22}) = \mathbf{w}.$

13.1 Probability Vectors and Matrices

Since all the components of **v** and **A** are nonnegative, then the components of **w** are nonnegative. Now, we need to see if the components of **w** add up to one. Let s = the sum of the components of **w**, then $s = v_1 a_{11} + v_2 a_{21} + v_1 a_{12} + v_2 a_{22}$. Regrouping the terms and factoring, we get $s = v_1(a_{11} + a_{12}) + v_2(a_{21} + a_{22})$. But the expressions in parentheses are the sums of the rows of the matrix. They each add up to one; therefore, $s = v_1 \cdot 1 + v_2 \cdot 1 = v_1 + v_2 = 1$. This shows that **w** is a probability vector.

d.
$$\mathbf{A} \cdot \mathbf{B} = \begin{bmatrix} 0.10 & 0.90 \\ 0 & 1 \end{bmatrix} \cdot \begin{bmatrix} 0.50 & 0.50 \\ 0.25 & 0.75 \end{bmatrix} = \begin{bmatrix} 0.275 & 0.725 \\ 0.250 & 0.750 \end{bmatrix}.$$

The product is a stochastic matrix because each row is a probability vector.

The multiplication of two stochastic matrices and the multiplication of a row vector times a stochastic matrix have a variety of uses, but before considering applications, we will state the general results.

1. A *one-by-n* probability row vector **v** times an *n-by-n* stochastic matrix **M** is a *one-by-n* probability vector, $\mathbf{v} \cdot \mathbf{M}$.
2. The product of two *n-by-n* stochastic matrices is a stochastic matrix.
3. For any positive integer m, the mth power of a stochastic matrix is a stochastic matrix.

Example 13.4

a. $(0.30, 0.50, 0.20) \cdot \begin{bmatrix} 1/2 & 1/3 & 1/6 \\ 1 & 0 & 0 \\ 1/2 & 1/2 & 0 \end{bmatrix} = (0.75, 0.20, 0.05).$

b. Let **A** be the stochastic matrix

$$\mathbf{A} = \begin{bmatrix} 1/3 & 2/3 \\ 1/4 & 3/4 \end{bmatrix},$$

then $\mathbf{A}^2$ and $\mathbf{A}^3$ are as indicated below:

$$\mathbf{A}^2 = \begin{bmatrix} 5/18 & 13/18 \\ 13/48 & 35/48 \end{bmatrix} \text{ and } \mathbf{A}^3 = \mathbf{A}^2 \mathbf{A} = \begin{bmatrix} 59/216 & 157/216 \\ 157/576 & 419/576 \end{bmatrix}.$$

Example 13.5 Refer to Example 13.1. Suppose the candidate finds that his stand on key issues has changed the probability distribution of the random variable X as follows. A given row in matrix **M** is the precinct's current voting probabilities for the candidate. The candidate's stand on the issues is likely to modify the voting probabilities by strengthening his vote among those who agree with him and weakening it among those who disagree with him. One way to express this modification of the current state is by *squaring* the matrix **M**. Other assumptions are possible, and actual empirical data are needed to determine the best-fitting assumption. But in each case, the matrix is an expression of one state and multiplication by another matrix represents the next state, etc. Let us take for granted the assumption that the second state of the voters' attitudes is $\mathbf{M}^2$ and find the new expected vote. First we find $\mathbf{M}^2$:

$$\mathbf{M}^2 = \begin{bmatrix} 0.50 & 0.15 & 0.35 \\ 0.48 & 0.24 & 0.28 \\ 0.60 & 0.30 & 0.10 \end{bmatrix}^2 = \begin{bmatrix} 0.532 & 0.216 & 0.252 \\ 0.523 & 0.214 & 0.263 \\ 0.504 & 0.192 & 0.304 \end{bmatrix}.$$

Now, the expected vote is

$$E(X) = \mathbf{M}^2 \cdot \mathbf{N} = \begin{bmatrix} 0.532 & 0.216 & 0.252 \\ 0.523 & 0.214 & 0.263 \\ 0.504 & 0.192 & 0.304 \end{bmatrix} \cdot \begin{bmatrix} 1000 \\ 3500 \\ 6500 \end{bmatrix} = (2926, 2981, 3152).$$

▲

EXERCISE 13.1

1. Find the value of the unknown component that will make the vector a probability vector.

   a. $\mathbf{v} = (v_1, 3/4, 1/8)$.

   b. $\mathbf{u} = (0.07, 0.15, u_3, 0.50, 0.02)$.

2. Find the value of the unknown entries that will make the matrix a stochastic matrix.

$$\mathbf{M} = \begin{bmatrix} 0.70 & 0.14 & a_{13} \\ a_{21} & 0.47 & 0.47 \\ 0.54 & a_{32} & 0 \end{bmatrix}.$$

3. Compute the product

$$\left(\frac{1}{3}, \frac{1}{6}, \frac{1}{2} \right) \cdot \begin{bmatrix} 1/2 & 1/2 & 0 \\ 0 & 4/5 & 1/5 \\ 1/7 & 2/7 & 4/7 \end{bmatrix}.$$

13.1 Probability Vectors and Matrices

4. Compute the product

$$(0.22, 0.41, 0, 0.37) \cdot \begin{bmatrix} 0.10 & 0 & 0.30 & 0.60 \\ 0.31 & 0.15 & 0.28 & 0.26 \\ 0.45 & 0.55 & 0 & 0 \\ 0 & 0.30 & 0.20 & 0.50 \end{bmatrix}.$$

5. Compute the product

$$\begin{bmatrix} 2/3 & 1/3 \\ 2/5 & 3/5 \end{bmatrix} \cdot \begin{bmatrix} 1/2 & 1/2 \\ 1/4 & 3/4 \end{bmatrix}.$$

6. If **A** is the stochastic matrix,

$$\mathbf{A} = \begin{bmatrix} 0 & 1 \\ 1/2 & 1/2 \end{bmatrix},$$

find $\mathbf{A}^2$ and $\mathbf{A}^3$.

▲ **Negotiation matrix** In Problems 7 through 13, refer to the following description. Two people are trying to reach a consensus in budgetary requests. Each one has a perception of what fraction he should get and what fraction the other person should get. This information can be represented by the matrix below, called a *negotiation matrix*.

	First Person's Needs	Second Person's Needs
Perception by the First Person	a_{11}	a_{12}
Perception by the Second Person	a_{21}	a_{22}

The a_{ij} entry equals the fraction of the budget that the ith person wants the jth person to have. For example, a_{11} = the fraction that the first person wants the first person to have, a_{12} = the fraction that the first person wants the second person to have, and so on.

The square of the negotiation matrix, $\mathbf{M}^2$, is the second-round matrix. It represents each person's modification of his proposal after seeing what the other person proposed. $\mathbf{M}^3$ represents the third round of negotiations. In general, $\mathbf{M}^k$ is the kth round of negotiations. If the sequence of matrices, $\mathbf{M}$, $\mathbf{M}^2$, $\mathbf{M}^3$, $\mathbf{M}^4$, ... approaches some definite matrix $\mathbf{T}$ (as k increases indefinitely), then $\mathbf{T}$ will be a matrix in which both rows are exactly the same. This means the two negotiators agree on the needs for each person; they have reached a consensus. This "final" matrix $\mathbf{T}$ is called the *consensus matrix*.

***7.** If the initial negotiation matrix is

$$\mathbf{M} = \begin{bmatrix} 2/3 & 1/3 \\ 2/3 & 1/3 \end{bmatrix},$$

a. Find M^2 and M^3.

b. Show that a consensus has been reached.

*8. If the initial matrix is

$$A = \begin{bmatrix} 0.60 & 0.40 \\ 0.60 & 0.40 \end{bmatrix},$$

a. Show that $A = A^2 = A^3$.

b. Show that a consensus has been reached.

*9. If the initial matrix is

$$B = \begin{bmatrix} 3/4 & 1/4 \\ 1/2 & 1/2 \end{bmatrix},$$

a. What fraction of the budget does the first negotiator perceive as needed by the second negotiator?

b. What fraction of the budget does the second negotiator perceive as needed by the second negotiator?

c. Compute the second-round negotiation matrix, B^2.

d. Compute the third- and fourth-round negotiation matrices, B^3 and B^4.

e. The matrix

$$T = \begin{bmatrix} 2/3 & 1/3 \\ 2/3 & 1/3 \end{bmatrix}$$

is the consensus matrix. Compute the differences $T - B$, $T - B^2$, and $T - B^3$.

f. Show that the vector $(2/3, 1/3)$ is a *fixed point* of B; that is, show that $(2/3, 1/3) \cdot B = (2/3, 1/3)$.

*10. Another committee of two people start with an initial negotiation matrix C:

$$C = \begin{bmatrix} 7/10 & 3/10 \\ 4/10 & 6/10 \end{bmatrix}.$$

a. Compute the second-round negotiation matrix, C^2.

b. Compute the third-round negotiation matrix, C^3.

c. The sequence of matrices, $C, C^2, C^3, \ldots$, approaches the matrix T:

$$T = \begin{bmatrix} 4/7 & 3/7 \\ 4/7 & 3/7 \end{bmatrix}.$$

Show that the entries in the third-round negotiation matrix differ from the entries in the consensus matrix by less than 0.05.

d. Show that the vector $(4/7, 3/7)$ is a fixed point for **C**; that is, show that $(4/7, 3/7) \cdot \mathbf{C} = (4/7, 3/7)$.

*11. **a.** Show that the negotiation matrix, $\mathbf{N} = \begin{bmatrix} 1 & 0 \\ 0 & 1 \end{bmatrix}$, will never reach a consensus.

b. Show that if $\mathbf{A} = \begin{bmatrix} 1 & 0 \\ 1/2 & 1/2 \end{bmatrix}$, then the sequence $\mathbf{A}^2$, $\mathbf{A}^3$, ... approaches the consensus $\begin{bmatrix} 1 & 0 \\ 1 & 0 \end{bmatrix}$.

*12. An easy way to find a consensus matrix starting with an initial matrix, **M** (with no zero entries), is to try to find a probability vector $(x, 1 - x)$, that will be a fixed point for **M**. (See Problem 9.f.) Use this method to find the fixed point for matrix **M** below:

$$\mathbf{M} = \begin{bmatrix} 3/5 & 2/5 \\ 1/3 & 2/3 \end{bmatrix}.$$

▲

*13. Let **M** be the two-by-two stochastic matrix $\begin{bmatrix} a & 1 - a \\ b & 1 - b \end{bmatrix}$ with no zero entries. Show that there is a fixed point $(x, 1 - x)$ for **M**. Hint: $(x, 1 - x) \cdot \mathbf{M} = (ax + b - bx, x - ax + 1 - b - x + bx)$. Now solve the vector equation $(ax + b - bx, -ax + 1 - b + bx) = (x, 1 - x)$ for x in terms of a and b and show that the vector $(x, 1 - x)$ is a fixed point for **M**.

13.2

STATES AND TRANSITIONS

In Example 13.5 we used stochastic matrices to describe changes in the state of a precinct's voting probabilities for a given political candidate. This is one of a general class of probability problems in which a given process changes from one state to another. Changes of state are called *transitions* and the matrices used to describe them are called **transition matrices.**

Any process in which a given transition occurs with a given probability is called a **Markov process** (Andrei A. Markov, 1856–1922).

By a "state" of a process, we mean a condition or circumstance that exists at some fixed time or place. In a specific problem, each state has a given probability that the process will either change to a different state or remain in that same state. If a_1 and a_2 are two states, a one-step transition from state a_1 to a_2 is denoted by $a_1 \to a_2$. If a process is in state a_i at some fixed time t and it *remains* in state a_i at the next time $t + 1$, then the process is said to have "changed" from state a_i to state a_i; that is, $a_i \to a_i$. The probability for this transition can be any number between 0 and 1 (including 0 and 1).

▲ Example 13.6   (Consumer loyalty) Customers shopping for a laundry detergent can be described as participating in a Markov process. At any one purchase they can continue to buy the same brand they bought before or they can switch to a new brand. We will consider a special case with just two states: x = the state that a customer buys Brand X (at a given purchase), and o = the state that a customer buys some other brand.

The one-step transition $o \to x$ means that the customer switches from buying another brand at some given time to buying Brand X on his very next purchase. The transition $x \to x$ means he bought Brand X two times in a row; $o \to o$ means he bought a brand other than Brand X two times in a row; and $x \to o$ means he switched from Brand X to another brand. Each of these four transitions is an event and can be assigned a probability value.

Suppose the manufacturer of Brand X knows the following information about consumer loyalty to Brand X: $P_{xx} = 0.30$, $P_{xo} = 0.70$, $P_{ox} = 0.20$, and $P_{oo} = 0.80$ [where the notation P_{xx} means the probability of the transition $x \to x$, $P_{xo} = P(x \to o)$, $P_{ox} = P(o \to x)$, and $P_{oo} = P(o \to o)$]. What other information can we obtain from these probabilities? Can we compute the probability that a customer who bought Brand X on one purchase, say at time $t = 0$, switches to another brand on the first purchase after that ($t = 1$), and then switches back to Brand X on the next purchase ($t = 2$)? This double transition $x \to o \to x$ is called a two-step transition. We can answer these and many other such questions by applying some of the probability formulas from Chapters 10 and 11. The double transition $x \to o \to x$ is the transition $x \to o$ followed by the transition $o \to x$. *We assume these two events are independent*; therefore, the probability of both of them occurring is the product of their probabilities P_{xo} times $P_{ox} = (0.70)(0.20) = 0.14$. What is the probability of the transition $x \to x \to x$? This double transition is $x \to x$ followed by $x \to x$, so the probability is P_{xx} times P_{xx} or $(0.30)(0.30) = 0.09$.

Suppose we wanted to know the total probability that a customer will buy Brand X at time $t = 0$ and that he will also buy Brand X at time $t = 2$, *without regard to what he did at time $t = 1$?* We solve this by observing that this is actually the union of the two disjoint events, $x \to x \to x$ and $x \to o \to x$. So, we want

13.2 States and Transitions

sums of the probabilities $P(x \to x \to x)$ and $P(x \to o \to x)$ or $0.09 + 0.14 = 0.23$. This is the probability that a two-step transition starts and ends at x, assuming all possible intermediate states. We denote such a transition as $x \xrightarrow{2} x$. Similarly, $x \xrightarrow{2} o$ represents all the ways that two-step transitions can start at x and end at o; $o \xrightarrow{2} x$ = all the two-step transitions from o to x; and $o \xrightarrow{2} o$ = all the two-step transitions from o to o. The probability of $P(x \xrightarrow{2} o)$ is

$$P(x \xrightarrow{2} o) = P(x \to x \to o) + P(x \to o \to o) = P_{xx}P_{xo} + P_{xo}P_{oo}$$
$$= (0.30)(0.70) + (0.70)(0.80) = 0.21 + 0.56 = 0.77.$$

Similarly, $P(o \xrightarrow{2} x) = 0.22$, and $P(o \xrightarrow{2} o) = 0.78$.

The manufacturer of Brand X might be interested in his long-run share of the market. He would need to know the three-step transitions, the four-step transitions, and, in general, the n-step transitions $x \xrightarrow{n} x$ for any positive integer n. The computations involved would get very complicated, especially in problems in which there are more than two states.

We can simplify the calculations by introducing some other ways of expressing the original data. In Figure 13.1 the data are expressed as tree diagrams.

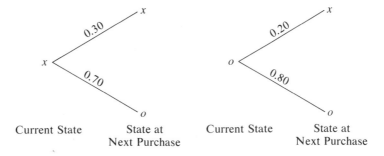

Figure 13.1
The numbers on the branches are the probabilities for the one-step transitions between the two states.

An alternative to a one-step tree diagram is a *directed graph*, or *transition diagram*, which connects a state to itself by a closed loop, and connects one state to another by an arc or line. These curves are labeled by numbers and arrows indicating the probabilities and the direction of the change in state. See Figure 13.2.

Transition diagrams are not practical for transitions of two or more steps. Tree diagrams are fine for two steps but become unwieldly for more than two. Fortunately,

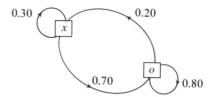

Figure 13.2
A transition diagram for the same one-step transitions as in the tree diagram in Figure 13.1.

there is another way to represent the data in the problem; the *transition matrix* is practical for any number of steps. Here is the matrix for these data:

$$\begin{array}{c} \text{State at} \\ \text{a Given} \\ \text{Purchase} \end{array} \begin{array}{c} \nearrow \\ x \\ o \end{array} \begin{array}{c} \overset{\text{State at the}}{\underset{\text{Next Purchase}}{}} \\ x \quad o \\ \begin{bmatrix} 0.30 & 0.70 \\ 0.20 & 0.80 \end{bmatrix} = \mathbf{M}. \end{array} \qquad (13.1)$$

Notice the arrow at the upper left. In this matrix, the numbers are the probabilities for the change in state indicated by the arrow. Explicitly, the entries in the matrix are the probabilities $P(x \to x) = P_{xx}$, $P(x \to o) = P_{xo}$, $P(o \to x) = P_{ox}$, and $P(o \to o) = P_{oo}$ as indicated below:

$$\mathbf{M} = \begin{array}{c} \nearrow \\ x \\ o \end{array} \begin{bmatrix} P_{xx} & P_{xo} \\ P_{ox} & P_{oo} \end{bmatrix}. \qquad (13.2)$$

There are eight two-step state changes and they are shown in Figure 13.3 below.

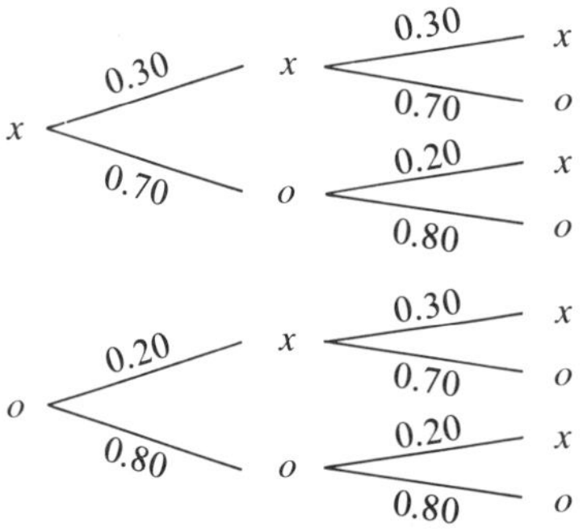

$$
\begin{array}{llll}
\text{(I)} \ x \to x \to x; & P(\text{I}) = P(x \to x)P(x \to x) = 0.09. \\
\text{(II)} \ x \to x \to o; & P(\text{II}) = P(x \to x)P(x \to o) = 0.21. \\
\text{(III)} \ x \to o \to x; & P(\text{III}) = P(x \to o)P(o \to x) = 0.14. \\
\text{(IV)} \ x \to o \to o; & P(\text{IV}) = P(x \to o)P(o \to o) = 0.56. \\
\text{(V)} \ o \to x \to x; & P(\text{V}) = P(o \to x)P(x \to x) = 0.06. \\
\text{(VI)} \ o \to x \to o; & P(\text{VI}) = P(o \to x)P(x \to o) = 0.14. \\
\text{(VII)} \ o \to o \to x; & P(\text{VII}) = P(o \to o)P(o \to x) = 0.16. \\
\text{(VIII)} \ o \to o \to o; & P(\text{VIII}) = P(o \to o)P(o \to o) = 0.64.
\end{array}
$$

Figure 13.3

13.2 States and Transitions

These eight disjoint chains of transitions can be grouped into four categories: $x \xrightarrow{2} x$; $x \xrightarrow{2} o$; $o \xrightarrow{2} x$; and $o \xrightarrow{2} o$. Each is composed of two disjoint events; therefore, their probabilities are

$$\left.\begin{aligned} P(x \xrightarrow{2} x) &= P(\text{I}) + P(\text{III}) = 0.09 + 0.14 = 0.23. \\ P(x \xrightarrow{2} o) &= P(\text{II}) + P(\text{IV}) = 0.21 + 0.56 = 0.77. \\ P(o \xrightarrow{2} x) &= P(\text{V}) + P(\text{VII}) = 0.06 + 0.16 = 0.22. \\ P(o \xrightarrow{2} o) &= P(\text{VI}) + P(\text{VIII}) = 0.14 + 0.64 = 0.78. \end{aligned}\right\} \quad (13.3)$$

We arrange these probabilities into a two-step transition matrix:

$$\mathbf{A} = \begin{matrix} & x & o \\ x & \\ o & \end{matrix} \begin{bmatrix} 0.23 & 0.77 \\ 0.22 & 0.78 \end{bmatrix}. \quad (13.4)$$

This matrix is actually the square of $\mathbf{M}$ in Equation (13.1), as shown below:

$$\mathbf{M}^2 = \begin{bmatrix} 0.30 & 0.70 \\ 0.20 & 0.80 \end{bmatrix} \begin{bmatrix} 0.30 & 0.70 \\ 0.20 & 0.80 \end{bmatrix} = \begin{bmatrix} 0.23 & 0.77 \\ 0.22 & 0.78 \end{bmatrix}.$$

It is no accident that the transition matrix for the two-step transition is the square of the one-step transition matrix. To see why this is the case, note that the two-step matrix in expression (13.4) is

$$\mathbf{A} = \begin{matrix} & x & o \\ x & \\ o & \end{matrix} \begin{bmatrix} P(\text{I}) + P(\text{III}) & P(\text{II}) + P(\text{IV}) \\ P(\text{V}) + P(\text{VII}) & P(\text{VI}) + P(\text{VIII}) \end{bmatrix}. \quad (13.5)$$

But according to the equations in Figure 13.3, expression (13.5) can be written

$$\begin{matrix} & x & o \\ x & \\ o & \end{matrix} \begin{bmatrix} P_{xx}P_{xx} + P_{xo}P_{ox} & P_{xx}P_{xo} + P_{xo}P_{oo} \\ P_{ox}P_{xx} + P_{oo}P_{ox} & P_{ox}P_{xo} + P_{oo}P_{oo} \end{bmatrix}. \quad (13.6)$$

Expression (13.6) is exactly the result that would be obtained by squaring $\mathbf{M}$, as written in Equation (13.2).

Any transitions with a higher number of steps can be obtained by raising the matrix to a higher power. This means that we can abandon the tree diagram and simply compute powers of $\mathbf{M}$. The three-step transitions are entries in the matrix $\mathbf{M}^3$:

$$\mathbf{M}^3 = \begin{matrix} & x & o \\ x & \\ o & \end{matrix} \begin{bmatrix} 0.223 & 0.777 \\ 0.222 & 0.778 \end{bmatrix}.$$

In other words, $P(x \xrightarrow{3} x) = 0.223$, $P(x \xrightarrow{3} o) = 0.777$, etc.

We can use powers of the transition matrix $\mathbf{M}$, $\mathbf{M}^2$, $\mathbf{M}^3$, $\mathbf{M}^4$, ..., to compute a manufacturer's share of the market after any number of transitions. Also we can compute the manufacturer's share of the market after some known *initial* market share. For example, suppose that a big advertising campaign (with coupons and price reductions) produced an initial market distribution of 75 percent of the consumers in state x and 25 percent in state o. Let $ denote the super sale. Then the probability distribution of the market at the time of the sale can be written as the initial vector $(0.75, 0.25)$. The tree diagram will have the two branches joined as shown in Figure 13.4.

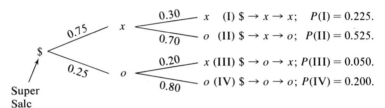

Figure 13.4

Now let us find the market distribution two purchases after the super sale; that is, we want the probabilities $P(\$ \xrightarrow{2} x)$ and $P(\$ \xrightarrow{2} o)$. They are

$$P(\$ \xrightarrow{2} x) = P(\text{I}) + P(\text{III}) = (0.75)(0.30) + (0.25)(0.20) = 0.275.$$

$$P(\$ \xrightarrow{2} o) = P(\text{II}) + P(\text{IV}) = (0.75)(0.70) + (0.25)(0.80) = 0.725.$$

If we interpret these computations strictly in terms of vectors and matrices, then the vector $\mathbf{u} = (0.75, 0.25)$ is the state of the market at the time of the sale. One step after the sale, the market state vector is the product of $\mathbf{u}$ and the transition matrix $\mathbf{M}$:

$$\mathbf{u} \cdot \mathbf{M} = (0.75, 0.25) \cdot \begin{bmatrix} 0.30 & 0.70 \\ 0.20 & 0.80 \end{bmatrix} = (0.275, 0.725).$$

The procedures discussed in the above example also apply to problems with three or more states.

▲ **Example 13.7** A Markov process may be used in biology to describe genetic combinations, or genotypes. A particular example of this application is as follows. Assume that the genotypes, AA, Aa, and aa, are three states whose probabilities are given in the tree diagram in Figure 13.5. The transition diagram for these data is given in Figure 13.6, where the branches with zero probability are omitted.

13.2 States and Transitions

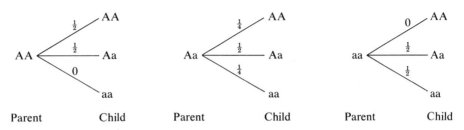

Figure 13.5
One-step transitions of genetic combinations.

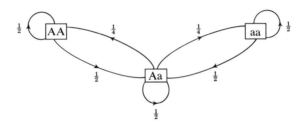

Figure 13.6
Transition diagram for the tree diagrams in Figure 13.5.

The transition matrix corresponding to the data in either diagram is the matrix **P**:

$$\mathbf{P} = \begin{array}{c} \\ \text{AA} \\ \text{Aa} \\ \text{aa} \end{array} \begin{array}{c} \text{AA} \quad \text{Aa} \quad \text{aa} \\ \left[\begin{array}{ccc} 1/2 & 1/2 & 0 \\ 1/4 & 1/2 & 1/4 \\ 0 & 1/2 & 1/2 \end{array} \right] \end{array}.$$

a. If, in some generation, the given population distribution is the initial vector $\mathbf{v} = (x, y, 1 - x - y)$ (sometimes called gene frequency), what will the vector be three generations later?

b. If the population distribution is the initial vector $\mathbf{u} = (1/10, 6/10, 3/10)$, answer part **a**.

Solution **a.** The answer to this question is $\mathbf{v} \cdot \mathbf{P}^3$.

$$\mathbf{P}^3 = \begin{bmatrix} 5/16 & 1/2 & 3/16 \\ 1/4 & 1/2 & 1/4 \\ 3/16 & 1/2 & 5/16 \end{bmatrix}.$$

Therefore, the gene frequency after three generations is

$$\mathbf{v} \cdot \mathbf{P}^3 = \left(\frac{2x + y + 3}{16}, \frac{1}{2}, \frac{-2x - y + 5}{16} \right).$$

b. If the initial distribution is $\mathbf{u} = (^1\!/_{10}, ^6\!/_{10}, ^3\!/_{10})$, then

$$\mathbf{u} \cdot \mathbf{P}^3 = \left(\frac{38}{160}, \frac{80}{160}, \frac{42}{160} \right) = (0.24, 0.50, 0.26). \quad ▲$$

EXERCISE 13.2

1. **a.** Find the matrix **M** that corresponds to the transition diagram on the right.

   b. Compute $\mathbf{M}^2$ and $\mathbf{M}^3$.

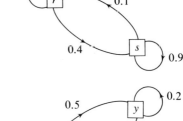

2. **a.** Find the matrix **P** that corresponds to the transition diagram on the right.

   b. Compute $\mathbf{P}^2$ and $\mathbf{P}^3$.

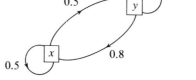

3. Use the data in Problem 1. If an initial-state vector is $\mathbf{v} = (0.7, 0.3)$, find the state after three steps.

4. Use the data in Problem 2. If an initial-state vector is $\mathbf{w} = (0,1)$, find the state after three steps.

5. In parts **a** and **b,** convert the tree diagram below

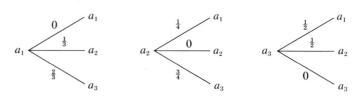

13.2 States and Transitions

a. To a transition diagram.

b. To a matrix **A**.

c. Find $\mathbf{A}^3$.

6. In parts **a** and **b**, convert the tree diagram below

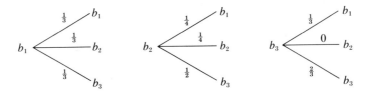

a. To a transition diagram.

b. To a matrix **B**.

c. Compute $\mathbf{B}^2$.

▲ 7. A stock room can be in two states at the end of each month: L = the state of a low inventory and H = the state of a high inventory. If, at the end of one month, the stock room is in state L, the stock clerk will order and receive more supplies before the end of the next month. Then the stock room has the probability 0.80 of being in state H (that is, the transition from L to H has the probability 0.80). If the stock room is in state H, the clerk will not order supplies; then the next month it has a probability 0.50 of being in state H.

a. Make up the transition diagram.

b. Make up the transition matrix.

8. A survey of eligible voters fits the following matrix of probabilities for voters changing from a voting or nonvoting state in one national election to a voting or nonvoting state in the next national election:

		Next Election	
		Voted	Did Not Vote
Previous Election	Voted	$\begin{bmatrix} 0.56 $	$ 0.44 \\$
	Did Not Vote	$ 0.88 $	$ 0.12 \end{bmatrix}$

If the initial vector in some election is (0.70, 0.30), what will be the voting probability vector two elections later?

9. Data collected from college students for intergenerational occupational movement from the father's occupation to the son's choice of occupation are given in the table below. (The only data available are for male ancestry and progeny.) Suppose that the distribution of these occupations at a given initial generation is given by the vector (0.40, 0.20, 0.20, 0.20). Find the distribution in the next generation. (See J. T. Mortimer's "Patterns of Intergenerational Movements: A Smallest Space Analysis." *American Journal of Sociology*, Vol. 79, no. 5, March 1974.)

Father's Occupation	Son's Occupational Choice			
	Professional Services	Businesses	Academic Artistic	Others
Professional Services (M.D., Law, etc.)	0.67	0.06	0.19	0.08
Businesses (All levels)	0.30	0.30	0.29	0.11
Academic, Artistic	0.28	0.30	0.32	0.10
All Others	0.28	0.17	0.27	0.28

10. Let A, B, C, and D be the four states:

A = high economic status, high social status.

B = high economic status, low social status.

C = low economic status, high social status.

D = low economic status, low social status.

The states and the probabilities for their transitions are shown in the transition diagram below. No connection between two states means a zero probability of transition from one to the other. For example $P(B \to C) = 0$.

a. Construct the transition matrix for these data.

13.2 States and Transitions

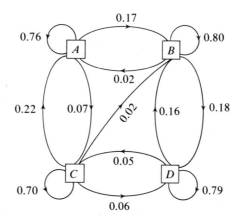

b. If the distribution in one generation is (0.25, 0.25, 0.25, 0.25), find the distribution in the next generation. (See Hunter's "Community Change: A Stochastic Analysis of Chicago's Local Communities, 1930-60." *American Journal of Sociology*, Vol. 79, no. 4, January 1974.)

11. In an "island" model consisting of four colonies, Y_1, Y_2, Y_3, and Y_4, the birthplaces of parents and offspring are represented by the intergenerational matrix shown below.

$$
\begin{array}{c} \text{Parents'} \\ \text{Birthplaces} \end{array}
\begin{array}{c} \\ Y_1 \\ Y_2 \\ Y_3 \\ Y_4 \end{array}
\begin{array}{c} \text{Offspring's Birthplaces} \\ \begin{array}{cccc} Y_1 & Y_2 & Y_3 & Y_4 \end{array} \\ \left[\begin{array}{cccc} 0.95 & 0.01 & 0.04 & 0.00 \\ 0.03 & 0.96 & 0.00 & 0.01 \\ 0.06 & 0.01 & 0.91 & 0.02 \\ 0.02 & 0.01 & 0.02 & 0.95 \end{array} \right] \end{array}
$$

a. If one generation has the distribution ($\frac{1}{4}$, $\frac{1}{4}$, $\frac{1}{4}$, $\frac{1}{4}$), what is the distribution in the next generation?

b. If **M** is the above matrix and **v** is the vector (1, 0, 0, 0), find the vector **v·M**. This vector is equivalent to confining all parental births to the first colony, while keeping the other colonies closed.

c. Use the vector **v·M** to find **v·M²** by multiplying (**v·M**)·**M**. Notice how the birthplace distribution slowly begins to spread to the other colonies. (See Bodmer and Cavalli-Sforza: "A Migration Matrix for the Study of Random Genetic Drift," *Genetics*, 59 (1968): 565-92.) ▲

13.3

THE MARKOV PROCESS—
THEORETICAL FOUNDATIONS

What would be the long-run consequences of a multiple-step transition if the Markov process was allowed to continue forever? What restrictions are necessary for a stochastic matrix to be a transition matrix for a Markov process?

In this section we present the definitions and theorems needed to answer these and other questions concerning Markov processes. Proofs are not given for all theorems; some proofs would require a level of mathematics much higher than that covered in this book.

Throughout this section we will use only stochastic matrices and probability vectors, unless otherwise indicated.

Definition (Regular stochastic matrix)
The statement that "the stochastic matrix $\mathbf{M}$ is regular" means that for some positive integer k, $\mathbf{M}^k$ has no zero entries. Of course, if $\mathbf{M}$ itself has no zero entries, then it is regular because for $k = 1$, $\mathbf{M}^1$ has no zero entries.

Example 13.8 **a.** If

$$\mathbf{M} = \begin{bmatrix} 1/2 & 1/3 & 1/6 \\ 2/5 & 2/5 & 1/5 \\ 1/10 & 7/10 & 1/5 \end{bmatrix},$$

is $\mathbf{M}$ regular?

b. If

$$\mathbf{P} = \begin{bmatrix} 1/2 & 1/4 & 1/4 \\ 1 & 0 & 0 \\ 0 & 1/2 & 1/2 \end{bmatrix},$$

is $\mathbf{P}$ regular?

c. If

$$\mathbf{Q} = \begin{bmatrix} 1 & 0 \\ 1/2 & 1/2 \end{bmatrix},$$

is $\mathbf{Q}$ regular?

13.3 The Markov Process—Theoretical Foundations

Solution **a.** Yes, since M^1 has no zero entries (and it is a stochastic matrix).

b. P has zero entries, but
$$P^2 = \begin{bmatrix} 1/2 & 1/4 & 1/4 \\ 1/2 & 1/4 & 1/4 \\ 1/2 & 1/4 & 1/4 \end{bmatrix}.$$

Since P^2 has no zero entries, P is regular.

c. $Q^2 = \begin{bmatrix} 1 & 0 \\ 3/4 & 1/4 \end{bmatrix}, Q^3 = \begin{bmatrix} 1 & 0 \\ 7/8 & 1/8 \end{bmatrix}, Q^4 = \begin{bmatrix} 1 & 0 \\ 15/16 & 1/16 \end{bmatrix}.$

Q is not regular because every power of Q has a zero entry in its first row, second column. It can be shown by mathematical induction that the nth power of Q is
$$Q^n = \begin{bmatrix} 1 & 0 \\ (2^n - 1)/2^n & 1/2^n \end{bmatrix}.$$

Definition **(Fixed point)**
If M is a regular n-by-n stochastic matrix, then the *one*-by-n vector v is called a *fixed point* for M if $v \cdot M = v$.

Theorem 13.1 If M is a regular stochastic matrix and v is a fixed point for M, then v is a fixed point for any positive integral power, M^k.

Proof If $v \cdot M = v$, then $v \cdot M^2 = v \cdot M \cdot M = (v \cdot M) \cdot M = (v) \cdot M = v$. If k is any positive integer such that $v \cdot M^k = v$, then
$$v \cdot M^k \cdot M = v \cdot M = v.$$
$$v \cdot M^{k+1} = v.$$

Therefore, there is no "last" integer n for which $v \cdot M^n = v$. (This is essentially the idea of a proof by mathematical induction.)

Example 13.9 **a.** Find a fixed point t for the matrix M:
$$M = \begin{bmatrix} 1/4 & 3/4 \\ 1/3 & 2/3 \end{bmatrix}.$$

b. Show that $t \cdot M^2 = t$.

Solution **a.** Let the components of the proposed fixed point $\mathbf{t}$ be $(t, 1 - t)$. Then,

$$\mathbf{t} \cdot \mathbf{M} = (t, 1 - t) \begin{bmatrix} 1/4 & 3/4 \\ 1/3 & 2/3 \end{bmatrix} = \left(\frac{t}{4} + \frac{1}{3} - \frac{t}{3}, \frac{3t}{4} + \frac{2}{3} - \frac{2t}{3} \right)$$

$$= \left(\frac{1}{3} - \frac{t}{12}, \frac{t}{12} + \frac{2}{3} \right).$$

Now set $\mathbf{t} \cdot \mathbf{M}$ equal to $\mathbf{t}$: $(t, 1 - t) = ((4 - t)/12, (t + 8)/12)$. The equality of the two vectors implies equality of their components $t = (4 - t)/12$ and $1 - t = (t + 8)/12$. Either of these equations is enough to determine the value of t, which is $t = 4/13$ and $1 - t = 9/13$. Therefore, the fixed point is $(4/13, 9/13)$. As a check, we compute the product

$$\left(\frac{4}{13}, \frac{9}{13} \right) \begin{bmatrix} 1/4 & 3/4 \\ 1/3 & 2/3 \end{bmatrix} = \left(\frac{1}{13} + \frac{3}{13}, \frac{3}{13} + \frac{6}{13} \right)$$

$$= \left(\frac{4}{13}, \frac{9}{13} \right).$$

b. We could show show that $\mathbf{t} \cdot \mathbf{M}^2 = \mathbf{t}$ as in the proof of Theorem 13.1: $\mathbf{t} \cdot \mathbf{M}^2 = \mathbf{t} \cdot \mathbf{M} \cdot \mathbf{M} = \mathbf{t} \cdot \mathbf{M} = \mathbf{t}$. But in order to show that we could actually square $\mathbf{M}$ first, then multiply by $\mathbf{t}$ and still get $\mathbf{t}$, we write the following equation:

$$\mathbf{t} \cdot \mathbf{M}^2 = \left(\frac{4}{13}, \frac{9}{13} \right) \cdot \begin{bmatrix} 5/16 & 11/16 \\ 11/36 & 25/36 \end{bmatrix} = \left(\frac{20}{208} + \frac{99}{468}, \frac{44}{208} + \frac{225}{468} \right)$$

$$= \left(\frac{4}{13}, \frac{9}{13} \right).$$

Theorem 13.2 If $\mathbf{T}$ is any n-by-n regular stochastic matrix in which all of the rows are identical, then every positive integral power of $\mathbf{T}$ is equal to $\mathbf{T}$, and multiplication of *any* probability vector by $\mathbf{T}$ produces one of the rows of $\mathbf{T}$. In symbols.

(I) For every positive integer k, $\mathbf{T}^k = \mathbf{T}$.

(II) If $\mathbf{v}$ is any *one*-by-n probability vector $(v_1, v_2, \ldots, v_n)$, then $\mathbf{v} \cdot \mathbf{T} = (t_1, t_2, \ldots, t_n)$.

Proof of (I) Squaring $\mathbf{T}$ we get $\mathbf{T}^2 = \mathbf{T} = n$ copies of the row vector:

$$\left(\sum_{i=1}^{n} t_1 t_i, \sum_{i=1}^{n} t_2 t_i, \ldots, \sum_{i=1}^{n} t_n t_i \right)$$

Now suppose there is some last positive integer $k > 1$ such that $\mathbf{T}$ to that power is $\mathbf{T}$, $(\mathbf{T}^k = \mathbf{T})$. Then multiplying both sides by $\mathbf{T}$: $\mathbf{T}^k \cdot \mathbf{T} = \mathbf{T} \cdot \mathbf{T} = \mathbf{T}^2 = \mathbf{T}$,

13.3 The Markov Process—Theoretical Foundations

or $\mathbf{T}^{k+1} = \mathbf{T}$. That is, $k + 1$ is a larger positive integer such that $\mathbf{T}$ to that power is $\mathbf{T}$, ($\mathbf{T}^{k+1} = \mathbf{T}$). Therefore, there is no end to the positive integers n for which $\mathbf{T}^n = \mathbf{T}$.

Proof of (II) If $\mathbf{v}$ is any probability vector $(v_1, v_2, \ldots, v_n)$, then,

$$\mathbf{v} \cdot \mathbf{T} = (v_1, v_2, \ldots, v_n) \begin{bmatrix} t_1 & t_2 & \cdots & t_n \\ t_1 & t_2 & \cdots & t_n \\ \vdots & \vdots & & \vdots \\ t_1 & t_2 & \cdots & t_n \end{bmatrix} = \left(\sum_{i=1}^{n} t_1 v_i, \ldots, \sum_{i=1}^{n} t_n v_i \right)$$

$$= \left(t_1 \sum_{i=1}^{n} v_i, t_2 \sum_{i=1}^{n} v_i, \ldots, t_n \sum_{i=1}^{n} v_i \right) = (t_1, t_2, \ldots, t_n).$$

Example 13.10 **a.** If

$$\mathbf{A} = \begin{bmatrix} 1/3 & 2/3 \\ 1/3 & 2/3 \end{bmatrix},$$

show that $\mathbf{A} = \mathbf{A}^2 = \mathbf{A}^3$. If $\mathbf{v} = (1/10, 9/10)$, show that $\mathbf{v} \cdot \mathbf{A} = (1/3, 2/3)$. If $\mathbf{w} = (x, 1 - x)$ for nonnegative x and $1 - x$, show that $\mathbf{w} \cdot \mathbf{A} = (1/3, 2/3)$.

b. Let $\mathbf{M}$ be the matrix

$$\mathbf{M} = \begin{bmatrix} 1/10 & 2/10 & 7/10 \\ 1/10 & 2/10 & 7/10 \\ 1/10 & 2/10 & 7/10 \end{bmatrix}$$

and let $\mathbf{v}$ be any one-by-three probability vector $\mathbf{v} = (x, y, 1 - x - y)$. Show that $\mathbf{v} \cdot \mathbf{M} = (1/10, 2/10, 7/10)$.

Solution **a.**

$$\mathbf{A}^2 = \begin{bmatrix} 1/3 & 2/3 \\ 1/3 & 2/3 \end{bmatrix} \cdot \begin{bmatrix} 1/3 & 2/3 \\ 1/3 & 2/3 \end{bmatrix} = \begin{bmatrix} 1/9 + 2/9 & 2/9 + 4/9 \\ 1/9 + 2/9 & 2/9 + 4/9 \end{bmatrix} = \mathbf{A}.$$

$\mathbf{A}^3 = \mathbf{A}^2 \cdot \mathbf{A} = \mathbf{A} \cdot \mathbf{A} = \mathbf{A}^2 = \mathbf{A}$. If $\mathbf{v} = (1/10, 9/10)$, then,

$$\mathbf{v} \cdot \mathbf{A} = \left(\frac{1}{10}, \frac{9}{10} \right) \cdot \begin{bmatrix} 1/3 & 2/3 \\ 1/3 & 2/3 \end{bmatrix} = \left(\frac{1}{10} \cdot \frac{1}{3} + \frac{9}{10} \cdot \frac{1}{3}, \frac{1}{10} \cdot \frac{2}{3} + \frac{9}{10} \cdot \frac{2}{3} \right)$$

$$= \left(\frac{1}{3}, \frac{2}{3} \right).$$

b.

$$\mathbf{v} \cdot \mathbf{M} = (x, y, 1 - x - y) \cdot \begin{bmatrix} 1/10 & 2/10 & 7/10 \\ 1/10 & 2/10 & 7/10 \\ 1/10 & 2/10 & 7/10 \end{bmatrix}$$

$$= \left[x \cdot \frac{1}{10} + y \cdot \frac{1}{10} + (1 - x - y)\frac{1}{10},\ x \cdot \frac{2}{10} + y \cdot \frac{2}{10} + (1 - x - y)\frac{2}{10},\ x \cdot \frac{7}{10} + y \cdot \frac{7}{10} + (1 - x - y)\frac{7}{10} \right].$$

Therefore, $\mathbf{v} \cdot \mathbf{M} = (1/10, 2/10, 7/10)$.

Proofs for Theorems 13.3 and 13.4 require transform analysis, including several topics from higher analysis. We are not going to even try to prove them here. The reader is referred to the proofs given in R. A. Howard's *Dynamic Probabilistic Systems, Volume I: Markov Models.* (New York: John Wiley, 1971).

Theorem 13.3 If $\mathbf{M}$ is a regular n-by-n stochastic matrix, then there is a unique n-by-n matrix $\mathbf{T}$ to which the sequence $\mathbf{M}, \mathbf{M}^2, \mathbf{M}^3, \ldots$ converges; that is, $\mathbf{M}^k$ approaches $\mathbf{T}$ as k increases indefinitely.

$\mathbf{T}$ is called the **terminal matrix** of this sequence. $\mathbf{T}$ is also called the **steady-state matrix** for the *chain of matrices*, $\mathbf{M}, \mathbf{M}^2, \mathbf{M}^3, \ldots$.

Theorem 13.4 If $\mathbf{T}$ is the terminal matrix of the sequence $\mathbf{M}, \mathbf{M}^2, \ldots$ for a regular stochastic matrix $\mathbf{M}$, then all of the rows of $\mathbf{T}$ are identical, and each row of $\mathbf{T}$ is *the* fixed point for $\mathbf{M}$.

The above theorem is very handy. We can find the steady-state matrix for a regular sequence $\mathbf{M}, \mathbf{M}^2, \mathbf{M}^3, \ldots$ by simply finding the fixed-point vector for $\mathbf{M}$.

Theorem 13.5 If $\mathbf{T}$ is the terminal matrix of the regular sequence $\mathbf{M}, \mathbf{M}^2, \mathbf{M}^3, \ldots$, then $\mathbf{T} \cdot \mathbf{M} = \mathbf{T}$ and $\mathbf{M} \cdot \mathbf{T} = \mathbf{T}$.

13.3 The Markov Process—Theoretical Foundations

Proof The general expression for the product of **M** and its fixed point **t** is

$$(t_1, \ldots, t_n) \begin{bmatrix} m_{11} & m_{12} & \cdots & m_{1n} \\ \vdots & \vdots & & \vdots \\ m_{n1} & m_{n2} & \cdots & m_{nn} \end{bmatrix} = \left(\sum_1^n t_i m_{i1}, \sum_1^n t_i m_{i2}, \ldots, \sum_1^n t_i m_{in} \right)$$

$$= (t_1, t_2, \ldots, t_n).$$

But multiplying the steady-state vector **T** times **M**, we get

$$\mathbf{T} \cdot \mathbf{M} = \begin{bmatrix} t_1 & t_2 & \cdots & t_n \\ \vdots & \vdots & & \vdots \\ t_1 & t_2 & \cdots & t_n \end{bmatrix} \cdot \begin{bmatrix} m_{11} & m_{12} & \cdots & m_{1n} \\ \vdots & \vdots & & \vdots \\ m_{n1} & m_{n2} & \cdots & m_{nn} \end{bmatrix}.$$

This is actually n copies of the multiplication of **M** by the fixed point **t**; or

$$\mathbf{T} \cdot \mathbf{M} = \begin{bmatrix} \sum_1^n t_i m_{i1} & \sum_1^n t_i m_{i2} & \cdots & \sum_1^n t_i m_{in} \\ \vdots & \vdots & & \vdots \\ \sum_1^n t_i m_{i1} & \sum_1^n t_i m_{i2} & \cdots & \sum_1^n t_i m_{in} \end{bmatrix} = \begin{bmatrix} t_1 & \cdots & t_n \\ \vdots & & \vdots \\ t_1 & \cdots & t_n \end{bmatrix} = \mathbf{T}.$$

Reversing the products gives a matrix whose rows are

$$\mathbf{M} \cdot \mathbf{T} = \begin{bmatrix} \sum_1^n t_1 m_{1i} & \sum_1^n t_2 m_{1i} & \cdots & \sum_1^n t_n m_{1i} \\ \sum_1^n t_1 m_{2i} & \sum_1^n t_2 m_{2i} & \cdots & \sum_1^n t_n m_{2i} \\ \vdots & \vdots & & \vdots \\ \sum_1^n t_1 m_{ni} & \sum_1^n t_2 m_{ni} & \cdots & \sum_1^n t_n m_{ni} \end{bmatrix} = \mathbf{T}.$$

Theorem 13.6 If **M** is a regular n-by-n stochastic matrix and **v** is *any* one-by-n probability vector, then the sequence of vectors $\mathbf{v} \cdot \mathbf{M}$, $\mathbf{v} \cdot \mathbf{M}^2$, $\mathbf{v} \cdot \mathbf{M}^3$, ... approaches the fixed-point vector **t** of **M**.

Discussion This theorem may be roughly argued as follows: Since $\mathbf{M}^k$ approaches **T**, then $\mathbf{v} \cdot \mathbf{M}^k$ approaches $\mathbf{v} \cdot \mathbf{T}$. Also, since every row of **T** is the same, then by Theorem 13.2, $\mathbf{v} \cdot \mathbf{T}$ = a row of $\mathbf{T}$. But, by Theorem 13.4, every row of $\mathbf{T}$ is the fixed point of $\mathbf{M}$. Therefore, $\mathbf{v} \cdot \mathbf{T} = \mathbf{t}$, or $\mathbf{v} \cdot \mathbf{M}^k$ approaches $\mathbf{t}$.

In each of the above theorems, we assumed that the transition matrices were regular. The Markov process associated with a regular transition matrix is called a *regular Markov chain*. There is also another type of Markov process called an *absorbing* Markov chain, which we will describe below. First, we state a definition of a Markov chain.

Definition (Markov chain)
A *Markov chain* is a probability space in which the simple events are state transitions. The probability function for the space is usually given as a transition matrix.

As mentioned above, the preceding examples of this section were all regular Markov chains; now let us look at an example of a transition matrix that is not regular.

Example 13.11 The transition matrix

$$\mathbf{M} = \begin{bmatrix} 1 & 0 & 0 & 0 \\ 0 & 1/2 & 1/4 & 1/4 \\ 1/5 & 1/5 & 2/5 & 1/5 \\ 0 & 1 & 0 & 0 \end{bmatrix}$$

is not regular. This can easily be seen by noticing that any power of $\mathbf{M}$ will have the same first row (1 0 0 0) as $\mathbf{M}$. Hence, all powers of $\mathbf{M}$ will contain zero entries. One important observation about this particular nonregular matrix is that if the Markov process associated with it ever gets into state a_1, then it will remain in that state forever. (Since the transition $a_1 \to a_1$ has the probability of one and $a_1 \to a_j$ (for any $j \neq 1$) has the probability zero, the process can never change out of state a_1 once it gets into that state.) For this reason, a_1 is called an *absorbing state* for this matrix.

Definition (Absorbing state)
A state a_i in a Markov chain is called *absorbing* if $P_{ii} = 1$ and $P_{ij} = 0$ for all $j \neq i$. (In other words, if a stochastic matrix has a one on the main diagonal, $a_{ii} = 1$, then the corresponding row is an absorbing state for that matrix.)

13.3 The Markov Process—Theoretical Foundations

Definition **(Absorbing Markov chain)**
A Markov chain is absorbing if its transition matrix has at least one absorbing state and if it is possible to get into some absorbing state from every nonabsorbing state.

The Markov chain with the transition matrix in Example 13.11 is absorbing. The process can get from any nonabsorbing state into a_1. For example, the process can get from a_4 to a_2 in one step, then from a_2 to a_3 in a second step, and finally from a_3 to a_1 in a third step.

The following application of an absorbing Markov chain is a favorite among probabilists. It serves to illustrate that even a slight bias toward losses can pull a gambler into the absorbing state of *certain ruin*. This type of application is called *the gambler's ruin* example.

Example 13.12 A gambler makes bets one dollar at a time. He has a probability p of winning and $q = 1 - p$ of losing on any one bet. We define a Markov chain with states $a_0, a_1, a_2, \ldots, a_n$ as follows: The gambler is in state a_i when he has i dollars left. Let a_0 and a_n be absorbing states ($P_{00} = 1$, and $P_{nn} = 1$). The game is over when he reaches one of the absorbing states.

a. Write the transition matrix for this Markov chain with $n = 5$.

b. In particular, write the matrix for $p = 0.40$ and $q = 0.60$ (and $n = 5$).

c. Draw a transition diagram for the matrix in part **b**.

d. If the gambler is in state a_2 at any given time, what is the probability that he will be in state a_0 three bets later? What is the probability that he will go from state a_2 to a_5 in three bets?

Solution **a.**

$$
\begin{array}{c c} & \begin{array}{cccccc} a_0 & a_1 & a_2 & a_3 & a_4 & a_5 \end{array} \\
\begin{array}{c} a_0 \\ a_1 \\ a_2 \\ a_3 \\ a_4 \\ a_5 \end{array} & \left[\begin{array}{cccccc} 1 & 0 & 0 & 0 & 0 & 0 \\ q & 0 & p & 0 & 0 & 0 \\ 0 & q & 0 & p & 0 & 0 \\ 0 & 0 & q & 0 & p & 0 \\ 0 & 0 & 0 & q & 0 & p \\ 0 & 0 & 0 & 0 & 0 & 1 \end{array}\right] = G.
\end{array}
$$

The matrix **G** satisfies the betting conditions. Notice, for example, that if the gambler has 3 dollars (state a_3) and he loses, he moves to state a_2 with probability q. If he wins, then he experiences the transition $a_3 \to a_4$, which occurs, according to the matrix **G**, with probability p, agreeing with the description above.

b.
$$\begin{bmatrix} 1 & 0 & 0 & 0 & 0 & 0 \\ 0.6 & 0 & 0.4 & 0 & 0 & 0 \\ 0 & 0.6 & 0 & 0.4 & 0 & 0 \\ 0 & 0 & 0.6 & 0 & 0.4 & 0 \\ 0 & 0 & 0 & 0.6 & 0 & 0.4 \\ 0 & 0 & 0 & 0 & 0 & 1 \end{bmatrix}$$

c.

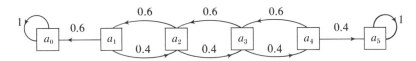

d. $P(a_2 \xrightarrow{3} a_0) = 0.36$, $P(a_2 \xrightarrow{3} a_5) = 0.064$. ▲

EXERCISE 13.3

In Problems 1 through 4, show that each matrix is regular.

1. $A = \begin{bmatrix} 0 & 1 \\ 1/2 & 1/2 \end{bmatrix}$.

2. $B = \begin{bmatrix} 2/3 & 1/3 \\ 2/5 & 3/5 \end{bmatrix}$.

3. $C = \begin{bmatrix} 0 & 1 & 0 \\ 0 & 0 & 1 \\ 1/2 & 1/2 & 0 \end{bmatrix}$.

4. $D = \begin{bmatrix} 1/2 & 1/4 & 1/4 \\ 1/2 & 0 & 1/2 \\ 0 & 1 & 0 \end{bmatrix}$.

5. Find the fixed point for matrix **C** in Problem 3.
6. Find the fixed point for matrix **D** in Problem 4.

13.3 The Markov Process—Theoretical Foundations

7. Write the terminal matrix **T** for integral powers of matrix **A** in Problem 1.

8. Write the terminal matrix **T** for integral powers of **B** in Problem 2.

9. a. If **v** is the vector $(3/7, 4/7)$, find $\mathbf{v} \cdot \mathbf{T}$ for **T** in Problem 7.

   b. What vector does $\mathbf{v} \cdot \mathbf{A}, \mathbf{v} \cdot \mathbf{A}^2, \mathbf{v} \cdot \mathbf{A}^3, \ldots$ approach? (**A** is the matrix in Problem 1.)

10. a. If $\mathbf{w} = (1/10, 9/10)$, find the vector $\mathbf{w} \cdot \mathbf{T}$ for **T** in Problem 8.

    b. What vector does $\mathbf{w} \cdot \mathbf{B}, \mathbf{w} \cdot \mathbf{B}^2, \mathbf{w} \cdot \mathbf{B}^3, \ldots$ approach?

▲ **Information theory problems** For Problems 11 and 12 refer to the following description. A telegraph company has a series of three messages: M_1 = Happy Birthday, M_2 = Come Home, and M_3 = Send Money. Each message is a state. The job of the company is to send a message from one city to the next. During the process of message sending, noise (interference and static) can enter the communication channel and distort the message, causing a transition from state M_i in one city to state M_j in the next. See the schematic diagram below:

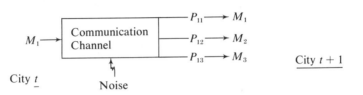

11. Assume that the Hopeless Telephone and Telegraph Company (HTT) has an extremely noisy communication system. It finds that when the office in city 1 sends Happy Birthday, the city 2 office receives Happy Birthday only 85 percent of the time. It receives Come Home 10 percent of the time and Send Money 5 percent of the time. Using the notation $P_{ij} = P(M_i \to M_j)$, this information is expressed as

$$P_{11} = 0.85, \ P_{12} = 0.10, \text{ and } P_{13} = 0.05.$$

Similarly, the other state changes are

$$P_{21} = 0.05, \ P_{22} = 0.90, \text{ and } P_{23} = 0.05.$$
$$P_{31} = 0.10, \ P_{32} = 0.10, \text{ and } P_{33} = 0.80.$$

a. Write the transition matrix **M** for this process.

b. Find $\mathbf{M}^2$ and interpret its meaning.

c. What would be the meaning of higher powers of **M**?

d. Find the steady state for **M**.

e. If you send M_1, what message is most likely to be received at some city far down the line.

12. Repeat parts **a** through **e** for the data given in the following diagrams:

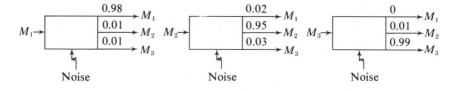

14
INTRODUCTION TO STATISTICS

14.1
ELEMENTARY CONCEPTS

The French population drinks about three billion gallons of wine each year. The U.S. population drinks only 0.3 billion gallons per year. So a Frenchman drinks about ten times as much wine as an American, right? Wrong! The population of France is only 60,000,000, whereas the U.S. population is 200,000,000; therefore, the *per capita* yearly wine consumption in France is 50 gallons and in America it is 1.5 gallons. Consequently, the Frenchman drinks, on the average, thirty-three times as much as an American!

In this example, the comparison of average gallons consumed provides more accurate information than the comparison of the total figures. There are some cases, however, in which an average is useless information. For example, the statistician who drowned while fording a stream whose average depth was only two feet would have been better off if he had known the maximum depth, not just the average. Knowing the average salaries in a corporation could be misleading if the salaries of the president and other high executives are included in the average. In this case, it would be more helpful to know the middle salary and the most frequent salary, as well as the average.

A collection of facts, such as each of those given above concerning wine consumption, stream depth, and corporate salaries, is called a set of **data**. Data are to statistics as simple events are to probability. They can both be expressed in terms of a random variable.

In statistics, the data are numbers that we consider as values of a random variable. The set of underlying data points $S = \{P_1, P_2, ..., P_n\}$ may map into a smaller set of measurements $\{x_1, x_2, ..., x_k\}$, since some of the measurements may correspond to more than one data point. Each measurement x_i has its frequency f_i depending upon how many points correspond to it. The sum of the frequencies

$\sum_{i=1}^{k} f_i$ is the number (n) of data points. The probability of selecting a point P_i at random from S is assumed to be $1/n$. That is, the underlying space is equiprobable, and the probability distribution for the set of measurements $\{x_j\}$ is found by using the set of frequencies $\{f_j\}$. The probability of x_j is

$$P(X = x_j) = \frac{f_j}{\sum_{j=1}^{k} f_j} = \frac{f_j}{n}.$$

This is equivalent to saying that each measurement x_j corresponds to a subset containing f_j points of S. Thus, for example, if only one point P_i corresponds to x_j, then $P(X = x_j) = 1/n$. If the frequency of x_m is two (exactly two points, P_i and P_j, correspond to x_m), then $P(X = x_m) = 2/n$, and so forth. The probability distribution $P(X = x_1)$, $P(X = x_2)$, ..., $P(X = x_k)$ is equivalent to the **relative frequency distribution** f_1/n, f_2/n, ..., f_k/n.

The **arithmetic mean** $\bar{x}$ of $\{P_1, P_2, ..., P_n\}$ is the expected value of the random variable X; that is,

$$\bar{x} = E(X) = \sum_{i=1}^{k} x_i \cdot P(X = x_i).$$

Expected value was defined in Chapter 12. An equivalent equation is

$$\bar{x} = \sum_{i=1}^{k} x_i \cdot \frac{f_i}{n} = \frac{1}{n} \sum_{i=1}^{k} x_i \cdot f_i.$$

Example 14.1 Consider the set of 25 data points, $P_1, P_2, ..., P_{25}$, below:

$P_1 = 3$	$P_8 = 6$	$P_{15} = 3$	$P_{22} = 3$
$P_2 = 4$	$P_9 = 4$	$P_{16} = 5$	$P_{23} = 5$
$P_3 = 7$	$P_{10} = 4$	$P_{17} = 4$	$P_{24} = 4$
$P_4 = 5$	$P_{11} = 7$	$P_{18} = 6$	$P_{25} = 7$
$P_5 = 6$	$P_{12} = 3$	$P_{19} = 6$	
$P_6 = 3$	$P_{13} = 3$	$P_{20} = 6$	
$P_7 = 4$	$P_{14} = 6$	$P_{21} = 5$	

The number n equals 25, but there are only five measurements in this set of data ($k = 5$); they are 3, 4, 5, 6, and 7. In the notation of the text, we have $x_1 = 3$, $x_2 = 4$, $x_3 = 5$, $x_4 = 6$, and $x_5 = 7$. Each of these five measurements appears more than once in the set of data. We tally their frequency by listing all five numbers, then going through the set of data and making a tally mark next to each measurement as we come across it in the data set. This produces the result in Figure 14.1. We write this as the frequency distribution table shown in Table 14.1.

14.1 Elementary Concepts

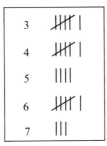

Figure 14.1

Table 14.1

x_i	3	4	5	6	7
f_i	6	6	4	6	3

The numbers f_i in Table 14.1 represent the frequencies with which the values x_i occur in the data; for example, $f_1 = 6$ which equals the frequency of x_1. The sum of the frequencies is the number n: $\Sigma_{i=1}^{5} f_i = 25 = n$. The relative frequencies for the five values are $6/25$, $6/25$, $4/25$, $6/25$, and $3/25$. These are the probabilities for the five values and they can be used to compute the mean $\bar{x}$. The mean is the expected value:

$$\bar{x} = E(X) = \frac{1}{n} \sum_{i=1}^{5} x_i \cdot f_i$$

$$= \frac{1}{25}(3 \cdot 6 + 4 \cdot 6 + 5 \cdot 4 + 6 \cdot 6 + 7 \cdot 3) = 4.76.$$

This same answer could have been obtained by adding up all 25 points, P_1, P_2, ..., P_{25}, and dividing by 25.

A frequency distribution table, such as Table 14.1, can be used to plot the distribution graph for the relative frequency distribution of the random variable. Figure 14.2 is the distribution graph for Table 14.1.

Sometimes, when the meaning is clear, the subscripts of x_i and f_i are dropped and the indices on the summation signs are omitted. Thus, the definition of the

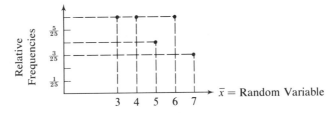

Figure 14.2
Distribution graph for the data in Example 14.1.

mean may be written $(1/n)\Sigma x \cdot f$. The mean is also called the *average* of the points in S.

Some other definitions that will be needed are for mode, median, and percentile.

Definition **(Mode)**
The value or values with the greatest probability (equivalently, the greatest frequency) are called the *modes* of the data. There may be more than one mode; that is, no unique or clearly defined mode.

Definition **(Median)**
If the measurements are arranged in order of size, the *median* is a number that precedes as many data points as it follows (counting the frequencies). In other words, if there is an odd number n of data points, then the median is the $(n+1)/2$ measurement (counting the frequencies). If n is an even number, the median is the average of the $n/2$ number and the $(n/2) + 1$ number.

Example 14.2 In Table 14.1, the frequency distribution table, we see that 5 is the median. (The first 5 precedes 12 points and follows 12 points. The 12 points it precedes are the other three 5's, the six 6's, and the three 7's. The 12 points it follows are the six 3's and the six 4's.) Alternatively, since $n = 25$, an odd number, we look for the thirteenth measurement, which is one of the 5's.

Definition **(Percentile)**
Assume the measurements are arranged in order of size. The kth percentile is the value preceded by k percent of the measurements (counting the frequencies). For example, the 37th percentile is the value greater than 37 percent of the data. The 50th percentile is the median. Another way to state this is in terms of the probability of an inequality of the random variable: "x_i is at the 37th percentile" means that $P(X \leq x_i) = 0.37$.

Suppose the data consist of n points that map into k distinct real numbers, $k \leq n$. These k values of the random variable are sometimes *grouped* into m *class intervals*; $m < k$. The midpoint of each class interval stands for all the data whose measurements are within that class. The end points of class intervals are called the *class boundaries*. For a given class, the *class frequency* is the number of data points whose measurements are in that class. The relative class frequency is the class frequency divided by n; it is also the probability that the random variable takes on the measurements within that class.

14.1 Elementary Concepts

Example 14.3 Given the frequency distribution in Table 14.2, group the data into the class intervals $[-2,0) = \{x_i$ such that $-2 \leq x_i < 0\}$, $[0,2) = \{x_i$ such that $0 \leq x_i < 2\}$, and so forth, including the left end point and excluding the right end point for each of the intervals $[2,4)$, $[4,6)$, and $[6,8)$. Find the class frequency for each class.

Table 14.2

x_i	-2	-1	0	1	2	3	4	5	6	7
f_i	4	5	7	6	5	2	1	2	1	1

Solution The measurements, $x_1 = -2$ and $x_2 = -1$, are both in the class interval $[-2,0)$, but $x_3 = 0$ is not in that class; it is in the next one. The frequency f_1 is 4 and f_2 is 5; therefore, the frequency for that class is 9, the sum of the frequencies of the points in the interval. Similarly, the class frequency for the interval $[0,2)$ is $7 + 6 = 13$. For $[2,4)$, f is 7; for $[4,6)$, f is 3; and for $[6,8)$, f is 2. We express this in Table 14.3.

Table 14.3

$[x_i, x_{i+1})$	$[-2,0)$	$[0,2)$	$[2,4)$	$[4,6)$	$[6,8)$
f	9	13	7	3	2

The class intervals create partitions of the set of measurements. They are a nonoverlapping collection of intervals filling up the range space of the random variable. A bar graph can be constructed over the class intervals as follows: Each class interval is the base of one bar of the bar graph. The height of the bar based on that interval is the relative frequency of the measurements in that class. If the class interval has class boundaries x_i and x_{i+1}, the base of the bar over this interval has length $x_{i+1} - x_i$ and height $P(x_i \leq X < x_{i+1})$. Note that we adopt the convention that only the left end point is in the interval; this means that a class interval is actually a half-closed, half-open interval $[x_i, x_{i+1})$.

The bar graph drawn in the manner described above is called a **histogram** for the data.

Example 14.4 Draw a histogram for the data in Example 14.3, using the class intervals given there.

Solution The sum of the frequencies $\Sigma f = 34$. From the table of class frequencies, we find that the relative frequency distribution is $9/34$, $13/34$, $7/34$, $3/34$, and $2/34$. These will be the heights of the bars as shown in Figure 14.3.

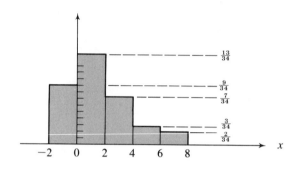

Figure 14.3
Histogram for the data in Example 14.3.

A table of cumulative frequencies can be obtained by assigning to any class interval the sum of all the frequencies in the preceding intervals. The cumulative frequencies are useful in computing the percentiles for any class interval.

Example 14.5 A table of cumulative frequencies for the grouped data in Example 14.3 is given in Table 14.4.

Table 14.4

$[x_i, x_{i+1})$ Cumulative Frequency	$[-2,0)$ 9	$[0,2)$ 22	$[2,4)$ 29	$[4,6)$ 32	$[6,8)$ 34

EXERCISE 14.1

1. Given the data

```
3 3 6 7 2 7 8 8 3 3
8 6 9 8 5 5 8 7 5 5
5 4 6 7 6 8 2 5 4 7
9 5 6 3 8 2 2 6 9
```

14.1 Elementary Concepts

  a. Tally the frequency of the data and construct the frequency distribution table.
  b. Draw the distribution graph.
  c. Compute the mean, mode, and median.
  d. For the four class intervals, [2,4), [4,6), [6,8), and [8,10), find the class frequencies and draw the histogram for the data.
  e. Make a cumulative frequency table for the data. At what percentile does 6 occur?

2. Given the data

$$
\begin{array}{cccccccc}
3.9 & 1.6 & 1.6 & 2.1 & 1.7 & 2.4 & 0.5 & 0.6 & 3.0 \\
1.2 & 1.3 & 4.2 & 1.5 & 2.2 & 0.5 & 2.0 & 3.4 & 1.5 \\
1.8 & 0.3 & 0.7 & 1.5 & 0.6 & 0.8 & 1.6 & 0.4 & 1.5 \\
0.7 & 1.3 & 2.5 & 1.3 & 2.6 & 1.0 & 3.9 & 2.2 \\
\end{array}
$$

  a. Tally the frequency of the data and construct the frequency distribution table.
  b. Draw the distribution graph.
  c. Compute the mean, mode, and median.
  d. For the five class intervals, [0.3,1.1), [1.1,1.9), [1.9,2.7), [2.7,3.5), and [3.5,4.3), find the class frequencies and draw the histogram for the data.
  e. Make a cumulative frequency table for the class intervals. At what percentile does 3.5 fall?

3. Given the following measurements and their frequencies

x	f	x	f	x	f	x	f	x	f
40	1	67	1	73	3	80	1	86	1
42	1	68	1	75	4	81	1	90	1
55	1	69	1	76	3	82	2	92	1
64	1	70	1	77	1	84	2	99	2
65	2	72	1	78	1	85	1		

  a. Find Σf.
  b. Find the mean $\bar{x}$.
  c. Using the class intervals

$$Z = [40,50) \qquad\qquad C = [70,80)$$

$$F = [50,60) \qquad B = [80,90)$$
$$D = [60,70) \qquad A = [90,100)$$

find the class frequencies and draw a histogram.

4. Given the following set of measurements and their frequencies

x	2	3	5	7	11	13	17	19	23	29
f	1	2	3	4	5	6	7	8	9	10

a. Find Σf.

b. Find the mean $\bar{x}$.

c. Using the class intervals, $[0,5), [5,10), \ldots, [25,30)$, find their class frequencies and draw the histogram.

5. In Problem 3, estimate the mean from the grouped data as follows: For each class interval, multiply the class frequency by the class midpoint and average these products. This gives the estimated mean:

$$\text{estimated } \bar{x} = \frac{\Sigma (\text{class midpoint})(\text{class frequency})}{n}.$$

(This method has the effect of assuming that all the data in the same class are located at the center of the class interval.)

6. In Problem 4, estimate the mean from the grouped data by the method described in Problem 5.

7. In Problem 3, between what percentiles does the class interval D fall?

8. In Problem 4, between what percentiles does the class interval $[10,15)$ fall?

▲ **Water quality control statistics** For Problems 9 through 14 use the following description. In the following data on water quality control, the random variable M is the monitoring variable. Its values are associated with the event that the water is checked for a given pollutant hourly, daily, weekly, monthly, quarterly, or annually. The frequency f represents the number of stations conducting these tests (f_{cb} = number of stations monitoring the water for *coliform bacillus*, etc.). Denote the events as follows: E_1 = the event of hourly testing, E_2 = event of daily testing, E_3 = weekly testing, E_4 = monthly testing, E_5 = quarterly testing,

14.1 Elementary Concepts

and E_6 = annual testing. The random variable M maps these events into the real numbers (representing hours) by the following equations: $M(E_1) = 1$, $M(E_2) = 24$, $M(E_3) = 168$, $M(E_4) = 730$, $M(E_5) = 2190$, and $M(E_6) = 8760$. The data [modified from the *C.R.C. Handbook of Environmental Control, Vol. III: Water Supply and Treatment* (Cleveland: C.R.C. Press, 1973)] are given in the accompanying table.*

	Monitoring Variable M (Hours)	Number of Stations			
		Suspended Solids f_{ss}	Coliform Bacillus f_{cb}	Pesticides f_p	Temperature f_t
Hourly	1	0	15	1	870
Daily	24	2	240	1	820
Weekly	168	50	425	7	530
Monthly	730	325	1260	70	2270
Quarterly	2190	210	635	10	830
Annually	8760	2	160	25	1060
Totals (n)		589	2735	114	6380

9. Find the mean number of hours between tests in monitoring the suspended solids.

10. Find the mean number of hours between tests in monitoring the pesticides.

11. a. Find the mode and median for monitoring the suspended solids.

    b. Find the mode and median for monitoring the *coliform bacillus*.

12. a. Find the mode and median for monitoring the pesticides.

    b. Find the mode and median for monitoring the temperature.

13. Suppose city X monitors the suspended solids weekly; between what two percentiles does this city fall?

14. Suppose city Y monitors the suspended solids quarterly; between what two percentiles does this city fall?

---

*Original Source: Federal Water Quality Administration, based on data provided by the U.S. Geological Survey, 1968.

15. **The prime commercial paper rates** The data points here are the average yearly prime commercial paper rates in percents for the 71 years from 1900 to 1970. Each number on the table below is the interest rate in percent taken from midyear points on a graph in the *Historical Chart Book* of the Federal Reserve System.

Year	Rate	Year	Rate	Year	Rate	Year	Rate	Year	Rate
1900	5.8	1915	4.0	1929	5.8	1943	0.8	1957	3.8
1901	5.6	1916	4.2	1930	3.6	1944	0.8	1958	2.8
1902	6.0	1917	5.6	1931	2.8	1945	0.8	1959	4.0
1903	6.0	1918	6.0	1932	2.6	1946	1.0	1960	4.0
1904	5.2	1919	5.6	1933	1.6	1947	1.2	1961	3.0
1905	5.8	1920	7.4	1934	1.0	1948	1.4	1962	3.4
1906	6.6	1921	7.2	1935	0.8	1949	1.4	1963	3.6
1907	6.4	1922	4.6	1936	0.8	1950	1.6	1964	3.8
1908	4.8	1923	4.8	1937	1.0	1951	2.2	1965	4.2
1909	5.2	1924	4.0	1938	0.8	1952	2.4	1966	5.2
1910	5.6	1925	4.0	1939	0.6	1953	2.6	1967	5.2
1911	4.8	1926	4.2	1940	0.6	1954	1.6	1968	5.8
1912	5.6	1927	4.4	1941	0.6	1955	2.2	1969	7.6
1913	6.4	1928	5.0	1942	0.8	1956	3.2	1970	7.8
1914	5.6								

a. List the numbers in ascending order and tally their frequencies.

b. Draw a graph of the relative frequency distribution.

c. Compute the arithmetic mean.

d. Find the mode, median, 75th percentile, and 25th percentile.

e. Group the data into ten class intervals [0.6,1.4), [1.4,2.2), ..., [7.8,8.6), and make a table of class frequencies.

f. Draw a histogram for the grouped data in part **e**.

g. Make a table of cumulative frequencies. At what percentile does the value 3.0 fall? 3.7? Between what two percentiles does the value 5.6 fall? ▲

14.2

DATA VARIABILITY

The mean and median of a set of data reveal central tendencies of the data; they are numbers around which the data are distributed. These numbers serve to represent the data as a single number. But, as we have seen, averages are not always the best information; sometimes it is essential to know something about the distribution.

A number related to the distribution (and to the mean) that describes the variability of the data would be a valuable statistical tool. From such a number, we could obtain a deeper statistical analysis of the data and even make some predictions.

Suppose, for example, a machine is known to be 99 percent reliable. Then from certain patterns of variability of data, we can determine what chance there is that the machine will have three failures in a trial run of ten attempts, or the chance that there will be zero failures, or one, or five, or nine, etc.

This predictive power of statistical analysis is based on the fact that large sets of measurements have certain characteristics that can be described by the way measurements tend to cluster around the mean. Two such measures of these population characteristics are the *variance*, and its square root, the *standard deviation*.

Before proceeding to the definition of variance, we need to understand the concept of *deviation from the mean*.

Definition (**Deviation from the mean**)
Given a set $\{x_1, x_2, ..., x_n\}$ of data points with mean $\bar{x}$, then the deviation from the mean of a given particular value x_i is the difference $x_i - \bar{x}$.

In compiling a list of deviations from the mean, it is important to *count the differences $x_i - \bar{x}$ as many times as x_i appears in the original data*; that is, the frequency of $x_i - \bar{x}$ is the same as the frequency of x_i.

Example 14.6 a. For the frequency distribution

x_i	2	3	4	5	6	7	10
f_i	1	3	1	1	1	2	1

compute the mean and the deviations from the mean for all points.

b. Compute the average deviation from the mean.

Solution **a.** The number $n = \Sigma f = 10$, and the mean is

$$\bar{x} = \frac{1}{10} \Sigma x \cdot f = \frac{1}{10}(2 + 9 + 4 + 5 + 6 + 14 + 10)$$

$$= \frac{50}{10} = 5.0.$$

The deviations from the mean are $2 - 5$, $3 - 5$, $4 - 5$, $5 - 5$, $6 - 5$, $7 - 5$, and $10 - 5$; or $-3, -2, -1, 0, 1, 2,$ and 5. The deviation -3 has frequency 1; -2 has frequency 3; -1 has frequency 1; and so on. A frequency distribution table is a convenient way to express the deviations and their frequencies. See Table 14.5.

Table 14.5

Deviations from the Mean	$x_i - \bar{x}$	-3	-2	-1	0	1	2	5
Frequency	f_i	1	3	1	1	1	2	1

b. The average deviation from the mean is

$$\text{Average}(x_i - \bar{x}) = \frac{1}{10} \Sigma (x - \bar{x}) f$$

$$= \frac{1}{10}(-3 + -6 + -1 + 0 + 1 + 4 + 5)$$

$$= \frac{1}{10} \cdot 0 = 0.$$

Surprisingly, the *average* deviation from the mean for *any* set of data is zero.

Note We wish to emphasize that it is necessary to make use of the frequencies in almost any work with statistical data (exceptions are for finding the range or finding the class intervals). For example, in computing the average of the deviations from the mean, it would be a mistake to simply add up the deviations, $-3, -2, -1, 0, 1, 2,$ and 5, and divide by 7.

A proof that the average deviation from the mean is always zero is as follows:

14.2 Data Variability

The sum of the deviations from the mean is

$$\Sigma(x - \bar{x})f = \Sigma xf - \Sigma \bar{x}f = n\bar{x} - \bar{x}\Sigma f = n\bar{x} - \bar{x}n = 0.$$

The average deviation = $(1/n)$ (sum of the deviations) = $(1/n) \cdot 0 = 0$.

Since this is true for any set of measurements, then the sum of the deviations is of no value in furnishing useful information about clustering. So, how are we going to find a number that expresses the magnitudes of the variations in a given set of data? The problem here is that the deviations from the mean are both positive and negative and the numbers balance perfectly; so they cancel each other out. What we need is a set of all positive (or zero) numbers. One way to get such nonnegative deviations is to square the differences $(x_i - \bar{x})^2$ before we add them together; these are the *squares of the deviations*.

$$\text{Total of squares of deviations} = \sum_{i=1}^{n} (x_i - \bar{x})^2 \cdot f_i.$$

Example 14.7 Find the sum of the squares of the deviations from the mean and find the average of these squared deviations for the data in Example 14.6.

Solution Use Table 14.5. The squares of the deviations are 9, 4, 1, 0, 1, 4, and 25; their respective frequencies are 1, 3, 1, 1, 1, 2, and 1. Therefore, the total of the squares of the deviations is

$$\Sigma(x - \bar{x})^2 \cdot f = 9 \cdot 1 + 4 \cdot 3 + 1 \cdot 1 + 0 \cdot 1 + 1 \cdot 1 + 4 \cdot 2 + 25 \cdot 1$$
$$= 56.$$

The average of the squared deviations is $56/10 = 5.6$.

In general, the sum of the squares of the deviations $\Sigma(x - \bar{x})^2 f$ is a numerical description of the *total variability* or total variation of the set of data. It seems natural to compute the *average of the squared deviations*, as was done in the preceding example. This average is called the **variance.**

Definition (Variance)

The variance of a set $\{x_1, x_2, \ldots, x_n\}$ of n measurements is the average squared deviation from the mean. In symbols, the variance, denoted by s^2, is defined by

$$s^2 = \frac{\sum_{i=1}^{n} (x_i - \bar{x})^2 f_i}{n}.$$

Since the square of a number distorts its magnitude (increases it if the number is greater than +1 or less than −1, and decreases it if the number is greater than −1 or less than +1), then the average of the squared deviations seems to be an exaggeration of the variability. We can get a truer picture of the variation of the data if we "undo" the squaring by computing the square root of the variance. In Example 14.7, the variance was 5.6 and its square root $\sqrt{5.6} = 2.366$. The square root of the variance is called the **standard deviation** of the data.

In symbols, the standard deviation is denoted by s:

$$s = \sqrt{\frac{\sum_{i+1}^{n}(x_i - \bar{x})^2 \cdot f_i}{n}}.$$

Example 14.8 For the following three sets of data, find the mean, variance, and standard deviation.

(I)

x	4	6	7	8	70
f	2	1	1	1	1

(II)

y	13	14	15	21	22
f	1	2	1	1	1

(III)

z	16.5
f	6

Solution In all three cases, $\Sigma f = 6$, and

(I) $\quad \bar{x} = \left(\dfrac{1}{6}\right)\Sigma xf = \left(\dfrac{1}{6}\right)(8 + 6 + 7 + 8 + 70) = 16.5.$

(II) $\quad \bar{y} = \left(\dfrac{1}{6}\right)\Sigma yf = \left(\dfrac{1}{6}\right)(13 + 28 + 15 + 21 + 22) = 16.5.$

(III) $\quad \bar{z} = \left(\dfrac{1}{6}\right)\Sigma zf = \left(\dfrac{1}{6}\right)(99) = 16.5.$

14.2 Data Variability

To find the variance and standard deviation, we construct the distribution table for the squared deviations, as in Table 14.6. For brevity, we compute only the squared deviations times their corresponding frequencies.

Table 14.6

(I) $(x - \bar{x})^2 f$	(II) $(y - \bar{y})^2 f$	(III) $(z - \bar{z})^2 f$
312.50	12.25	0
110.25	12.50	
90.25	2.25	
72.25	20.25	
2862.25	30.25	
3447.50	77.50	0

The sums of the squared deviations as shown in the table are

(I) $$\Sigma (x - \bar{x})^2 f = 3447.5.$$

(II) $$\Sigma (y - \bar{y})^2 f = 77.5.$$

(III) $$\Sigma (z - \bar{z})^2 f = 0.$$

The variance and standard deviation for each case are

(I) $$s^2 = \frac{(3447.5)}{6} = 574.6, \quad \text{standard deviation} = \sqrt{574.6} = 23.97.$$

(II) $$s^2 = \frac{(77.5)}{6} = 12.9, \quad \text{standard deviation} = \sqrt{12.9} = 3.59.$$

(III) $$s^2 = 0, \quad \text{standard deviation} = 0.$$

The large variability in (I) is because the point 70 is so extreme compared to the other points. In (II), the variability of the data is not large. In (III), there is no variability since all six points have exactly the same value. It is clear that the standard deviation reveals something about the variability of the data.

A useful feature of the standard deviation is that in any set of *normally distributed* measurements (those whose histograms approximate a bell-shaped curve), 67 percent of the measurements fall within ±1 standard deviation of the mean, 95 percent fall within ±2 standard deviations, and 99 percent fall within ±3 standard deviations of the mean.

Cases (I) and (II) are not normally distributed, and case (III) could be considered as a perverse example of a normal distribution.

Finally, we note the following short method for computing the variance and, hence, the standard deviation.

Theorem 14.1 Given the set $x_1, x_2, \ldots, x_n$ of n measurements with frequencies $f_1, f_2, \ldots, f_n$, and mean $\bar{x}$, then

$$s^2 = \frac{\sum_{i=1}^{n} x_i^2 f_i}{n} - (\bar{x})^2.$$

Proof (As usual, the reader who abhors proofs may skip this and accept the theorem.)

(I) $\sum_{i=1}^{n} (x_i - \bar{x})^2 f_i = \sum_{i=1}^{n} (x_i^2 - 2\bar{x}x_i + \bar{x}^2) f_i.$ (I) Square of a binomial.

(II) $\sum_{i=1}^{n} (x_i - \bar{x})^2 f_i = \sum_{i=1}^{n} x_i^2 f_i$ (II) Linear properties of sums.
$\qquad - 2\bar{x} \sum_{i=1}^{n} x_i f_i + \bar{x}^2 \sum_{i=1}^{n} f_i.$

(III) $\sum_{i=1}^{n} (x_i - \bar{x})^2 f_i = \sum_{i=1}^{n} x_i^2 f_i - 2\bar{x}n\bar{x} + n\bar{x}^2.$ (III) Definition of $\bar{x}$ and sum of f_i.

(IV) $\sum_{i=1}^{n} (x_i - \bar{x})^2 f_i = \sum_{i=1}^{n} x_i^2 f_i - n\bar{x}^2.$ (IV) Combining terms.

(V) $s^2 = \frac{1}{n} \sum_{i=1}^{n} x_i^2 f_i - \bar{x}^2.$ (V) Dividing both sides in Step 4 by n, and by the definition of s^2.

Example 14.9 a. Find the mean, variance, and standard deviation for the following set of data:

$x_1 = 8$ $x_5 = 24$ $x_9 = 14$
$x_2 = 10$ $x_6 = 8$ $x_{10} = 24$
$x_3 = 50$ $x_7 = 40$ $x_{11} = 14$
$x_4 = 18$ $x_8 = 30$

b. Draw a histogram for these data using the three intervals, $[0,20)$, $[20,40)$, and $[40,60)$.

c. Does $\bar{x} \pm s$ cover 67 percent of the data? Would you say the histogram for these data approximates a bell-shaped curve?

14.2 Data Variability

Solution a. First, we convert these data to a frequency distribution table:

y	8	10	14	18	24	30	40	50
f	2	1	2	1	2	1	1	1

Now, to compute the mean:

$$\bar{x} = \sum_{i=1}^{8} y_i f_i = \frac{240}{11} \approx 21.8.$$

Variance s^2, by the short method, is

$$s^2 = \frac{\sum_{i=1}^{8} y_i^2 f_i}{11} - (21.8)^2 = \frac{7096}{11} - 475.2 = 169.8.$$

Therefore, the standard deviation is (approximately)

$$s = 13.$$

b. See Figure 14.4 for the histogram.

c. No, $\bar{x} \pm s = 21.8 \pm 13$ contains seven of the eleven or approximately 64 percent of the data points. No, the histogram is not a bell-shaped curve (high in the middle and low at the two ends.)

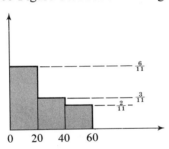

Figure 14.4

The variance and the standard deviation can be computed from data grouped into class intervals by using class frequencies and the midpoints of the classes.

Example 14.10 For the grouped data in Table 14.7,

a. Find the mean, variance, and standard deviation.

b. Draw a histogram for the given class intervals.

c. Find the approximate percentage of the numbers that are within one standard deviation from the mean.

Table 14.7

Class $[x_i, x_{i+1})$	[2,5)	[5,8)	[8,11)	[11,14)	[14,17)	[17,20)
Class Frequency, f_i	4	7	17	12	9	3

Solution

a. First, we compute the number n of data points: $n = \Sigma f = 52$. Now to get the mean, we use the midpoints 3.5, 6.5, 9.5, 12.5, 15.5, and 18.5 of the class interval. For example, 3.5 is the midpoint of the interval [2,5). In general, the midpoint of $[x_i, x_{i+1})$ is $m_i = (x_{i+1} + x_i)/2$. The mean is

$$\bar{x} = \frac{1}{n} \sum_{i=1}^{6} m_i f_i = \frac{1}{52} [(3.5)4 + (6.5)7 + (9.5)17 + (12.5)12 + (15.5)9 + (18.5)3]$$

$$= 10.9.$$

Variance:

$$s^2 = \frac{\sum_{i=1}^{6}(m_i - \bar{x})^2 f_i}{n} = \frac{1}{52}[(54.76)4 + (19.36)7 + (1.96)17 + (2.56)12$$

$$+ (21.16)9 + (57.76)3] = 15.04.$$

Standard deviation:

$$s = 3.88.$$

b. For the histogram, see Figure 14.5

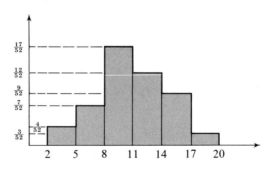

Figure 14.5

14.2 Data Variability

c. $\bar{x} + s = 10.9 + 3.88 = 14.78$, and $\bar{x} - s = 10.9 - 3.88 = 7.02$. Here, 14.78 is in the interval $[14,17)$, which exceeds $40/52 \approx 77$ percent of the measurements; 7.02 is in the class interval $[5,8)$, which exceeds $4/52 \approx 8$ percent of the measurements. Therefore, approximately $77 - 8$, or 69 percent of the data lies within one standard deviation from the mean.

EXERCISE 14.2

1. For the data $x_1 = -5$, $x_2 = -1$, $x_3 = 3$, $x_4 = 4$, $x_5 = 6$, $x_6 = 7$, $x_7 = 9$, and $x_8 = 11$,

   a. Find the mean, variance, and standard deviation.

   b. At what percentile does the number 11 occur?

2. For the data $y_1 = -3$, $y_2 = -2$, $y_3 = -1$, $y_4 = 10$, and $y_5 = 14$,

   a. Find the mean, variance, and standard deviation.

   b. At what percentile does the number -1 occur?

3. For the following frequency distribution, find the mean, variance, and standard deviation.

x	4	6	8	10	12
f	5	2	1	3	6

4. For the following frequency distribution, find the mean, variance, and standard deviation.

x	-3	-2	-1	0	1	2	3
f	3	6	8	16	8	6	3

5. For the following grouped data,
   a. Find the mean, variance, and standard deviation.
   b. Draw a histogram for the given classes.
   c. Find the percentiles between which the interval [70,80) falls.

Class $[x_i, x_{i+1})$	[30,40)	[40,50)	[50,60)	[60,70)	[70,80)	[80,90)	[90,100)
Class Frequency	2	1	4	6	15	10	5

6. For the following grouped data,
   a. Find the mean, variance, and standard deviation.
   b. Draw a histogram for the given classes.
   c. Between what two percentiles does the mean fall?

Class $[x_i, x_{i+1})$	[0,5)	[5,10)	[10,15)	[15,20)	[20,25)	[25,30)
Class Frequency	15	8	5	2	1	1

▲ 7. The litter size of swine has the distribution shown in the table below. For the purpose of computation, count the 16 or more group as 17; also, in the other groups choose the middle number as the representative for the group.

Litter Sizes	1,2, or 3	4,5, or 6	7,8, or 9	10,11, or 12	13,14, or 15	16 or more
Frequency	13	44	32	4	3	4

   a. Find the mean litter size.
   b. Find the variance and standard deviation.
   c. If a sow has a litter of 14, at which percentile is she ranked?

14.2 Data Variability

d. If a litter is 1 ½ standard deviations greater than the mean, at what percentile does it fall?

8. From the Uniform Crime Reports in 1967 for 49 states (from the U.S. National Prisoner Statistics, U.S. Department of Justice), the distribution table below is derived. Use the number 7000 as the representative for the over 6000 group.

Number of Prison Admissions	Less than 1000	1000 to 2000	2000 to 3000	3000 to 4000	4000 to 5000	5000 to 6000	over 6000
Frequency	19	14	7	2	3	1	3

a. Find the mean number of prison admissions. (Hint: The calculations are easier if you drop the zeros from all the numbers; that is, use [1,2) instead of [1000,2000), then restore the zeros at the end of the solution.)

b. Find the variance and standard deviation.

c. Assume the U.S. population was 200,000,000 at the time these data were taken. Find the number of prisoners admitted per 100,000 of the U.S. population.

9. Black lung benefit program payments to coal miners' widows are as listed in the following table.*

Monthly Payments (Dollars), Class Intervals	[160,165)	[165,170)	[170,175)	[175,180)
Numbers of Widows Receiving Them, Frequency	6,000	49,000	10,000	23,000

a. Find the mean number of dollars.

b. Find the variance and standard deviation.

*Source: U.S. Social Security Administration, data simplified from the U.S. Bureau of the Census, *Statistical Abstract of the United States*, 1974 (95th edition).

10. The following table gives the frequency distribution of the heights of entering freshmen at a given college. Compute the mean, variance, and standard deviation. (Hint: Use a calculator.) ▲

Height in Inches	Number of Students
80.5 to 78.5	2
78.5 to 76.5	4
76.5 to 74.5	11
74.5 to 72.5	36
72.5 to 70.5	76
70.5 to 68.5	207
68.5 to 66.5	219
66.5 to 64.5	57
64.5 to 62.5	40
62.5 to 60.5	4
60.5 to 58.5	1

11. Explain how you could find the mean and standard deviation if you were given a *probability* distribution instead of a *frequency* distribution. Find the mean and standard deviation for the following probability distribution:

x	0	1	2	3	4
$P(X=x)$	$(1/3)^4$	$4(1/3)^3(2/3)$	$6(1/3)^2(2/3)^2$	$4(1/3)(2/3)^3$	$(2/3)^4$

14.3

BINOMIAL PROBABILITY DISTRIBUTION

Random experiments with only two outcomes (or with several outcomes grouped into only two categories) are called **binomial probability experiments.** Such experiments arise in statistical sampling problems in which the statistician needs to determine the frequency of occurrence of a certain trait in the population by testing a random sample of the population.

14.3 Binomial Probability Distribution

Example 14.11 A manufacturer believes that his production process is 99 percent reliable, but he wants to keep a quality control check on it to protect his good reputation. Suppose he cannot test every item produced, so he must be contented with a sampling technique. The testing of an item is an experiment that has two outcomes: good or defective, called success and failure. (Which is the *success* outcome and which is the *failure* depends upon how the problem is set up.) It is assumed that the manufacturing process is operating at a 99 percent success rate. A given sample of n items could have from zero to n failures. The manufacturer is satisfied that the process is performing at the hypothesized rate when the number of failures is below a certain level.

Example 14.12 A biologist receives a shipment of 1000 newborn white rats, 75 percent of which are supposed to have a certain organ disorder, X. He wants to raise them on a special diet designed to eliminate X. The only way he can find out if a rat has X is to kill it and examine the organ. He calls having the disorder a success and not having it a failure. Before starting his experiment, he wants to be sure that his population actually does have 75 percent successes. Since he has to kill the animal to test for success, he can't test the entire population before the experiment begins. He decides to select a sample of n rats for testing. If the number of successes is above a certain number k, then he has no reason to assume that the population frequency of success is misrepresented by the biology supply laboratory that sent him the rats.

We will take up these two problems (Examples 14.11 and 14.12) again later and derive probabilities for various samples, but first we need to examine a single binomial experiment and repetitions of the experiment.

Let the two outcomes in a binomial experiment be denoted by S and F (for success and failure). Let the random variable Y = the number of successes in any binomial experiment. If the experiment is repeated n times, then Y has the range $0, 1, \ldots, n$, since there could be from zero to n successes.

In a single experiment, if S occurs, then $Y = 1$ and if F occurs, then $Y = 0$. Let p denote the probability for a success: $P(S) = P(Y = 1) = p$; and let q denote the probability for a failure: $P(F) = P(Y = 0) = q$.

For an experiment repeated twice, Table 14.8 gives the outcomes in terms of

Table 14.8 Probabilities for a Two-Trial Binomial Experiment

S and F Events	SS	SF or FS	FF
Random Variable $Y = y$	$Y = 2$	$Y = 1$	$Y = 0$
Probability $P(Y = y)$	p^2	$2pq$	q^2

both combinations of S's and F's and the random variable Y. Also shown in the table is the probability for these events.

The two trials are independent; therefore, the probabilities for the combinations of successes and failures can be expressed as products of the probabilities for the single trial.

Similarly, Table 14.9 shows the combinations of successes and failures, their corresponding random variable value, and their probabilities for a three-trial experiment.

Table 14.9 Probabilities for a Three-Trial Binomial Experiment

S and F events	SSS	SSF, SFS, or FSS	SFF, FSF, or FFS	FFF
Random Variable $Y = y$	$Y = 3$	$Y = 2$	$Y = 1$	$Y = 0$
Probability $P(Y = y)$	p^3	$3 \cdot p^2 \cdot q$	$3 \cdot p \cdot q^2$	q^3

In general, for n trials, the distribution table is shown Table 14.10.

Table 14.10 Probabilities for an n-Trial Binomial Experiment

$Y = y$: Number of Successes	n	$n - 1$	$n - 2$	...	$n - k$	...	0
$P(Y = y)$	p^n	$np^{n-1}q$	$\dfrac{n(n-1)}{2} p^{n-2} q^2$	...	$\dfrac{n!}{(n-k)!k!} p^{n-k} q^k$	...	q^n

Notice that the probabilities are exactly the same as the terms in expansion of $(p + q)^n$:

$$(p + q)^n = p^n + np^{n-1}q + \frac{n(n-1)}{2} p^{n-2} q^2 + \ldots + \frac{n!}{(n-k)!k!} p^{n-k} q^k + \ldots + q^n.$$

Tables 14.8, 14.9, and 14.10 also represent the number of successes in a sample of size 2, 3, and n, respectively. (In other words, a repetition of n trials is equivalent to the selection of n elements in a sample.)

14.3 Binomial Probability Distribution

Example 14.13 Refer to the manufacturer's problem in Example 14.11. If the success probability p is assumed to be 0.99, then $q = 0.01$.

  a. Find the probability that a sample of size three will have exactly two failures.
  b. Find the probability that a sample of size three will have no failures.
  c. Find the probability that a sample of size four will have exactly three successes.
  d. Find the probability that a sample of size four will have three or more successes.

Solution
  a. Here the number of trials is $n = 3$, and the number of successes is $Y = 1$; from Table 14.9 $P(Y = 1) = 3pq^2 = 0.000297$. This means that if the manufacturing process is 99 percent reliable, there is less than a $3/100$ of one percent chance that two of the three items selected at random will be defective.

  b. Zero failures in three trials means $n = 3$ and $Y = 3$. From Table 14.9, $P(Y = 3) = p^3 = (0.99)^3 \approx 0.970$. If the process is 99 percent reliable, then there is a 97 percent chance that of the three items selected at random, none will be defective.

  c. Here $n = 4$ and $Y = 3$; so from Table 14.10, $P(Y = 3) = 4p^3 q = 4(0.99)^3 (0.01) \approx 0.0388$.

  d. Here we want $P(Y \geq 3) = P(Y = 3 \text{ or } Y = 4) = P(Y = 3) + P(Y = 4)$. $P(Y = 3) = 0.0388$ from part c. Now, $P(Y = 4) = (0.99)^4 \approx 0.96$. Therefore, $P(Y \geq 3) = 0.0388 + 0.96 = 0.9988$. ▲

Example 14.14
  a. Write the binomial distribution for a four-trial experiment.
  b. Find the mean $\bar{y}$ for the distribution of the random variable Y in the four-trial experiment.
  c. Find the variance for the distribution.

Solution a.

$Y = y$	$Y = 4$	$Y = 3$	$Y = 2$	$Y = 1$	$Y = 0$
$P(Y = y)$	p^4	$4p^3 q$	$6p^2 q^2$	$4pq^3$	q^4

  b. The mean $\bar{y}$ is the expected value of the random variable:
  $$E(Y) = \Sigma yP(Y = y) = 4p^4 + 12p^3 q + 12p^2 q^2 + 4pq^3 + 0q^4$$
  $$= 4p(p + q)^3 = 4p.$$

c. The variance s^2 is the expected value of $(Y - \bar{y})^2$: $E(Y - \bar{y})^2 = \Sigma(y - \bar{y})^2 P(Y = y)$, which, by Theorem 14.1, is

$$\begin{aligned} s^2 &= E(Y - \bar{y})^2 = \Sigma y^2 P(Y = y) - (\bar{y})^2 \\ &= 16p^4 + 36p^3 q + 24p^2 q^2 + 4pq^3 - 16p^2 \\ &= 36p^3 q + 24p^2 q^2 + 4pq^3 - 16p^2(1 - p^2) \\ &= 4pq[9p^2 + 6pq + q^2 - 4p(1 + p)] \\ &= 4pq[(3p + q)^2 - 4p - 4p^2] \\ &= 4pq[(3p + 1 - p)^2 - 4p - 4p^2] \\ &= 4pq. \end{aligned}$$

Thus, for this example, the mean is $4p$ and the variance is $4pq$.

In general, for any binomial distribution of n trials, the mean is np, the variance is npq, and the standard deviation is $\sqrt{npq}$.

▲ **Example 14.15** Refer to the biologist's problem in Example 14.12. There the probability of a success is $p = 0.75$ and $q = 0.25$.

a. Find the probability of seven successes in a sample of eight rats.

b. What is the probability for two or fewer successes in a sample of ten rats?

c. For a sample of four elements, show that the mean is three and find the standard deviation.

d. If a sample of ten rats has two or fewer successes, should the biologist be suspicious of the population of his shipment? (That is, should he suspect that it is not 75 percent successes?)

Solution   a. For $n = 8$, $P(Y = 7) = 8p^7 q = 8(0.75)^7 (0.25) \approx 0.267$. (Pocket calculator accuracy.)

b. $P(Y \leq 2) = P(Y = 0) + P(Y = 1) + P(Y = 2) = (0.25)^{10} + 10(0.75)(0.25)^9 + 45(0.75)^2 (0.25)^8 \approx 0.0004$.

c. By the formula derived in Example 14.14, $\bar{y} = np = 4(0.75) = 3$. Also $s^2 = npq$; so the variance is $4(0.75)(0.25) = 0.75$, and the standard deviation = $\sqrt{0.75} \approx 0.866$.

14.3 Binomial Probability Distribution

d. Yes, if a sample of ten rats had two or fewer successes, the shipment is suspected of being not as claimed. If the biologist assumes that the shipment contains 75 percent successes, then he knows that a sample of ten will contain two or fewer successes with a probability of only 0.0004. He can decide that this is too rare an occurrence to make him confident that the shipment actually is at the 75 percent level. If his test *did* result in only two successes in ten trials, he should not believe the claim of 75 percent successes. ▲

What *would* it take to convince him that the shipment is as claimed? The answer to this is a decision-making problem, and we look to the traditional method of solution.

The shipper and the receiver come to an agreement about the rejection level (or acceptance level) concerning the number of successes in a given random sample. Suppose the biologist had decided before testing that if the random sample of ten rats has a result with so few successes that such a result occurs less than five percent of the time in shipments that are supposed to be 75 or more percent successes, then he has decided on a *five percent rejection level*. In other words, if an actual experiment with ten rats had k successes and $P(Y = k) \leq 0.05$, then he rejects the shipment (that is, he rejects the assumption that there are actually 75 percent or more successes in the shipment). If $P(Y = k) \geq 0.05$, then he accepts the lot because he has already decided that in testing ten rats he expects the given result at least five percent of the time when the shipment *is* as claimed.

Example 14.16 A biologist orders 1000 rats in which 60 percent or more are supposed to have a certain property X. He decided he will test the shipment at the 10 percent acceptance level. He selects 12 rats at random and finds that only four of them have property X. Should he accept the shipment?

Solution Here $n = 12$. He will accept if $P(Y \leq 4) \geq 0.10$, otherwise he will reject.

$$P(Y \leq 4) = P(Y = 0) + P(Y = 1) + P(Y = 2) + P(Y = 3) + P(Y = 4)$$
$$= (0.40)^{12} + 12(0.40)^{11}(0.60) + 66(0.40)^{10}(0.60)^2$$
$$\quad + 220(0.40)^9 (0.60)^3 + 495(0.40)^8 (0.60)^4$$
$$= 0.0000 + 0.0003 + 0.0025 + 0.0125 + 0.0420$$
$$= 0.0573.$$

The result of 0.0573 is less than 10 percent and, according to his prior decision, too rare to verify a 60 or more percent success rate. He rejects the shipment. ▲

EXERCISE 14.3

In Problems 1 and 2, let S = success and F = failure in the binomial experiments; their probabilities are $P(S) = p$ and $P(F) = q$.

1. Suppose $p = 0.2$ and $q = 0.8$,

    a. Find the probability that there will be at least three successes in four trials.

    b. Find the probability of exactly two successes in five trials.

    c. In a sample of 100 trials, find the mean number of successes and the standard deviation. (Hint: Use the formulas $\bar{y} = np$ and $s^2 = npq$.)

2. Suppose $p = 0.4$ and $q = 0.6$,

    a. Find the probability that there will be at least three successes in four trials.

    b. Find the probability of exactly three failures in five trials.

    c. In a sample of 150 trials, find the mean number of successes and the standard deviation.

3. Write the binomial expansion for $(p + q)^5$.

4. Write the binomial expansion for $(p + q)^6$.

▲ 5. A baseball player has a lifetime batting average of 0.350. What is the probability that he will not have a hit in five times at bat? (Assume the batting average does not change during the process.)

6. A certain dominant gene appears in a population of rats with a probability of 0.685. What is the probability that in a sample of six of those rats, only two will have that gene?

7. In Problem 5, what is the probability that the baseball player will get at least two hits in five times at bat?

8. In Problem 6, what is the probability that a sample of six rats will have three or more rats with the dominant gene?

14.3 Binomial Probability Distribution

9. The postal service claims that 98 percent of the mail is delivered accurately. You mail three letters. What chance is there that at least two of them make it to their destination?

10. A manufacturer believes his production process is 90 percent reliable. In a sample of six items, he finds that four of them are faulty. What is the probability that such a sample could occur? Should he be suspicious of the estimated percent of reliability? ▲

*11. Let p be the probability for a success in a single trial of a binomial experiment, and let $P(Y = n - k)$ be the probability for $n - k$ successes in n trials with

$$P(Y = n - k) = \frac{n!}{(n-k)!k!} p^{n-k}(1-p)^k.$$

Prove that the mean number of successes in the n trials is np.

12. Explain how you can find the percentile if you are given a *probability* distribution instead of a *frequency* distribution. In the following probability distribution, find the percentile of $Y = 3$ and of $Y = 1$.

$Y = y$	$Y = 0$	$Y = 1$	$Y = 2$	$Y = 3$	$Y = 4$
$P(Y = y)$	$(1/3)^4$	$4(1/3)^3(2/3)$	$6(1/3)^2(2/3)^2$	$4(1/3)(2/3)^3$	$(2/3)^4$

▲ 13. A drug is hypothesized to be effective 85 percent of the time. Suppose 170 patients are given the drug; what is the mean number of patients for which this drug is effective? What is the standard deviation?

14. It is estimated that 30 percent of American households have two or more television sets. Find the mean number of households with two sets in a survey of 450 households; find the standard deviation. ▲

15. Write the expansion of $(p + q)^9$.

16. Write the expansion of $(p + q)^{10}$.

▲ 17. A manufacturer believes his production process is at least 80 percent reliable. He decides to make a random sample test choosing nine items and a five percent rejection level. He finds that he has only four successes.

   a. What is the probability $P(Y \leq 4)$?

   b. Should he reject the assumption that the process is at least 80 percent reliable?

18. In 1972, 37 percent of the U.S. population was under 19 years of age. An advertising agency assumed that the audience of a given television news broadcast had at most 37 percent of the under 19 age group in it. They decided to test this assumption with a 10 percent rejection level. In a sample of ten people watching the show, they found that seven of them were under 19.

   a. What is the probability $P(Y \geq 7)$?

   b. Should they have accepted the assumption?

14.4

NORMAL DISTRIBUTION

Narrow old stairs are often worn in a bell-shaped pattern. (See Figure 14.6) This happens because the normal tendency is to step in the middle of the stair most frequently and to step on the outer edges with less frequency.

In general, when the frequencies for any set of measurements are distributed

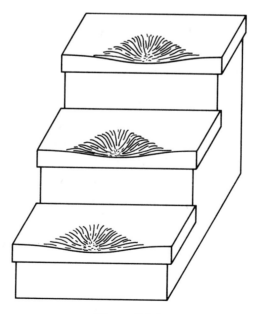

Figure 14.6
The bell-shaped curve.

14.4 Normal Distribution

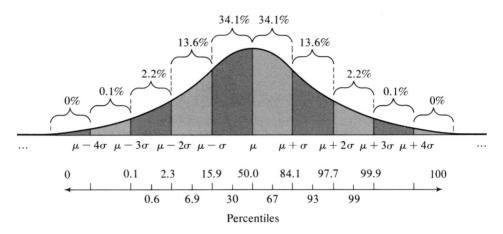

Figure 14.7
Normal probability distribution.

in such a bell-shaped curve, they are said to be **normally distributed.** A normal distribution is one that has a graph similar to Figure 14.7 above. The center of the graph is at the mean, μ, for the set of measurements; σ is the standard deviation.

Normal distributions have been studied closely and their features are well known. The domain of the curve is the set of real numbers (a *continuum* rather than just a finite, discrete set). This means that the measurements are assumed to be a connected, infinite, and continuous set of variables. This assumption implies that the measurements are continuously spread out over the interval with no gaps.

The normal distribution has a known relationship between the values of the variable and the percentiles for those values. This information is useful in the case of the discrete random variable. For example, when a set of discrete variables $\{x_1, x_2, \ldots, x_n\}$ has a distribution that approximates the normal distribution depicted in Figure 14.7, then the normal distribution can be used to approximate the percentiles of the values x_i.

The percentile line at the bottom of Figure 14.7 gives some idea of the approximation, but the relationship between an approximately normal distribution of discrete variables and the percentiles is more clearly revealed in the *z-scores* given in Table 14.11.

The z-score of x_i is defined as

$$z_i = \frac{x_i - \mu}{\sigma}, \tag{14.1}$$

where μ is the population mean and σ is the standard deviation. This equation gives z_i as the number of standard deviations from the mean. Computation of

Table 14.11 Table of Z-scores and Cumulative Percentiles in a Normal Distribution

z_i	Percentile	z_i	Percentile	z_i	Percentile	z_i	Percentile
−3.50	0.01	−0.63	26.43	0.00	50.00	0.83	79.67
−3.00	0.13	−0.59	27.76	0.03	51.20	0.87	80.78
−2.42	0.78	−0.55	29.12	0.06	52.39	0.91	81.86
−2.10	1.79	−0.51	30.50	0.09	53.59	0.95	82.89
−1.91	2.11	−0.48	31.56	0.12	54.78	1.00	84.13
−1.77	3.84	−0.45	32.64	0.15	55.96	1.05	85.31
−1.65	4.95	−0.42	33.72	0.18	57.14	1.10	86.43
−1.55	6.06	−0.39	34.83	0.21	58.32	1.15	87.49
−1.46	7.21	−0.36	35.94	0.24	59.48	1.20	88.49
−1.39	8.23	−0.33	37.07	0.27	60.64	1.26	89.62
−1.32	9.34	−0.30	38.32	0.30	61.79	1.32	90.66
−1.26	10.38	−0.27	39.36	0.33	62.93	1.39	91.77
−1.20	11.51	−0.24	40.52	0.36	64.06	1.46	92.79
−1.15	12.51	−0.21	41.68	0.39	65.17	1.55	93.94
−1.10	13.57	−0.18	42.86	0.42	66.28	1.65	95.05
−1.05	14.69	−0.15	44.04	0.45	67.36	1.77	96.16
−1.00	15.87	−0.12	45.22	0.48	68.44	1.91	97.19
−0.95	17.11	−0.09	46.41	0.51	69.50	2.10	98.21
−0.91	18.14	−0.06	47.61	0.55	70.88	2.42	99.22
−0.87	19.22	−0.03	48.80	0.59	72.24	3.00	99.87
−0.83	20.33	0.00	50.00	0.63	73.57	3.50	99.99
−0.79	21.48			0.67	74.86		
−0.75	22.64			0.71	76.11		
−0.71	23.89			0.75	77.36		
−0.67	25.14			0.79	78.52		

a z-score is a way to "standardize" *any* set of data. Standardized scores always have a mean of zero and a standard deviation of one.

In Table 14.11, the percentile column gives the percent *having the corresponding z-score or less*. Thus, for example, 93.94 percent of the measurements have a z-score of +1.55 or less and the rest, 6.06 percent, have a z-score of more than 1.55.

Example 14.17 Suppose that the ages of U.S. citizens voting in a recent election are normally distributed. The mean age is 50 and the standard deviation is 14. Answer the following questions.

14.4 Normal Distribution

a. At what percentile did a voter fall whose age is 57?

b. At what percentile did a voter fall whose age is 36?

c. What percent of the voters were under 30?

d. What percent of the voters were over 22 *and* under 78?

e. What percent of the voters were over 75?

Solution

a. We need to see how many standard deviations the age 57 is above the mean. That is, what is the z-score for $x = 57$? By Equation (14.1), $z = (57 - 50)/14 = 7/14 = 0.5$. In other words, 57 is 7 more than the mean, and since the standard deviation is 14, then 57 is 0.5 of one standard deviation above the mean. According to Table 14.11, this value of z corresponds to the 69.16 percentile (approximately). This value is obtained by "reading between the lines" in the table entries. There were 69.16 percent of the voters younger than 57 years old.

b. Here $x = 36$; so $z = (36 - 50)/14 = -14/14 = -1.00$, and according to Table 14.11, $z = -1$ corresponds to the 15.87th percentile.

c. For ages under 30, let $x = 30$, then $z = -1.43$. Approximately 7.65 percent of the voters were under 30.

d. For ages between 22 and 78: For $x = 22$, $z = -2.0$; and for $x = 78$, $z = +2.0$. Now $z = -2.0$ is at the two percent level (approximately) and $+2.0$ is at the 98 percent level (approximately). Therefore, 98 percent were under 78 and two percent were under 22, so $98 - 2 = 96$ percent were between 78 and 22.

e. For ages over 75, $x = 75$ and $z = (75 - 50)/14 = 1.78$, corresponding to the 96.2 percentile. That is, 96.2 percent of the voters were *under* 75 (Table 14.11 is a table of percentiles corresponding to a certain z-score or less); so $100 - 96.2$ or 3.8 percent of the voters were over 75 years of age. ▲

Suppose the z-score for certain data is known, then can the original score x be found? Yes. If the data are in a normal distribution and z is known, then x can be determined from Equation (14.1) by solving for x in terms of z:

$$x = \sigma \cdot z + \mu. \tag{14.2}$$

Example 14.18 For normally distributed data with mean $\mu = 10$ and standard deviation $\sigma = 3$, find the value x of the data that corresponds to:

a. A z-score of -1.5.

b. A z-score of 2.1.

Solution   **a.** Here $z_i = -1.5$. Then $x_i = 3(-1.5) + 10 = 5.5$.

b. Here $z_i = 2.1$. Then $x_i = 3(2.1) + 10 = 16.3$.

Finally, we sometimes need to find the value of the variable x that corresponds to a given percentile. This is done in two steps. First use Table 14.11 to find the z-score for the given percentile, then use the z-score to find the value of x by Equation (14.2) above.

Example 14.19   In the voting example where the mean is 50 and the standard deviation is 14, we will find the age that corresponds to the 25th percentile; that is, what is the age of the youngest 25 percent of the voters? According to Table 14.11, the z-score corresponding to the 25th percentile is $z = -0.67$ (approximately). Now, by Equation (14.2), since $z = -0.67$, we get

$$x = 14(-0.67) + 50 = -9.3 + 50 = 40.7.$$

EXERCISE 14.4

In these problems suppose that the data are approximately normal.

1. If the mean is 73 and the standard deviation is 12 for a set of data, find the percentiles of the following quantities:

   a. $x = 53$,   **b.** $x = 79$,   **c.** $x = 92$,   **d.** $x = 40$.

▲ 2. The mean IQ is 100 and the standard deviation is 16.4. Find the percentiles for the following IQs:

   a. IQ = 110,   **b.** IQ = 89,   **c.** IQ = 138,   **d.** IQ = 70.

3. In experiments with rats, the normal weight of the adrenal gland tissue is 37.5 milligrams with a standard deviation of 1.1. Some rats under stress were found to have heavier adrenal glands. Rats subjected to 12 hours of stress had adrenal glands weighing 45.5 milligrams and those subjected to 60 hours had glands weighing 55 milligrams. Find the percentiles of these two weights as if they were part of the normal population. (What percentage of the normal population of rats would have adrenal glands weighing this much or more?) Is it safe to assume that such great weights are too rare (less than one percent) to occur in the normal distribution? Does it seem that stress must be a factor

14.4 Normal Distribution

in their larger size? (See Knigge: *American Journal of Physiology,* 169 (1959): 579-82.)

4. Records are kept of public attendance at city council meetings. The mean is 25 and the standard deviation is 10.

   a. How often can the council expect crowds of 35 or more?

   b. How often can they expect crowds of forty or more?

   c. How often can they expect crowds of fifteen or less?

   d. The council will build a larger hall if there is a five percent chance that 50 or more people attend. Should they build the hall? ▲

5. A set $\{x_1, x_2, \ldots, x_n\}$ has mean = 20 and standard deviation = 4.

   a. Find the value x corresponding to the z-score of -1.6.

   b. Find the value x corresponding to $z = 3.1$.

6. If the mean is 62 and the standard deviation is 12,

   a. Find the value x corresponding to a z-score of -1.17.

   b. Find the value x corresponding to $z = 0.50$.

7. If the mean of a set of data is 50 and the standard deviation is 3.6, find the value x that occurs at the

   a. 37th percentile   c. 98th percentile

   b. 54th percentile   d. 10th percentile

▲ 8. An aptitude test has a mean of 23 and a standard deviation of 4.9. Persons receiving scores below the 15th percentile are refused admittance to a given training program. Persons receiving scores between the 15th and 75th percentiles are admitted on probation. Persons between the 75th and 90th percentiles are given partial scholarships and persons in the 90th percentile or higher are given full scholarships. Find the test scores corresponding to each of these key percentiles.

For Problems 9 through 12, use the following data. Two rival cities advertise their climates. City X boasts of an average of 200 days of sunshine per year with a standard deviation of 8. City Y claims a mean of 225 days and standard deviation of 10.

9. Last year, both cities had 225 days of sunshine. What percentage of all years have had less than that for city X? For city Y?

10. Approximately what chance is there that each city will have more than 170 days of sunshine in a year?

11. What percent of the years have less than 210 days of sunshine for each city?

12. What percent of the years have more than 250 days of sunshine for each city?

A nose for statistics In Problems 13 through 18 refer to the following description. Anthropometric features in a given population can be measured for the purpose of identifying regional differences, if any. For example, head breadth, body weight, nasal bone length, stature, etc., are useful indices in anthropological studies. For a given population, the nasal bone length mean is 26.4 millimeters and the standard deviation is 3.1 millimeters. Assume a normal population distribution and solve the following problems.

13. At what percentile does an individual fall with a nasal bone length of 28.8 mm?

14. At what percentile does an individual fall with a nasal bone length of 20.5 mm?

15. In each of four villages, A, B, C, and D, a random sample of 400 persons was selected. The mean nasal bone length for each sample is given in the table below. Find the percentile in the population for each village.

Village	Sample Size	Sample Mean
A	400	26.5
B	400	26.2
C	400	26.3
D	400	26.6

16. Find the mean of the four means for the four villages. Find the percentile in the population for this mean of the means.

17. a. Assume that in a distribution of all 400-person samples, the mean is 26.38 and the standard deviation is 0.16. At what percentile (in the population of 400-person samples) will a sample from another village, E, fall if the sample from village E has a mean of 28.8 mm?

   b. Explain the difference in the percentile for the *individual* with a nasal bone length of 28.8 mm in Problem 13 and the percentile of the village E with a mean nasal bone length of 28.8 mm in part **a** of this problem.

18. a. What nasal bone length must an *individual* have to be in the upper 5 percent of the population?

   b. What mean must a *village* of 400 people have for this village to be in the upper 5 percent of such villages?

15

CORRELATION

15.1

LINEAR REGRESSION AND CORRELATION

In a given statistical experiment, a researcher may collect data on two or more characteristics of the population. This means that he will study several different random variables $X, Y, Z, \ldots$, each one with the same sample space S as its domain: $X:S \to R$, $Y:S \to R$, $Z:S \to R$, ...

Example 15.1 Blood samples are collected from 100 individuals. Let the random variable X = percent concentration of cholesterol; X has a range $\{x_1, x_2, \ldots, x_{k_1}\}$. Let Y = percent concentration of uric acid; Y has a range $\{y_1, y_2, \ldots, y_{k_2}\}$. Let Z = percent of blood sugar; Z has a range $\{z_1, z_2, \ldots, z_{k_3}\}$. Let W = percent concentration of protein; W has a range $\{w_1, w_2, \ldots, w_{k_4}\}$. Each of these is a random variable with a definite range, an observable frequency, and a computable mean and standard deviation. These four random variables, X, Y, Z, and W, have the same underlying sample space—the 100 individual blood samples.

Sometimes it is useful to determine whether or not there is any linear connection between the different random variables. It frequently happens that two of these variables will increase at the same time (a positive correlation) or one will increase while the other decreases (a negative correlation). On other occasions, neither of these trends may be discernible (a zero correlation).

Furthermore, it may appear that there is some way that a change in one of the random variables forces a change in the other (a causal relation) or the correlation may be false (a spurious correlation). Often, a high positive correlation (close to +1) calls for a deeper investigation of the underlying causes, if any. It could be that two highly correlated variables are both influenced by the same hidden variable, but not by each other. A classic example is the high correlation between the number of days and the number of nights in a given time interval. Days do not cause nights (or the reverse) but both are caused by a hidden variable, the spin of the earth.

Although *causality* cannot necessarily be established by high correlation, it is possible to use a consistently high correlation as a *predictive* or *inferential* device. Days do not cause nights, but if a cruise trip takes 15 days, we can infer that it will also take 14, 15, or 16 nights.

The simplest correlation to investigate involves the *linear regression* between two random variables. The best-fitting line through a set of points that appear to have a linear relation is found by the *method of least squares*. The correlation coefficient can then be found, as will be demonstrated later.

Example 15.2 Two points completely determine a unique line, but suppose you wanted to find the best-fitting line for three points that were not all on the same line? (We shall assume that the best-fitting line is the one with the least *sum of the squared errors*, defined below.) Let *A*, *B*, and *C* be three points whose coordinates are given by the table of data in Figure 15.1. Each of the three lines, *AC*, *AB*, and *BC*, fits two of the points, but not all three. The line *AC* misses *B* by a large amount. (The amount by which the line misses *B* is called the *error or deviation of B* from the line, and it is defined as the vertical distance from *B* to the line.) The line *AB* misses *C* by a large amount, and the line *BC* misses *A* by an even larger amount.

What we want is a line with a minimum total amount by which the line misses

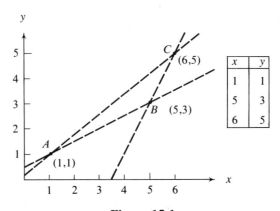

Figure 15.1

15.1 Linear Regression and Correlation

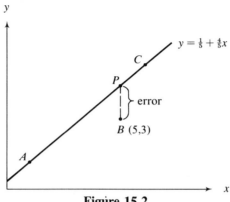

Figure 15.2

the data points. This line is called the best-fitting line; it does not necessarily go through any one of the three points.

Let us find the equations of a few lines and compute the errors. First, the line AC through points $A = (1,1)$ and $C = (6,5)$ has equation

$$y - 1 = \frac{(5-1)}{(6-1)}(x - 1) \quad \text{or} \quad y = \frac{1}{5} + \left(\frac{4}{5}\right)x.$$

Its graph is shown in Figure 15.2. The error (or deviation) of B from the line AC is the vertical distance from the point B to the point P of AC with the same abscissa as B. Since P has coordinates satisfying the line equation $y = 1/5 + (4/5)x$, then for $x = 5$, $y = 21/5$. The distance $\overline{PB}$ is $21/5 - 15/5 = 6/5$. To eliminate the trouble caused by positive and negative signs in computing the total error, we compute the squared errors. Here, the *squared error from B to the line AC is* $36/25 = 1.44$.

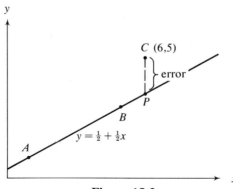

Figure 15.3

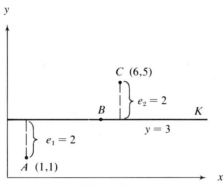

Figure 15.4

Another line, AB, is shown in Figure 15.3. It has equation $y = 1/2 + (1/2)x$. The deviation of C from AB is the distance $\overline{PC} = 5 - 7/2 = 3/2$. *The squared error is $9/4 = 2.25$.* Another line that may be thought of as fitting the data would be a horizontal line through the average ordinate of the three points. It is the line K passing through point B, as shown in Figure 15.4. This mean line K misses the two points A and C by a *total squared error of 8*.

Without any further trial and error, we present the line with the smallest squared deviation. It is the line L depicted in Figure 15.5. The equation of L is $y = 1/7 + (5/7)x$. This line does not contain any of the three points, A, B, or C, yet the sum of the three squared errors is the least possible amount.

The three errors are $e_1 = (6/7 - 1) = {}^-1/7$, $e_2 = (26/7 - 3) = 5/7$, and $e_3 = (31/7 - 5) = {}^-4/7$. The sum of the squared errors (SSE) is

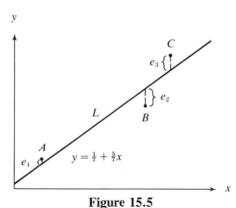

Figure 15.5
Line L is the line of the least squared errors.

15.1 Linear Regression and Correlation

$$\text{SSE} = \sum_{i=1}^{3} e_i^2 = (^-1/7)^2 + (^5/7)^2 + (^-4/7)^2 = {}^{42}/_{49} \approx 0.86.$$

If we tried to fit any other line for the three points, A, B, and C, we would find that none have an SSE less than 0.86.

But how did we get this best-fitting line? The actual derivation of the least squares line equation requires calculus. We will simply state the general formula here.

FORMULAS FOR THE COEFFICIENTS IN THE LEAST SQUARES LINE

Given a set of n points (x_1, y_1), (x_2, y_2), ..., (x_n, y_n), and the slope-intercept line equation

$$Y = a + bX,$$

then the sum of the squared errors (SSE) defined by

$$\text{SSE} = \Sigma(Y - a - bX)^2$$

is minimum if and only if the numbers a and b satisfy the equations

$$\left. \begin{array}{l} a\Sigma X + b\Sigma X^2 = \Sigma XY \\ na + b\Sigma X = \Sigma Y. \end{array} \right\} \quad (15.1)$$

To solve for b, multiply the first equation by n and the second by ΣX. Then,

$$\left. \begin{array}{l} na\Sigma X + nb\Sigma X^2 = n\Sigma XY \\ na\Sigma X + b(\Sigma X)(\Sigma X) = (\Sigma Y)(\Sigma X). \end{array} \right\} \quad (15.2)$$

Subtracting one equation from the other and solving for b yields

$$b = \frac{n\Sigma XY - \Sigma X \Sigma Y}{n\Sigma X^2 - (\Sigma X)^2}. \quad (15.3)$$

This number, b, is the slope of the least squares line. Once it is found, the intercept a can be found in terms of b by the second equation in system (15.1):

$$a = \frac{\Sigma Y - b\Sigma X}{n}. \quad (15.4)$$

Equation (15.3) suggests how we should arrange the data. The given points can be tabulated into columns that exhibit the quantities (sums) required by the equation. We return to Example 15.2 and arrange the data in the suggested form.

Example 15.3 The three points (x_i, y_i) are tabulated in Table 15.1. Notice that the column headings correspond to the various terms in the formulas for computing a and b. Also, note the sums at the bottoms of each column.

Table 15.1

X	Y	XY	X²
1	1	1	1
5	3	15	25
6	5	30	36
$\Sigma X = 12$	$\Sigma Y = 9$	$\Sigma XY = 46$	$\Sigma X^2 = 62$
$(\Sigma X)(\Sigma Y) = 108$			

$$b = \frac{3\,\Sigma XY - (\Sigma X)(\Sigma Y)}{3\,\Sigma X^2 - (\Sigma X)^2}$$

$$= \frac{3(46) - 108}{3(62) - (12)^2} = \frac{30}{42}$$

$$= \frac{5}{7}.$$

$$a = \frac{\Sigma Y - b\,\Sigma X}{3} = \frac{9 - \left(\frac{5}{7}\right)12}{3}$$

$$= \frac{\frac{3}{7}}{3} = \frac{3}{21}$$

$$= \frac{1}{7}.$$

As shown in Figure 15.5, the equation for the best-fitting line is $y = 1/7 + 5/7\,x$.

15.1 Linear Regression and Correlation

We now take a slightly larger example, tabulate the data, and find the least squares line.

Example 15.4 Find the least squares line for the ten points (2,2), (3,2), (3,3), (5,4), (6,5), (6,6), (6,7), (7,6), (7,7), and (7,8) plotted in Figure 15.6. (Such a graph is called a "scatter diagram.")

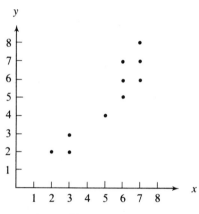

Figure 15.6

The data are tabulated in Table 15.2.

Table 15.2

X	Y	XY	X^2	Y^2
2	2	4	4	4
3	2	6	9	4
3	3	9	9	9
5	4	20	25	16
6	5	30	36	25
6	6	36	36	36
6	7	42	36	49
7	6	42	49	36
7	7	49	49	49
7	8	56	49	64
$\Sigma X = 52$	$\Sigma Y = 50$	$\Sigma XY = 294$	$\Sigma X^2 = 302$	$\Sigma Y^2 = 292$

Using Equations (15.3) and (15.4),

$$b = \frac{10(294) - (52)(50)}{10(302) - (52)(52)}$$

$$= \frac{2940 - 2600}{3020 - 2704} = \frac{340}{316}$$

$$\approx 1.076;$$

$$a = \frac{50 - (1.076)(52)}{10}$$

$$= -0.59.$$

The line equation is $y = -0.59 + 1.076x$. The Y^2 column in Table 15.2 was not needed in the computation of a and b, but it will be used later.

The sign of the slope of the least squares line indicates whether the two random variables tend to increase at the same time or whether one increases when the other decreases. This tendency is formalized by referring to the *least squares line* as the **regression line** for Y on X (where regression means reasoning back from effect to cause). The slope b is called the **regression coefficient.**

We could also have found another line—the regression line for X on Y, which would have minimized the squared errors in the X direction between the given data points and the line.

In either case, we can define a **correlation coefficient** r, which is a measure of the closeness of the two variables. The correlation coefficient is defined below in terms of the slope of the regression line for Y on X, but this does not imply that the correlation "favors" this order in the regression. The correlation coefficient is symmetric in X and Y and would be the same for the regression line for X on Y. The equation for the correlation coefficient is

$$r = b \frac{S_X}{S_Y}, \tag{15.5}$$

where b is the slope of the regression line Y on X and S_X and S_Y are the standard deviations for X and Y, respectively.

The correlation coefficient is not greater than $+1$ and not less than -1, and it has the sign of the slope of the regression line b. If $b > 0$, then $r > 0$, and if $b < 0$, then $r < 0$. A zero correlation between the variables means they are not *linearly* related (although they may be related by a *nonlinear* equation). If r is close to $+1$, the variables are said to have a *high positive correlation*; if r is close to -1, the variables are said to have a *high negative correlation*.

15.1 Linear Regression and Correlation

For computational purposes, we rewrite Equation (15.5) to get r in terms of the basic data. Recall that

$$S_X = \sqrt{\frac{1}{n} \cdot \Sigma X^2 - (\bar{X})^2} = \frac{1}{n}\sqrt{n\Sigma X^2 - (\Sigma X)^2}.$$

Similarly,

$$S_Y = \frac{1}{n}\sqrt{n\Sigma Y^2 - (\Sigma Y)^2}.$$

From the definition of b in Equation (15.3) and the definition of r, we get

$$r = \frac{n(\Sigma XY) - (\Sigma X)(\Sigma Y)}{\sqrt{n(\Sigma X^2) - (\Sigma X)^2}\sqrt{n(\Sigma Y^2) - (\Sigma Y)^2}}. \tag{15.6}$$

Notice that in this form the symmetry* of the variables X and Y in defining r is clearer than it is in Equation (15.5).

Example 15.5 Find the correlation coefficients for the data in

a. Example 15.3.

b. Example 15.4.

Solution a. For Example 15.3, we need Y^2 and ΣY^2. These are $y_1^2 = 1$, $y_2^2 = 9$, $y_3^2 = 25$, and $\Sigma Y^2 = 35$. Now from Equation (15.6), the correlation coefficient for a linear regression is

$$r = \frac{3(46) - 108}{\sqrt{3(62) - (12)^2}\sqrt{3(35) - (9)^2}} \approx 0.94,$$

meaning X and Y have a high positive linear correlation.

b. In Example 15.4, the Y^2 column is already computed. $\Sigma Y^2 = 292$; hence, $n\Sigma Y^2 = 2920$, so

$$r = \frac{340}{\sqrt{316}\sqrt{2920 - (50)^2}} \approx 0.93.$$

---

*The formula is the same if X and Y are interchanged.

Example 15.6 For the data below,

X	2	3	4	5	3½	4½	6	4	3½	7	9	9
Y	12	11	12	5	8½	9½	7	7	5½	3	5	2

a. Graph the points in a coordinate plane (that is, draw the scatter diagram for these data).

b. Write the equation for the least squares line.

c. Use the least squares line to estimate y when $x = 8.0$.

d. Find the correlation coefficient r.

Solution **a.** See Figure 15.7 and Table 15.3. The sums computed from Table 15.3 are $\Sigma X = 60.5$, $\Sigma Y = 87.5$, $\Sigma XY = 375.75$, $\Sigma X^2 = 361.75$, and $\Sigma Y^2 = 762.75$.

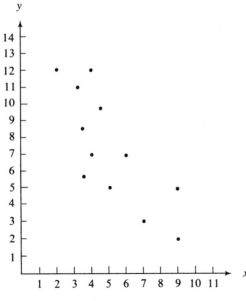

Figure 15.7

Table 15.3

X	Y	XY	X^2	Y^2
2.0	12.0	24.00	4.00	144.00
3.0	11.0	33.00	9.00	121.00
3.5	8.5	29.75	12.25	72.25
3.5	5.5	19.25	12.25	30.25
4.0	7.0	28.00	16.00	49.00
4.0	12.0	48.00	16.00	144.00
4.5	9.5	42.75	20.25	90.25
5.0	5.0	25.00	25.00	25.00
6.0	7.0	42.00	36.00	49.00
7.0	3.0	21.00	49.00	9.00
9.0	5.0	45.00	81.00	25.00
9.0	2.0	18.00	81.00	4.00

15.1 Linear Regression and Correlation

b. From the above data, we obtain the regression coefficient

$$b = \frac{12(375.75) - (60.5)(87.5)}{12(361.75) - (60.5)(60.5)} = \frac{-784.75}{680.75}$$

$$= -1.152 \text{ (pocket calculator accuracy)}.$$

The intercept a, is

$$a = \frac{87.5 - (-1.152)(60.5)}{12} = \frac{157.08}{12} = 13.09.$$

The *line equation* is $y = 13.09 - 1.152x$.

c. For $x = 8.0$, $y = 13.09 - (1.152)(8) = 13.09 - 9.22 = 3.87$.

d. The correlation coefficient is

$$r = \frac{12(375.75) - (60.5)(87.5)}{\sqrt{12(361.75) - (60.5)^2}\sqrt{12(762.75) - (87.5)^2}} \approx -0.778.$$

Finally, we state alternatives to Equations (15.3) and (15.6)

$$b = \frac{\Sigma(X - \bar{X})(Y - \bar{Y})}{\Sigma(X - \bar{X})^2}$$

and

$$r = \frac{\Sigma(X - \bar{X})(Y - \bar{Y})}{\sqrt{\Sigma(X - \bar{X})^2 \Sigma(Y - \bar{Y})^2}}.$$

EXERCISE 15.1

1. For the data below,

X	2.2	3.0	3.2	3.6	3.8	4.5
Y	4.0	1.4	2.6	2.8	2.0	3.0

a. Plot the scatter diagram.

b. Write the equation of the least squares line.

c. Use the *line equation* to obtain y when $x = 4.0$.

d. Find the correlation coefficient.

2. For the data below,

X	29	31	40	44	46	51	52	53	58	60	66
Y	37	54	36	52	32	40	40	27	24	44	22

a. Plot the scatter diagram.

b. Write the equation of the least squares line.

c. Use the *line equation* to obtain y when $x = 35$.

d. Find the correlation coefficient.

▲ 3. Ten college students were in a study to determine the linear relationship between the number of television football games they watched on New Year's weekend and the grade point average made in the spring semester. X = number of football games watched, $X \leq 8$; and Y = the spring grade point average, $Y \leq 4$. The following data were collected: $n = 10$, $\Sigma X = 30$, $\Sigma Y = 24$, $\Sigma XY = 75$, $\Sigma X^2 = 100$, and $\Sigma Y^2 = 72$.

a. Write an equation for the least squares line.

b. Predict the grade point average of the student who watched five television football games that weekend.

c. Find the correlation coefficient.

d. Would a high correlation of these data be considered valid in constructing a causal model?

4. Let X = the number of dollars spent per week on advertising and Y = number of dollars in sales per week. Over a ten-week period, the company collects the following data: $n = 10$; $\Sigma X = 6630$; $\Sigma Y = 14{,}390$; $\Sigma XY = 9{,}574{,}800$; $\Sigma X^2 = 4{,}511{,}600$; and $\Sigma Y^2 = 20{,}973{,}900$.

a. Write an equation for the least squares line.

b. Find the correlation coefficient.

15.1 Linear Regression and Correlation

   c. Does the correlation justify an assumption that increased advertising cost causes increased sales?

5. One method of identifying cells in the microscope depends on the scattering of light intensity. A study by Brunsting and Mullaney in *Biophysical Journal* (Vol. 14, no. 6, 1974, pp 439–53) uses information correlating, by linear regression, nuclear diameters X on whole cell diameters Y as a means of identification. The data are given in the table below. Both measurements are in microns.

X	Y	X	Y	X	Y
6.6	9.3	7.4	11.2	8.5	13.4
6.8	10.1	7.7	10.5	8.6	11.4
7.0	10.8	7.8	10.6	8.8	12.3
7.1	10.6	8.1	12.3	9.1	11.8
7.2	10.6	8.2	12.0	9.2	12.9
7.3	10.3	8.2	11.1	9.7	12.7
7.4	10.4	8.3	10.8	11.3	14.5

   a. Find the regression line (least squares line) equation.

   b. Find the correlation coefficient.

6. In an elementary school experiment, students were allowed to evaluate their own work on a scale from zero to seven. The teachers evaluated the same work on the same scale. Let X = student self-evaluation and Y = teacher's evaluation. The following data were collected:

X	Y	X	Y	X	Y
1.5	2.0	3.0	3.0	4.8	4.7
1.8	1.0	3.3	1.2	5.0	3.5
2.0	1.0	3.5	2.2	5.3	4.0
2.3	1.0	3.8	2.7	5.5	4.7
2.5	2.0	4.0	3.5	6.3	5.0
2.5	1.0	4.5	3.5		

   a. Find the linear regression equation.

   b. Find the correlation coefficient. ▲

15.2

MULTIPLE REGRESSION AND PATH MODELS

In the previous section, the linear regression for Y on a single variable X was obtained from Equations (15.1). In this section, we give a similar system for the general case of the linear regression of Y on k variables, $X_1, X_2, ..., X_k$.

In the single variable case, the regression coefficient was denoted by b and the correlation coefficient by r. In anticipation of the generalizations to be made, we introduce the notation b_{YX} as the coefficient for the regression of Y on X; that is, b_{YX} = the slope of the least squares line minimizing the SSE for the errors in the Y direction. We denote by b_{XY} the coefficient for the regression of X on Y; that is b_{XY} is the slope of the least squares line minimizing the SSE for the errors in the X direction. By analogy to the derivation for b_{YX} we have

$$b_{XY} = \frac{n(\Sigma XY) - (\Sigma X)(\Sigma Y)}{n(\Sigma Y^2) - (\Sigma Y)^2}.$$

The correlation coefficient r may be denoted by either r_{XY} or r_{YX} since $r = \sqrt{(b_{XY})(b_{YX})}$. Also, $r = b_{XY}(s_Y/s_X)$.

If Y is the dependent variable and X_1 is the independent variable, then sometimes b_{XY} and b_{YX} are written as b_{1Y} and b_{Y1}, respectively. When Y is denoted by X_2, then these are b_{12} and b_{21}.

The coefficients a and b defined in Equation (15.1) furnish the minimum of the sum of the squared errors, $\text{SSE} = \Sigma(Y - a - bX)^2$, by a method of calculus. The same calculus method can be used to find the coefficients a, b_1, and b_2 in the two-variable linear regression equation for the random variable Y on the two random variables X_1 and X_2:

$$Y = a + b_1 X_1 + b_2 X_2. \tag{15.7}$$

The best-fitting linear equation will be one for which

$$\text{SSE} = \Sigma(Y - a - b_1 X_1 - b_2 X_2)^2 \tag{15.8}$$

will be minimum. SSE will be least if and only if the coefficients a, b_1, and b_2 satisfy the system

$$\left. \begin{aligned} a \Sigma X_1 + b_1 \Sigma X_1^2 + b_2 \Sigma X_1 X_2 &= \Sigma X_1 Y \\ a \Sigma X_2 + b_1 \Sigma X_2 X_1 + b_2 \Sigma X_2^2 &= \Sigma X_2 Y \\ a \cdot n + b_1 \Sigma X_1 + b_2 \Sigma X_2 &= \Sigma Y. \end{aligned} \right\} \tag{15.9}$$

15.2 Multiple Regression and Path Models

The solutions to Equations (15.9) are lengthy, but can be simplified if we translate the coordinate axes by expressing each variable as a deviation from the mean; that is, if we use the equations $y = Y - \bar{Y}$, $x_1 = X_1 - \bar{X}_1$, and $x_2 = X_2 - \bar{X}_2$.

$$\left. \begin{array}{l} b_1 \Sigma x_1^2 + b_2 \Sigma x_1 x_2 = \Sigma x_1 y \\ b_1 \Sigma x_1 x_2 + b_2 \Sigma x_2^2 = \Sigma x_2 y \end{array} \right\} \quad (15.10)$$

These may be solved for b_1 and b_2:

$$b_1 = \frac{\Sigma x_2^2 (\Sigma x_1 y) - (\Sigma x_1 x_2)(\Sigma x_2 y)}{\Sigma x_1^2 \Sigma x_2^2 - (\Sigma x_1 x_2)^2} \quad (15.11)$$

$$b_2 = \frac{\Sigma x_1^2 (\Sigma x_2 y) - (\Sigma x_1 x_2)(\Sigma x_1 y)}{\Sigma x_1^2 \Sigma x_2^2 - (\Sigma x_1 x_2)^2}. \quad (15.12)$$

The above values for b_1 and b_2 can be substituted into the third equation in system (15.9) and a can be found:

$$a = \frac{\Sigma Y - b_1 \Sigma X_1 - b_2 \Sigma X_2}{n}. \quad (15.13)$$

Example 15.7 Let the variable Y represent a product index for manufactured goods and the variables X_1 and X_2 represent two products of an agricultural economy. A model assuming a linear regression equation

$$Y = a + b_1 X_1 + b_2 X_2$$

is proposed and the data for a six year period are gathered and tabulated in Table 15.4 below.

Table 15.4

Year	X_1	X_2	Y
1	3	4	4
2	2	5	4
3	5	5	7
4	3	6	9
5	3	4	4
6	3	4	3

a. Find the means $\bar{X}_1$, $\bar{X}_2$, and $\bar{Y}$.

b. Find the standard deviations S_1, S_2, and S_Y for the three variables X_1, X_2, and Y, respectively.

c. Find the coefficients a, b_1, and b_2.

d. Predict the value of Y in the seventh year if in that year $X_1 = 6$ and $X_2 = 5$.

Solution **a.** $\bar{X}_1 = 3.167$, $\bar{X}_2 = 4.667$, $\bar{Y} = 5.167$.

b. To get the standard deviations we use the short method described in Theorem 14.1:

$$S_1 = \sqrt{\frac{\Sigma(X_1^2)}{6} - (\bar{X}_1)^2}$$

$$= \sqrt{10.833 - 10.029}$$

$$= \sqrt{0.804}$$

$$= 0.897.$$

Similarly, $S_2 = 0.743$ and $S_Y = 2.114$.

c. To find a, b_1, and b_2, we will use Equations (15.11), (15.12), and (15.13). From the translation equations $y = Y - \bar{Y}$, $x_1 = X_1 - \bar{X}_1$, and $x_2 = X_2 - \bar{X}_2$, and from Table 15.4, we get a new table in terms of x_1, x_2, and y (see Table 15.5 below).

Table 15.5

x_1	x_2	y	x_1^2	x_2^2	y^2	$x_1 x_2$	$x_1 y$	$x_2 y$
−0.167	−0.667	−1.167	0.028	0.445	1.362	0.111	0.195	0.778
−1.167	0.333	−1.167	1.362	0.111	1.362	−0.389	1.362	−0.389
1.833	0.333	1.833	3.360	0.111	3.360	0.610	3.360	0.610
−0.167	1.333	3.833	0.028	1.777	14.692	−0.223	−0.640	5.109
−0.167	−0.667	−1.167	0.028	0.445	1.362	0.111	0.195	0.778
−0.167	−0.667	−2.167	0.028	0.445	4.696	0.111	0.362	1.445

From this table we get the sums

$$\Sigma x_1^2 = 4.834$$

$$\Sigma x_2^2 = 3.334$$

$$\Sigma x_1 x_2 = 0.331$$

$$\Sigma x_1 y = 4.834$$

$$\Sigma x_2 y = 8.331.$$

15.2 Multiple Regression and Path Models

We use these values in Equations (15.11) and (15.12) to find b_1 and b_2:

$$b_1 = 0.834$$
$$b_2 = 2.416.$$

To find a, we use the values of b_1 and b_2 above in Equation (15.13):

$$a = \frac{31 - (0.834)(19) - (2.416)(28)}{6}$$

$$= -8.749.$$

(Here, the sums ΣX_1, ΣX_2 and ΣY were obtained from Table 15.4.)

d. The regression equation is

$$Y = -8.749 + 0.834\, X_1 + 2.416\, X_2$$

So, if $X_1 = 6$, and $X_2 = 5$, then $Y = 8.335$. ▲

The two coefficients, b_1 and b_2, in Equations (15.11) and (15.12) are usually denoted by $b_{Y1.2}$ and $b_{Y2.1}$, where the numbers following the dot indicate the subscripts of the other variables in the same equation. That is, $b_{Y1.2}$ denotes the coefficient of X_1 in an equation whose only other independent variable is X_2. This could be significant in a system of equations in which, for example, the coefficient $b_{Y3.12}$ and the coefficient $b_{Y3.1}$ would play different roles. The former is the weight of the X_3 variable in a regression equation that has both X_1 and X_2 variables, but the latter is the weight given to the X_3 variable in an equation that has X_1 but not X_2. This occurs in a hierarchical or recursive system of causal equations. The coefficients subscripted by the dot notation are called *partial regression coefficients*.

Similar dot notation subscripts are used for the correlation coefficients, which would be called *partial correlation coefficients*, namely $r_{Y1.2}$ and $r_{Y2.1}$. The notation r_{12} may be used instead of $r_{12.Y}$.

The definitions of these partial correlation coefficients are

$$\left.\begin{aligned} r_{12} &= \frac{\Sigma x_1 x_2}{\sqrt{(\Sigma x_1^2)(\Sigma x_2^2)}} \\ r_{Y1.2} &= \frac{\Sigma y x_1}{\sqrt{(\Sigma y^2)(\Sigma x_1^2)}} \\ r_{Y2.1} &= \frac{\Sigma y x_2}{\sqrt{(\Sigma y^2)(\Sigma x_2^2)}}. \end{aligned}\right\} \quad (15.14)$$

Example 15.8 Compute the partial correlation coefficients for the production data in Example 15.7.

Solution We use Table 15.5 and Equation (15.14).

$$r_{12} = \frac{0.331}{\sqrt{(4.834)(3.334)}} = 0.082$$

For $r_{Y1.2}$ and $r_{Y2.1}$ we need the sum $\Sigma y^2 = 26.834$. Now,

$$r_{Y1.2} = 0.424$$

and

$$r_{Y2.1} = 0.881.$$

In Equations (15.14), $x_1 = X_1 - \bar{X}_1$, $x_2 = X_2 - \bar{X}_2$, and $y = Y - \bar{Y}$.

Equations (15.11) through (15.14) yield the following equations for $b_{Y1.2}$ and $b_{Y2.1}$ in terms of the correlation coefficients and the standard deviations, $S_{X_1}(= S_1)$ and $S_{X_2}(= S_2)$, and S_Y.

$$\left.\begin{array}{l} b_{Y1.2} = \dfrac{r_{Y1.2} - r_{Y2.1}\, r_{12}}{1 - (r_{12})^2} \cdot \dfrac{S_Y}{S_1} \\[2ex] b_{Y2.1} = \dfrac{r_{Y2.1} - r_{Y1.2}\, r_{12}}{1 - (r_{12})^2} \cdot \dfrac{S_Y}{S_2} \end{array}\right\} \quad (15.15)$$

In many applications, the variables may be in different units (such as X_1 = rainfall in inches, X_2 = fertilizer in pounds, and Y = wheat in bushels). Before any comparisons can be made, the units must be converted to "pure" numbers whose units are standard deviations. The method for doing this is given below.

From Equation (15.13) we get

$$a = \bar{Y} - b_{Y1.2}\bar{X}_1 - b_{Y2.1}\bar{X}_2.$$

Now, Equation (15.7) can be written as

$$Y = \bar{Y} - b_{Y1.2}\bar{X}_1 - b_{Y2.1}\bar{X}_2 + b_{Y1.2}X_1 + b_{Y2.1}X_2,$$

which is the same as

$$Y - \bar{Y} = b_{Y1.2}(X_1 - \bar{X}_1) + b_{Y2.1}(X_2 - \bar{X}_2). \quad (15.16)$$

If we divide both sides of Equation (15.16) by S_Y and multiply the X_1 term by S_1/S_1 and the X_2 term by S_2/S_2, we get

15.2 Multiple Regression and Path Models

$$\frac{Y - \bar{Y}}{S_Y} = b_{Y1.2} \frac{(X_1 - \bar{X}_1)}{S_1} \cdot \frac{S_1}{S_Y} + b_{Y2.1} \cdot \frac{(X_2 - \bar{X}_2)}{S_2} \cdot \frac{S_2}{S_Y}. \qquad (15.17)$$

Let $V = (Y - \bar{Y})/S_Y$. The variable V is in standard form; its mean is zero and its standard deviation is one.* Similarly, the variables $U_1 = (X_1 - \bar{X}_1)/S_1$ and $U_2 = (X_2 - \bar{X}_2)/S_2$ are in standard form. ($\bar{U}_1 = 0$, $S_{U_1} = 1$, etc.) Denote the coefficients $b_{Y1.2}(S_1/S_Y)$ by $\beta_{V1.2}$ and $b_{Y2.1}(S_2/S_Y)$ by $\beta_{V2.1}$. Then Equation (15.17) becomes

$$V = \beta_{V1.2} U_1 + \beta_{V2.1} U_2. \qquad (15.18)$$

By virtue of Equations (15.15), the β coefficients can be written

$$\beta_{V1.2} = \frac{r_{Y1.2} - r_{Y2.1} r_{12}}{1 - (r_{12})^2}$$

$$\beta_{V2.1} = \frac{r_{Y2.1} - r_{Y1.2} r_{12}}{1 - (r_{12})^2}. \qquad (15.19)$$

The β coefficients are called the **path coefficients** in the path analysis causal models used in social and biological sciences. Equation (15.18) can be represented graphically by a *path diagram* as in Figure 15.8.

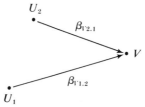

Figure 15.8

If the equation showing the regression of U_2 on U_1 was included in a system with Equation (15.18), such as, for example,

$$U_2 = b_{21} U_1,$$

then the equations

---

*V is a multiple of the deviation from the mean. As shown on page 421 the average deviation from the mean is zero; so $\bar{V} = 0$. Also from Theorem 14.1 and the definition of variance, the variance of V is one. Therefore the standard deviation of V is also one. Similar remarks hold for U_1 and U_2.

$$\left.\begin{array}{l} U_2 = \beta_{21} U_1 \\ V = \beta_{V1.2} U_1 + \beta_{V2.1} U_2 \end{array}\right\} \quad (15.20)$$

would describe a three-point causal model. The coefficient $\beta_{21} = b_{21}(S_1/S_2)$. The path diagram for system (15.20) is shown in Figure 15.9. Notice that in the diagram the connection between U_2 and U_1 shows up as a result of the first equation in system (15.20).

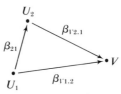

Figure 15.9

Example 15.9 For the production data in Example 15.7 and the correlation coefficients in Example 15.8 compute the path coefficients and draw a path diagram as in Figure 15.8.

Solution We use Equations (15.19) and the known values of r_{12}, $r_{Y1.2}$ and $r_{Y2.1}$.

$$\beta_{V1.2} = \frac{0.424 - (0.881)(0.082)}{1 - (0.082)^2} = 0.354.$$

Similarly,

$$\beta_{V2.1} = 0.852.$$

Now let $V = y/s_Y$, $U_1 = x_1/s_1$, and $U_2 = x_2/s_2$. Then,

$$V = 0.354\, U_1 + 0.852\, U_2.$$

The path diagram is as shown in Figure 15.10 below.

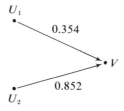

Figure 15.10

15.2 Multiple Regression and Path Models

The general problem of finding the linear regression for Y on k random variables, $X_1, X_2, \ldots, X_k$, each consisting of n values, is solved by methods exactly analogous to the case for one and two independent variables.

In $(k + 1)$-dimensional space, the coefficients $a, b_1, b_2, \ldots, b_k$ of the linear regression equation

$$Y = a + b_1 X_1 + b_2 X_2 + \ldots + b_k X_k$$

minimize the SSE,

$$\text{SSE} = \Sigma(Y - a - b_1 X_1 - \ldots - b_k X_k)^2,$$

if and only if they satisfy the system

$$\begin{aligned}
a\Sigma X_1 + b_1 \Sigma X_1^2 + b_2 \Sigma X_2 X_1 + \ldots + b_k \Sigma X_k X_1 &= \Sigma X_1 Y \\
a\Sigma X_2 + b_1 \Sigma X_1 X_2 + b_2 \Sigma X_2^2 + \ldots + b_k \Sigma X_k X_2 &= \Sigma X_2 Y \\
\vdots \qquad \vdots \qquad \vdots \qquad \qquad \vdots \qquad & \quad \vdots \\
a\Sigma X_j + b_1 \Sigma X_1 X_j + b_2 \Sigma X_2 X_j + \ldots + b_k \Sigma X_k X_j &= \Sigma X_j Y \\
\vdots \qquad \vdots \qquad \vdots \qquad \qquad \vdots \qquad & \quad \vdots \\
a\Sigma X_k + b_1 \Sigma X_1 X_k + b_2 \Sigma X_2 X_k + \ldots + b_k \Sigma X_k^2 &= \Sigma X_k Y \\
a \cdot n + b_1 \Sigma X_1 + b_2 \Sigma X_2 + \ldots + b_k \Sigma X_k &= \Sigma Y.
\end{aligned}$$

These coefficients can be converted to β coefficients as in the case of two variables. A causal model consists of a hierarchical system of equations in standardized variables:

$$\begin{aligned}
U_1 &= e_1 \\
U_2 &= \beta_{21} U_2 + e_2 \\
U_3 &= \beta_{31.2} U_1 + \beta_{32.1} U_2 + e_3 \\
&\vdots \\
V &= \beta_{V1.23\ldots k} U_1 + \beta_{V2.13\ldots k} U_2 + \ldots \beta_{Vk.12\ldots(k-1)} U_k + e_V
\end{aligned}$$

The β's are the path coefficients and the e's are exogenous variables affecting the model from the outside. A path diagram for a four-variable system is suggested by Figure 15.11. (The path diagram does not exhibit the dot notation since the graph usually shows all the variables involved.)

Researchers working in the social sciences recognize the value of path models in establishing theoretical and practical relationships between variables. In the article, "Path Analysis: Sociological Examples" (*The American Journal of Sociology* Vol. 72, no. 1, July 1966, pp 1-16), O. D. Duncan gives a clear expression of the value of path analysis:

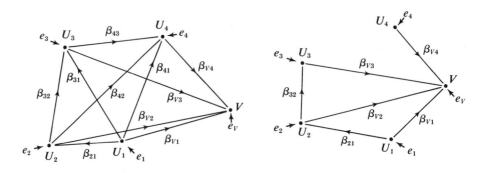

All Variables Correlated   Some Variables with a Zero Correlation

Figure 15.11
Some path models. The e's are exogenous variables.

The great merit of the path scheme, then, is that it makes the assumptions explicit and tends to force the discussion to be at least internally consistent, so that mutually incompatible assumptions are not introduced surreptitiously unto different parts of an argument extending over scores of pages. With the causal scheme made explicit, moreover, it is in a form that enables criticism to be sharply focused and hence potentially relevant not only to the interpretation at hand but also, perchance, to the conduct of future inquiry.

EXERCISE 15.2

*1. Read and prepare an oral report on material from one of the following references:

 a. Simon, Herbert A. *Models of Man.* New York: John Wiley, Inc., 1957. Especially Chapter 1: "Causal Ordering and Identifiability" and Chapter 2: "Spurious Correlation: A Causal Interpretation."

 b. Duncan, Otis D. "Partials, Partitions, and Paths." In *Sociological Methodology 1970,* edited by George W. Borgatta, pp. 38–47. San Francisco: Jossey-Bass, Inc., 1970.

 c. Gibson, Frank K., Prather, J. E., and Taylor, G. A. "A Path Analytic Treatment of Correction Outputs." *Social Science Quarterly* 54 (September 1973): 281-91.

15.2 Multiple Regression and Path Models

   d. Schussler, Karl. "Ratio Variables and Path Models." In *Structural Equation Models in the Social Sciences,* edited by Arthur S. Goldberger and O. D. Duncan, pp 201-228. New York: Seminar Press, 1973.

   e. Wright, Sewall. *Genetic and Biometric Foundations.* Vol. 1, Chicago: University of Chicago Press, 1968. Especially Chapter 13: "Path Analysis: A Theory" and Chapter 14: "Interpretation by Path Analysis."

*2. Read and prepare an oral report on material from one of the following references:

   a. Blalock, Hubert M. *Causal Inference in Nonexperimental Research.* Chapel Hill: University of North Carolina Press, 1964. Chapter 2: "Mathematical Representations of Causal Models" and Chapter 3: "Evaluating Causal Models."

   b. Stokes, Donald E. "Compound Paths: An Expository Note." *American Journal of Political Science* 18 (February 1974): 191-214.

   c. Crow, James F., and Kimura, M. *An Introduction to Population Genetics.* New York: Harper and Row, 1970. Especially Chapter 3: "Inbreeding."

   d. Duncan, Otis D. "Unmeasured Variables in Linear Models for Panel Analysis." In *Sociological Methodology 1972,* edited by Herbert L. Costner, pp 36-82. San Francisco: Jossey-Bass, Inc., 1972.

   e. Duncan, Otis D. "Path Analysis: Sociological Examples." *American Journal of Sociology* 72 (July 1966): 1-16.

*3. Given the data in the table below, for a linear regression $Y = a + b_1 X_1 + b_2 X_2$:

X_1	X_2	Y
3	5	3
4	5	4
4	4	4
2	3	3
5	6	4
4	4	3

   a. Find the partial regression coefficients.

   b. Find the partial correlation coefficients.

c. Put the data in standard form and compute the path coefficients.

d. Draw a path diagram for the data.

*4. Given the data in the table below, for the linear regression equation $Y = a + b_1 X_1 + b_2 X_2$:

X_1	X_2	Y
4	1	7
7	2	12
9	5	17
12	8	20

a. Find the partial regression coefficients.

b. Find the partial correlation coefficients.

c. Put the data into standard form and compute the path coefficients.

d. Draw a path diagram for the data.

▲ ***5. Immunology** Experiments designed to study body functions for the purpose of developing specific vaccines are often performed as follows. A known amount of a given chemical is introduced into the body, samples of cells are removed and examined, and the percentage of changed cells is noted. The data obtained from such experiments can effectively be analyzed by multiple regression. A linear correlation between the various types of changed cells would be useful in describing the immunological reaction.

In this problem we use data from David G. Haegert and Robin R. A. Coombs' "Immunoglobulin-Positive Mononuclear Cells in Human Peripheral Blood." *Journal of Immunology* 116 (May 1976): 1426-30. The numbers in the table below represent percentages of the indicated cells in the samples. Use the notation X_1 = lymphocyte yield, X_2 = mixed antiglobulin rosettes, and Y = direct antiglobulin rosettes.

a. Find the partial regression coefficients a, b_1, and b_2.

b. Find the partial correlation coefficients r_{12}, $r_{Y1.2}$, and $r_{Y2.1}$.

c. Find the path coefficients $\beta_{Y1.2}$ and $\beta_{Y2.1}$.

15.2 Multiple Regression and Path Models

X_1	X_2	Y
71	11	13
71	13	13
91	8	11
76	13	10
70	10	10
70	8	9
87	7	6
82	8	10
86	6	16
72	9	8

For convenience, we present the following calculations, which you can verify for yourself:

$\Sigma X_1 = 776$, $\Sigma X_2 = 93$, $\Sigma Y = 106$, $\bar{X}_1 = 77.6$, $\bar{X}_2 = 9.3$, $\bar{Y} = 10.6$.

The standard deviations are

$$S_1 = 7.7, \ S_2 = 2.3, \ \text{and} \ S_Y = 2.7.$$

For the translated variables $x_1 = X_1 - \bar{X}_1$, $x_2 = X_2 - \bar{X}_2$, $y = Y - \bar{Y}$, we have the sums

$$\Sigma x_1^2 = 594.4, \ \Sigma x_2^2 = 52.1, \ \Sigma y^2 = 72.4,$$

$$\Sigma x_1 x_2 = -107.8, \ \Sigma x_1 y = 5.4, \ \text{and} \ \Sigma x_2 y = 6.2. \quad \blacktriangle$$

16

INTRODUCTION TO GAME THEORY

16.1

MATRIX GAMES

In everyday language, the term *game* may be applied to a large variety of human activities involving conflict, negotiation, competition, etc., between two or more parties; but here we will confine our studies to a more limited concept. We consider only two-person (or two-player) games. Each of the two players is given an opportunity to choose a single alternative from among a given set of alternatives, while the opposing player is also making a similar choice. Each side then reaps a reward or suffers a loss according to some prearranged scheme.

In this sense, playing a game is simply deciding which alternative to select. For this reason, game theory is often called *decision theory*, a term that should help to dispel the onus of frivolity associated with games.

There are several differences between real-life decisions and game-theory decisions. The fundamental assumptions for playing game-theory games are simple enough, but the real-life games, which the game-theory games are supposed to represent, do not have such nice crisp features.

Listed below are some elements of real-game play with the corresponding theoretical assumptions for theory-game play.

REAL GAMES VERSUS THEORY GAMES

1. Assignment of numerical values to the decisions.

   a. Real games: In most situations, it is not clear what "value" a decision has. For example, in selecting a vice-presidential candidate, can a political party assign a value of +5 to choosing a person from one geographical section of the country and a value of −3 to choosing a person from another section? Does making your friend wait in the rain have a value of −3, while the act of driving around the block in heavy traffic on a dark, rainy night has a value of −1?.

b. Theory games: All the outcomes (or payoffs) have definite unambiguous numbers assigned to them; the numbers can be compared and manipulated by the rules of arithmetic. This is called the **postulate of well-defined utility.**

2. Play of the game.

    a. Real games: Swindling, bluffing, soul-searching, cheating, and otherwise "psyching-out" the opponent are all part of a real game.

    b. Theory games: Straightforward play; pick a strategy and use it.

3. Clarity of goal.

    a. Real games: The player may have a fuzzy, unclear, or even contradictory motivation to win a "unit of utility." A player may wish to "win by losing." In playing a weaker opponent or a child, the stronger player may let the weaker one win; a player may not try to take every point; and so on.

    b. Theory games: Each decision-maker always tries to optimize his gain. This idea is important to the method of solution. This is the **desire-to-win postulate.**

4. Knowledge of available strategies.

    a. Real games: A player may not be completely aware of the opponent's possible moves, or he may make the mistake of failing to assume his opponent will make the best move.

    b. Theory games: Players are aware of each other's possible strategies and make use of that information. (As we shall see, both players may choose a probability strategy, in which case neither will know exactly which decision will be made on a given play.) This is the **intelligent-player postulate.**

5. Form of the game.

    a. Real games: The game is a sequence of moves made on alternate turns by the two players. The number of strategies is enormous even in simple games. These games are illustrated by tree diagrams and they are said to be in *extensive* form.

    b. Theory games: The games are in *normal* or *matrix* form; each game consists of a single play (the game may be repeated as often as desired). The number of strategies is small, usually just two or three.

16.1 Matrix Games

The above contrasts between real and theory games seem to indicate that game theory is not very useful in real life, but implicit in the list is a case in favor of game theory. *Game theory provides a method of approach to decision-making problems.*

In the Foreword to *Readings in Game Theory and Political Behavior* (edited by Martin Shubik, New York: Doubleday & Co., 1954), Richard Snyder writes:

All students of politics are interested in how and why decision-makers make mistakes in judgment, in how and why strategies fail, and in how and why coalitions become unstable. Game theory directs our attention to such relevant factors as the possible effects: of the rules of the game on the range of choice, of the amount of information known to decision-makers and to opponents, of the different ways of calculating the distribution of gains and losses, and of the kinds of strategies in fact available. It is obvious that political decisions are not made in a vacuum, are not always made under ideal circumstances, and are not always the result of free choice in the conventional sense. Game theory offers a way of thinking systematically about the limiting conditions *under which choices must be made.*

Professor Snyder's statement that *"game theory directs our attention to the relevant factors"* is the key to learning how real games can be "converted" to theory games for the purpose of solving the real-game decision problems with theory-game techniques. Converting real into theory games is a difficult and even controversial task. One area of controversy lies in utility theory and its use in assigning numerical payoffs in real games.

Fortunately, there are well-developed, widely accepted, and easily learned methods for constructing and solving some of the simplest theory games.

Example 16.1 We begin with Game 1 in Table 16.1, which represents a typical matrix game. It is a three-by-four, two-person, zero-sum game. The table is called a *payoff matrix*. It exhibits the results of any decision by either player. Both players see the entire payoff matrix. The row player secretly selects one of the three rows, and the column player secretly selects one of the four columns, then the two choices are simultaneously revealed. The outcome of the game is shown in the box that is the intersection of the chosen row and the chosen column. This outcome is the payoff and what one player wins the other loses (the sum of the payoffs equals zero).

In Game 1, if the row player chooses row 1 (r_1) and the column player chooses column 4 (c_4), then the intersection (r_1,c_4) shows that C wins 3 points (and R loses 3 points). If R plays r_1 and C plays C_1, then according to the intersection (r_1,c_1), R wins 1 point. If R plays r_2 and C plays c_4, then R wins 10 points.

Although neither player knows what the other will play, they can easily figure out what the other should play because they both see the entire payoff board.

Table 16.1 Table of Payoffs for a Matrix Game

C = Column Player

R = Row Player

	c_1	c_2	c_3	c_4
r_1	R wins 1 point	C wins 2 points	C wins 1 point	C wins 3 points
r_2	R wins 5 points	C wins 4 points	C wins 2 points	R wins 10 points
r_3	R wins 7 points	C wins 1 point	R wins 10 points	R wins 5 points

Example 16.2   a. Show that there is a column that C can play and never lose.

b. Show that R does not have a no-lose row.

c. Find the *safest row* for R to play (the one that will cut losses to a minimum).

Solution   a. If C plays column c_2, then no matter what R plays, C wins either 2, 4, or 1 point.

b. Every row has some win for C in it, so regardless of what R plays, C could play a winning column (for example, C could always play c_2).

c. R's safest play is r_3. Then if C plays the no-lose column (c_2), R would lose only one point; if C played some other column, R would win something.

This game is biased in favor of C.

According to the *intelligent-player postulate*, R should not try to benefit by taking a chance on another row in hope that C will play some column other than c_2; this would be hoping that C is not intelligent. For example, if C ever plays c_3 while R is playing r_3, then C will lose 10 points, at (r_3, c_3). If C assumes R is stupid enough to play r_1, and C tries to "catch" R by playing c_4, hoping to win 3 points at (r_1, c_4), then he is in for a surprise when he finds that R

16.1 Matrix Games

is intelligent and sticking to r_3 (to minimize his losses), in which case it is C who is "caught" at (r_3, c_4) with a 5-point loss.

Usually the payoff matrix is written in more compact form by using "signed" numbers instead of spelling out the winner in words (such as "R wins" and "C wins"). The convention is to let a positive number in the payoff matrix represent a win for the row player (and a loss for the column player in zero-sum games), while a negative entry represents a loss for the row player (and a win for the column player).

Example 16.3 The payoff matrix in Game 1 can be expressed in terms of signed numbers as in Table 16.2.

Table 16.2 Game 1

		Column Player				
		c_1	c_2	c_3	c_4	Row Minima
Row Player	r_1	1	−2	−1	−3	−3
	r_2	5	−4	−2	10	−4
	r_3	7	−1	10	5	(−1)
Column Maxima		7	(−1)	10	10	

We have added a column of "row minima" and a row of "column maxima"; these are to be used in the method of solution.

The minimum in a given row is the least amount R can win in that row. Of course, if he "wins" a −3, then he really loses 3, but if −3 is the minimum in its row, then he cannot lose more than 3 by playing that row. The row minima are the minimum amounts R can win.

Similarly, the maximum in any given column is the largest amount C can lose by playing that column. Again, if C "loses" a −2, then he really wins 2, and if −2 is the largest number in the column, then C cannot win less than 2. The column maxima are the maximum amounts C can lose.

Now, we are ready to find R's best play; he would like to maximize the smallest amount he can win. He wants the maximum of the minima, or the *maxi-min* (written maximin). R selects the row with the largest row minimum.

C's best play is to minimize the largest amount he can lose. He wants the minimum of the maxima, or the *mini-max* (written minimax). C selects the column with the smallest column maximum.

In Game 1, the maximin is the -1 at the (r_3, c_2) position and the minimax is the -1 in the same position. When a number $x_{ij} =$ both the maximin and the minimax, then the corresponding position $(r_i c_j)$ is called a *saddle point* for the matrix; the formal definition is given below.

Definition **(Saddle point)**
In a matrix game, a *saddle point* is a position (r_i, c_j) in which there appears an entry that is both the smallest in its row and the largest in its column.

Note that "smallest" is not used in the strict sense; there may be other entries in the same row that are equally small, but none is smaller. Also, "largest" is not used in the strict sense. Games with saddle points are called *strictly determined*.

Figure 16.1 illustrates the term *saddle point*. It is the lowest point in the X-direction and the highest point in the Y-direction. It makes a perfect seat for the old-time movie cowboy, Tom Minimax.

In Game 1, the position (r_3, c_2) is a saddle point for the matrix since it contains the minimax solution -1 (no number in row r_3 is smaller than -1 and no number in column c_2 is larger than -1).

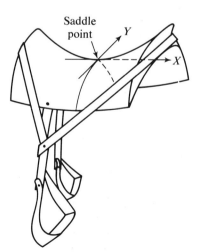

Figure 16.1

16.1 Matrix Games

SOLUTION OF A MATRIX GAME BY MINIMAX STRATEGY

Solving a matrix game is an orderly process of searching out the best row for the row player and the best column for the column player. More often it involves the selection of the best combination of rows and columns for both players. The process is called the *minimax strategy* and consists of two branches, which are explained below. (See Figure 16.2.)

Branch 1 Check for a saddle point.

a. If there are one or more saddle points, play for any of them.

b. If there is no saddle point, go to Branch 2.

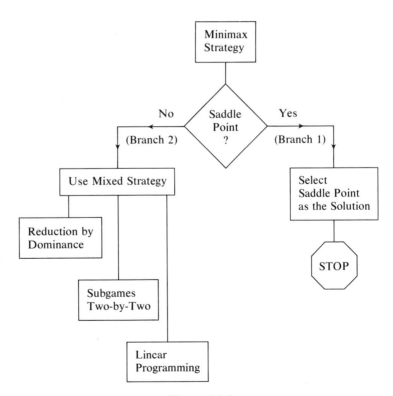

Figure 16.2

Branch 2 In the absence of saddle points, play for a mixed strategy. This is a probabilistic strategy that is used in a run of several plays of the game. Each player makes his choices of rows or columns in a certain proportion, to be determined by the payoffs, but in a random pattern. This will insure a minimax solution that allows R to maximize the minimum he can win and C to minimize the maximum he can lose. The mixed strategy will be discussed in later sections.

Note In Branch 1, if a saddle point exists, *neither player can improve on the saddle point solution* if both are trying to obtain their optimal value from the game. The saddle point, when it exists, is called the **value** of the game, and it is denoted by V. The game favors the row player when V is positive, and it favors the column player when V is negative. In Branch 2, the value of the game is the expected value of the mixed strategy.

▲ **Example 16.4** Find the best move for the amateur chess player against the former world champion, Tigran Petrosian, in Table 16.3, a matrix representation of a chess game.

Table 16.3

		Petrosian				
		Preparation	Defense	Counter-attack	Siege	Row Minima
	Preparation	0	2	−1	1	−1
	Attack	−2	−20	−2	3	−20
	Diversion	−1	−6	−4	−4	−6
Amateur	Sacrifice	−3	0	−1	−10	−10
	Flight	−2	−1	−6	−2	−6
	Siege	1	−2	−3	0	−3
	Defense	0	0	−1	0	−1
Column Maxima		1	2	−1	3	

16.1 Matrix Games

Solution According to the row minima and the column maxima, the minimax solutions are the -1's occurring in two places (two saddle points). This is the best the amateur can hope to do against Petrosian. Therefore, the amateur should play either *preparation* or *defense* and he can last longest against Petrosian. Note that the -1's at (r_1, c_3) and (r_7, c_3) are in the saddle points, but the position (r_4, c_3) is not a saddle point even though there is a -1 in that position. (It is not the smallest number in its row.) ▲

Example 16.5 Find the saddle points, if any, in the following matrix games.

a. $\begin{bmatrix} 3 & 0 & 4 \\ 1 & 2 & -1 \\ -2 & 3 & 2 \end{bmatrix}$ b. $\begin{bmatrix} 1 & 2 & -3 & -2 \\ 5 & 6 & 7 & 5 \\ -1 & 0 & -3 & 2 \\ 5 & 7 & 8 & 5 \end{bmatrix}$ c. $\begin{bmatrix} -3 & 4 & -2 \\ 3 & -5 & -1 \\ 1 & 3 & 0 \\ 2 & 2 & 1 \end{bmatrix}$

Solution

a. The row minima are 0, -1, and -2; the column maxima are 3, 3, and 4. None is the same, so there is no saddle point.

b. The row minima are -3, 5, -3, and 5; the column maxima are 5, 7, 8, and 5. Since 5 is both a row minimum and a column maximum, it is a minimax and the four positions containing the four 5's all happen to be saddle points. The 5 at (r_2, c_1) is the smallest in its row and the largest in its column, etc.

c. The 1 at (r_4, c_3) is the smallest number in its row and the largest number in its column; therefore, it is a saddle point for this game.

Two useful properties concerning matrix games are multiplication invariance and translation invariance of a saddle point.

Property 1 **(Multiplication invariance)**
If all the entries in the matrix game A are multiplied by the same positive number, then the new matrix game B will have the same position for its saddle point as A did (provided A has a saddle point) or the mixed strategy on B will be the same as it is for A. That is, multiplication of a matrix by a positive number does not change the strategy (it just changes the payoffs).

Property 2 **(Translation invariance)**
Addition or subtraction of the same constant to all the entries of a matrix game does not change the strategy.

Example 16.6 Given the matrix game **G**:

$$G = \begin{bmatrix} -1/15 & 1/6 \\ 1/10 & 2/15 \end{bmatrix}$$

a. Convert the game to one with all whole numbers by multiplying the matrix by an appropriate number.

b. Find the saddle point for the new matrix.

c. Add 3 to each entry and find the saddle point of the result.

Solution a. Multiply each term in **G** by 30 to obtain

$$30\mathbf{G} = \begin{bmatrix} -2 & 5 \\ \text{③} & 4 \end{bmatrix}.$$

b. The saddle point 3 is circled.

c. Adding 3 to each entry of 30**G** is

$$\begin{bmatrix} 1 & 8 \\ \text{⑥} & 7 \end{bmatrix}.$$

The saddle point is in the same relative position.

We now propose a method for constructing a theory game from a real-life situation.

CONSTRUCTING A GAME PLAN

First, identify the two sides. Next, partition each side into a set of *factors* (alternatives that represent different possible disjoint choices). Third, arrange the choices into a matrix pattern. Fourth, and this is the difficult part, assign numerical values to the factors as they appear in the matrix. The numbers must reflect the empirical evidence, value judgments, and personal biases toward any of the factors. Furthermore, since each entry in the matrix is the intersection of a row and a column, the number in that entry must reflect the competition between the factors represented by that row and that column, Finally, apply the minimax game-theory technique: look for a saddle point, and in the absence of one apply the mixed strategy.

In most cases, of course, it is possible to execute the above plan only to a small degree, but it does offer a way of thinking systematically and this has value.

16.1 Matrix Games

Example 16.7 **a.** Convert the following decision problem to a theory game.

b. Give numerical values to the decisions and test the game for saddle points.

A college student is trying to decide among three colleges that have invited her to apply. She sets up a matrix game in which the row player is the set of three colleges and the column player is the state of her future interest, based on her present self-assessed inclinations and aptitudes. The row player is partitioned into

r_1: Technological Institute

r_2: Business College

r_3: Liberal Arts College

The student's future job interest, the column player, is partitioned into

c_1: Engineer

c_2: Mathematician

c_3: Musician

c_4: Business Manager

c_5: Journalist

c_6: Stockbroker

c_7: Computer Programmer

Solution **a.** The first two steps in the construction of the game plan (determining the sides and partitioning them into factors) has been completed in the statement of the problem. The matrix arrangement is given below. We need to assign numbers to the boxes in the matrix, but before filling in the numbers, let us examine the meanings of the positions. The (r_i, c_j) position should be filled with a number that represents the student's evaluation of obtaining her education for the c_j profession at the r_i college. For example, a number in the (r_2, c_3) position

Future State of Job Interest

		c_1	c_2	c_3	c_4	c_5	c_6	c_7
	r_1							
Colleges	r_2							
	r_3							

will reflect her desire to take music education at the business college, and it reflects her desire to major in music given she is in the business college. The utility of these decisions are usually unique to the person filling in the matrix.

b. An example of some numerical values that could be assigned is given below.

	c_1	c_2	c_3	c_4	c_5	c_6	c_7
r_1	13	6	2	4	4	2	15
r_2	5	4	1	11	6	12	8
r_3	7	⑦	8	7	10	8	7

The saddle point is circled. It should be clear that a saddle point is a compromise between the absolutely best possible outcome and the absolutely worst possible outcome; that is, the saddle point is a *safe* goal. ▲

EXERCISE 16.1.

1. Solve the matrix games:

   a. $\begin{bmatrix} 5 & 6 \\ 3 & -1 \end{bmatrix}$ **b.** $\begin{bmatrix} -2 & 4 \\ -4 & -3 \end{bmatrix}$ **c.** $\begin{bmatrix} 5 & 10 & -2 \\ 3 & -5 & -6 \end{bmatrix}$

2. Solve the matrix games:

   a. $\begin{bmatrix} 4 & 3 \\ -2 & 1 \end{bmatrix}$ **b.** $\begin{bmatrix} -7 & -6 \\ 5 & -1 \end{bmatrix}$ **c.** $\begin{bmatrix} 5 & 7 \\ -6 & 1 \\ 2 & -5 \end{bmatrix}$

3. Convert the following matrix games to ones with whole numbers by suitable multiplication, and find the saddle points, if any.

   a. $\begin{bmatrix} -1/2 & 2/3 \\ -1 & -1/3 \end{bmatrix}$ **b.** $\begin{bmatrix} -1/2 & -1/4 & 1/8 & 5/8 & 9/8 \\ -1/4 & -1/8 & 0 & 1/2 & 1 \end{bmatrix}$ **c.** $\begin{bmatrix} 1/2 & 2/3 \\ -3/4 & 4/5 \end{bmatrix}$

16.1 Matrix Games

4. Convert the following matrix games to ones with whole numbers by suitable multiplication, and find the saddle points, if any.

   a. $\begin{bmatrix} 3/8 & 5/8 & 7/4 \\ -7/4 & 11/8 & 1/4 \\ -1/4 & 0 & -7/8 \end{bmatrix}$ b. $\begin{bmatrix} -1/5 & 2/5 \\ 3/5 & 1 \end{bmatrix}$ c. $\begin{bmatrix} 1/3 & -3/5 \\ 5/7 & 2/3 \end{bmatrix}$

▲ 5. A factory finds that if it increases its control of air pollution, production will suffer. But if it doesn't control pollution, boycotts, fines, and lawsuits will be presented and the company's sales and image (which is really bad) will suffer. The company's study of the situation implies the assignment of utilities in the table below, where the factors J_1, J_2, and J_3 are three judgment states against the company. They are the *composite variables* consisting of:

   J_1 = Consumer boycotts of the company's products and their effects on the possible production gain from not controlling pollution.

   J_2 = Production function of pollution minus lawsuits and boycotts.

   J_3 = Production function of pollution minus court fines for lack of controls.

		Public Judgments Against the Company		
		J_1	J_2	J_3
Company Production and Profit	Pollution Control	4	1	2
	No Control	−1	−7	5

   a. Solve for the saddle point.

   b. Suppose that after one year the company improves its image and thereby increases each payoff by two units. Write the new payoff matrix. Can the company safely return to the no-control decision?

6. The war game illustrated by the table below is between an Earth spaceship and a spaceship from Planet III in Alpha Centauri's planetary system.

   a. Find the best action for the Earth spaceship.

Earth Spaceship

		Defend	Attack	Make Friendly Contact	Self-Destruct
Ship from Planet III of Alpha Centauri	Attack	3	−10	15	20
	Defend	5	0	0	50
	Make Friendly Contact	−3	−50	0	100
	Self-Destruct	−10	−35	0	0

    b. What happens if the Planet III ship tries to do something other than defend? Especially, what happens to the Planet III ship if it tries to make friendly contact?

7. By filling in the blank matrix below with numerical utilities based upon your own circumstance and values, create a decision game and see if you can find a saddle point. To construct the composite index consider: chance to get a good job, chance to get financial aid, prospects of getting the major in a short time (the units you have already completed in the subject will count here), and other components you may think of. ▲

Composite Index

		High	Medium	Low
Major Subject in College Curriculum	Management			
	Mathematics			
	Chinese			
	Physical Education			
	Psychology			

16.2
MIXED STRATEGIES

What is the best way to play a game that does not have a saddle point? This question has a practical answer only if one anticipates playing the game several times, since the best strategy is to compute probabilities for playing the rows and columns.

Example 16.8 A door-to-door salesman devises the following two-person game. His *sales pitch* is the row player and his *customers* constitute the column-playing team. He partitions his sales pitch into two factors: r_1 = glad-hand approach and r_2 = serious approach. His customers are of two types: c_1 = naive customer and c_2 = wary customer. His matrix diagram for the game is shown below.

		Customers	
		Naive	Wary
Sales	Glad-Hand		
Pitch	Serious		

He finds that when he applies the glad-hand approach and the customer turns out to be naive, he makes an $11 sale. If the glad-hand approach is used against a wary customer, he makes a $3 sale. A serious approach to a naive customer yields a $4 sale and a serious approach to a wary customer yields an $8 sale. He uses these sales as the utilities to fill in the matrix, as shown below.

	Naive	Wary
Glad-Hand	11	3
Serious	4	8

He never knows before he gives his sales pitch whether he is facing a naive or a wary customer. What strategy should he adopt to get a consistently decent sale? He could mix his approaches in hopes of getting something between an $11 sale and a $3 sale.

He assumes that there is some optimal expected value V to his contacts, *regardless* of what type of customer he is contacting. He wants to find out what strategy he can use to insure him *at least that value*. He will try a strategy that will provide him with exactly the same expected value (in the long run) against either type of customer.

Suppose he decides to use the glad-hand approach with probability p and the serious approach with probability $1 - p$. Then, against naive customers, the expectation is $11 \cdot p + 4 \cdot (1 - p) = E_1$. (Recall the definition of expected value.) Against wary customers the expectation is $3 \cdot p + 8 \cdot (1 - p) = E_2$. He wants the values E_1 and E_2 to equal the optimal value V. That is,

$$E_1 = E_2 = V, \text{ or } 11p + 4(1 - p) = 3p + 8(1 - p).$$

Solving for p yields $p = 1/3$, and $1 - p = 2/3$.

If he uses the glad-hand approach $1/3$ of the time and the serious approach $2/3$ of the time, his expected values will be

Against naive customer: $\quad (11)\left(\dfrac{1}{3}\right) + (4)\left(\dfrac{2}{3}\right) = \$6.33,$

Against wary customer: $\quad (3)\left(\dfrac{1}{3}\right) + 8\left(\dfrac{2}{3}\right) = \$6.33.$ ▲

In general, any two-by-two game *without saddle points* can be solved for the row player by computing the expected values against each of the two columns.

Example 16.9 Find the optimal play for the row player in the two-by-two game in the matrix below.

$$\begin{array}{c} & \begin{array}{cc} c_1 & c_2 \end{array} \\ \begin{array}{c} r_1 \\ r_2 \end{array} & \left[\begin{array}{cc} 3 & -7 \\ -5 & 10 \end{array}\right] \end{array}$$

Solution Let x = the probability for playing r_1 and let $1 - x$ = the probability for playing r_2. Then the expected values for the row player are

Assuming C plays c_1: $\quad 3 \cdot x + (-5)(1 - x) = E_1,$

Assuming C plays c_2: $\quad (-7) \cdot x + (10)(1 - x) = E_2.$

16.2 Mixed Strategies

To get the optimal value V, independent of what the column player plays, we must have

$$V = E_1 = E_2, \text{ or } 3x + (-5)(1 - x) = (-7)x + 10(1 - x).$$

Thus, $x = 3/5$ and $1 - x = 2/5$.

The value of the game is the expected value when C plays either column c_1 or c_2. That is, the game value is $V = E_1 = E_2$: $V = 3(3/5) + (-5)(2/5) = -(1/5)$. This game favors the column player.

The same procedure may be used by the column player for finding his optimal strategy. In this example, suppose the column player C plays c_1 with probability y and plays c_2 with probability $1 - y$. Then the expected values for C are

Assuming R plays r_1: $\quad 3y + (-7)(1 - y) = F_1$,

Assuming R plays r_2: $\quad (-5)y + 10(1 - y) = F_2$.

When both F_1 and F_2 equal the optimal value V, we get

$$3y + (-7)(1 - y) = (-5)y + 10(1 - y).$$

Solving for y, $y = 17/25$ and $1 - y = 8/25$. Thus, the best play for the column player C is to select column c_1 for $17/25$ of the time and to select column c_2 for $8/25$ of the time.

Neither player should let the other know when he is about to select one or the other of his alternatives. To insure the secrecy of the selection, some randomizing device is needed. The row player could have a spinner with a dial that gives the spinner a $3/5$ chance to stop on r_1 and a $2/5$ chance to stop on r_2. Or R could have a box with 5 marbles in it, 3 red and 2 white. After shaking the box, R can reach in and take out a marble; if he draws out a red marble, he plays r_1 and a white marble means r_2. Similarly, to insure a random choice of $17/25$ c_1 and $8/25$ c_2, the column player could have a box with 25 marbles, 17 red and 8 white. After each player secretly draws a marble, they simultaneously reveal the color that indicates which row and column are to be played. The marbles are then returned to the boxes and the process is repeated at the next play of the game. In the long run, C wins an average of $1/5$ per game.

Theorem 16.1 (Optimal strategy in a nonstrictly determined two-by-two game)
Let a matrix game $\mathbf{G}$ be

$$\begin{array}{c} \begin{array}{cc} c_1 & c_2 \end{array} \\ \begin{array}{c} r_1 \\ r_2 \end{array} \left[\begin{array}{cc} a & b \\ c & d \end{array} \right] = \mathbf{G} \end{array}$$

and suppose *there is no saddle point* (each of the elements of one diagonal is greater than each of the elements in the other diagonal). Then optimal strategy for the row player is to

$$\text{Play } r_1 \text{ with probability } \frac{d-c}{a-b+d-c},$$

$$\text{Play } r_2 \text{ with probability } \frac{a-b}{a-b+d-c}.$$

Optimal strategy for the column player is to

$$\text{Play } c_1 \text{ with probability } \frac{d-b}{d-b+a-c},$$

$$\text{Play } c_2 \text{ with probability } \frac{a-c}{d-b+a-c}.$$

The proof of Theorem 16.1 is delayed until the section in which linear programming is applied to solving matrix games.

For convenience, we present the following *short method* for computing strategies of a two-by-two game without saddle points. (See Figure 16.3.)

Example 16.10 Find the mixed strategy for R and C in the following two-by-two matrix game:

$$\begin{bmatrix} -2 & 1 \\ 2 & 0 \end{bmatrix}$$

Solution There are no saddle points; therefore, Theorem 16.1 applies. We use the short method of Figure 16.3.

$$\begin{bmatrix} -2 & 1 \\ 2 & 0 \end{bmatrix} \begin{matrix} 2 \\ 3 \end{matrix} \quad \text{and} \quad \begin{bmatrix} -2 & 1 \\ 2 & 0 \end{bmatrix} \begin{matrix} 2/5 \\ 3/5 \end{matrix}$$
$$\quad\quad 1 \quad 4 \quad\quad\quad\quad\quad\quad\quad 1/5 \quad 4/5$$

In other words, R plays r_1 with probability $2/5$ and r_2 with probability $3/5$. Also, C plays c_1 with probability $1/5$ and c_2 with probability $4/5$. The value V of the game may be computed from either R's strategy or C's strategy. The following scheme computes the value four times; the student may take his choice and use any one of the four computations shown here.

16.2 Mixed Strategies

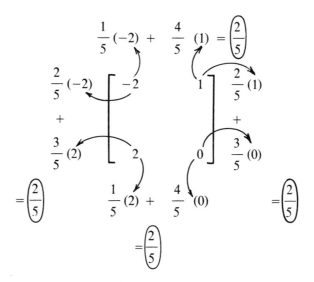

$V = 2/5$.

If there is no saddle point then:

$$\begin{bmatrix} a & b \\ c & d \end{bmatrix} \begin{matrix} d-c \\ a-b \end{matrix}$$

In each row, subtract the smaller element from the larger and write down the answers, interchanging rows as indicated. (Here we assume both a and d are greater than each of c and b.)

$$\begin{bmatrix} a & b \\ c & d \end{bmatrix} \begin{matrix} \dfrac{d-c}{a-b+d-c} \\ \dfrac{a-b}{a-b+d-c} \end{matrix}$$

Divide each of the two numbers just found by their sum. This gives the probabilities with which R should play his rows.

$$\begin{bmatrix} a & b \\ c & d \end{bmatrix}$$
$$ d-b \quad a-c$$

In each column, subtract the smaller element from the larger element and write down the answers, interchanging columns as indicated.

$$\begin{bmatrix} a & b \\ c & d \end{bmatrix}$$
$$\dfrac{d-b}{d-b+a-c} \qquad \dfrac{a-c}{d-b+a-c}$$

Divide each of the two numbers just found by their sum. This gives the probabilities with which C should play his columns.

Figure 16.3

Example 16.11 Find the strategies for each player and the value for the game in each of the following matrix games. (Look out for saddle points.)

a. $\begin{bmatrix} 2 & -2 \\ -4 & 5 \end{bmatrix}$ b. $\begin{bmatrix} 6 & 3 \\ -4 & 5 \end{bmatrix}$ c. $\begin{bmatrix} 6 & 3 \\ 2 & -2 \end{bmatrix}$

Solution
a. R's strategy is to play r_1 with probability $9/13$ and r_2 with probability $4/13$. C's strategy is to play columns c_1 and c_2 with probabilities $7/13$ and $6/13$, respectively. The game value is $2/13$.

b. R's strategy: r_1 probability is $3/4$, r_2 probability is $1/4$. C's strategy: c_1 probability is $1/6$, c_2 probability is $5/6$. Game value is $7/2$.

c. There is a saddle point in the (r_1, c_2) position. The game value is 3.

We can use the short method just discussed to solve some larger games (especially two-by-n and n-by-two games) through a technique called the *subgame strategy*. For example, in a two-by-three game, the column player may choose to ignore one of the columns and play the game as if it were a two-by-two game. He can compute every strategy available to R for every such two-by-two subgame. Of course, if C can determine that there is a column he should ignore, then R can also see that C will not play that column and he can also play the same two-by-two game.

Example 16.12
a. Find the three two-by-two subgames for the two-by-three game:

$$\begin{bmatrix} 5 & -3 & 2 \\ 0 & 4 & -6 \end{bmatrix}$$

Find the strategies and values of each subgame.

b. Which one subgame would C rather play?

c. Does R have anything to say about which two-by-two subgame will be played?

Solution
a. Three subgames are obtained by having C, in turn: 1. ignore c_1, leaving c_2, c_3; 2. ignore c_2, leaving c_1, c_3; or 3. ignore c_3, leaving c_1, c_2.

$$\begin{array}{c} c_2 c_3 \\ \begin{matrix} r_1 \\ r_2 \end{matrix} \begin{bmatrix} -3 & 2 \\ 4 & -6 \end{bmatrix} \end{array} \qquad \begin{array}{c} c_1 c_3 \\ \begin{matrix} r_1 \\ r_2 \end{matrix} \begin{bmatrix} 5 & ② \\ 0 & -6 \end{bmatrix} \end{array} \qquad \begin{array}{c} c_1 c_2 \\ \begin{matrix} r_1 \\ r_2 \end{matrix} \begin{bmatrix} 5 & -3 \\ 0 & 4 \end{bmatrix} \end{array}$$

1. 2. 3.

16.2 Mixed Strategies

In the subgame 1, C has value $-2/3$. The subgame 2 has a saddle point in the (r_1, c_2) position. The game value is the minimax entry 2 in the saddle point position. In the subgame 3, C has the strategy of playing c_1 with probability $7/12$ and playing c_2 with probability $5/12$. The value of subgame 3 is $5/3$.

b. The column player would rather play subgame 1 because it favors C the most (value = $-2/3$).

c. The row player cannot determine which two-by-two subgame is to be played since the act of ignoring a column is C's option, not R's.

When playing a two-by-n game, the column player's best course is to follow the strategy of the two-by-two subgame. That is, C should reduce the game to a set of two-by-two games and play only the subgame that has the best value from C's viewpoint—the one with the smallest value.

In Example 16.12, the column player's best choice in playing the given two-by-three game is to ignore the first column and play the mixed strategy on the remaining subgame. A similar strategy exists in playing an n-by-two game, when the row player is the one who has the option of ignoring rows and creating two-by-two subgames.

Example 16.13 a. Find the six two-by-two subgames for the four-by-two game **G**.

$$\mathbf{G} = \begin{array}{c} \\ r_1 \\ r_2 \\ r_3 \\ r_4 \end{array} \begin{array}{c} c_1 \quad c_2 \end{array} \left[\begin{array}{cc} -8 & 2 \\ 0 & 5 \\ -1 & 10 \\ 9 & -6 \end{array} \right]$$

b. Find R's strategy for each subgame and the value of each subgame.

c. What should R's strategy be in the original game **G**?

Solution a. Each of the six subgames comes from selecting two (out of four) rows to ignore. Dropping two rows at a time, the six subgames are

$$\mathbf{H}_1 = \begin{bmatrix} -8 & 2 \\ \boxed{0} & 5 \end{bmatrix}, \quad \mathbf{H}_2 = \begin{bmatrix} -8 & 2 \\ \boxed{-1} & 10 \end{bmatrix}, \quad \mathbf{H}_3 = \begin{bmatrix} -8 & 2 \\ 9 & -6 \end{bmatrix} \begin{array}{c} 15/25 \\ 10/25 \end{array},$$

$$\mathbf{H}_4 = \begin{bmatrix} \boxed{0} & 5 \\ -1 & 10 \end{bmatrix}, \quad \mathbf{H}_5 = \begin{bmatrix} 0 & 5 \\ 9 & -6 \end{bmatrix} \begin{array}{c} 15/20 \\ 5/20 \end{array}, \quad \mathbf{H}_6 = \begin{bmatrix} -1 & 10 \\ 9 & -6 \end{bmatrix} \begin{array}{c} 15/26 \\ 11/26 \end{array}$$

b. Of these subgames, three have saddle points as indicated by the circled values. The strategies in those cases is to play the saddle point row only. In each of these games, the entry at the saddle point is the value of the game. In games H_3, H_5, and H_6, the mixed strategy found by the short method is indicated next to each matrix. H_3 has value $-6/5$, H_5 has value $9/4$, and H_6 has value $42/13$.

c. The best subgame for R is H_6, which has the largest value. Therefore, R's strategy for the original game is to force the game into the H_6 subgame by never playing rows r_1 or r_2, and to play r_3 with probability $15/26$ and r_4 with probability $11/26$. Then regardless of what C does, R is insured of the payoff $42/13$ in the long run.

$$\begin{matrix} 0 \\ 0 \\ 15/26 \\ 11/26 \end{matrix} \begin{bmatrix} -8 & 2 \\ 0 & 5 \\ -1 & 10 \\ 9 & -6 \end{bmatrix}$$

For example, if C plays c_1, then R's expected payoff is

$$(0)(-8) + (0)(0) + \left(\frac{15}{26}\right)(-1) + \left(\frac{11}{26}\right)(9) = \frac{84}{26} = \frac{42}{13}.$$

EXERCISE 16.2

1. Write the mixed strategy solutions to the two-by-two games:

    a. $\begin{bmatrix} 3 & 2 \\ 0 & 7 \end{bmatrix}$ b. $\begin{bmatrix} 1/2 & -1/7 \\ -2/3 & 1/3 \end{bmatrix}$ c. $\begin{bmatrix} -0.01 & 0.99 \\ 0.99 & -0.01 \end{bmatrix}$

2. Write the mixed strategy solutions to the two-by-two games:

    a. $\begin{bmatrix} -6 & 7 \\ 3 & -4 \end{bmatrix}$ b. $\begin{bmatrix} 3 & 1/3 \\ 1/2 & 2 \end{bmatrix}$ c. $\begin{bmatrix} -1/2 & -1/5 \\ -1/10 & -1/4 \end{bmatrix}$

3. Find the value of each of the games in Problem 1.
4. Find the value of each of the games in Problem 2.

16.2 Mixed Strategies

5. How many two-by-two subgames does a two-by-n game have, given n is a positive integer greater than two.

6. Write out the six subgames of the two-by-four game
$$\begin{bmatrix} 3 & -2 & 10 & -1 \\ -1 & 2 & 1 & 5 \end{bmatrix}.$$
Find the value for each one. Which one should C play?

7. Given the three-by-two game,
$$\begin{bmatrix} 6 & -1 \\ -2 & 4 \\ 1/2 & -1 \end{bmatrix}$$
write out the three subgames; find the value of each and indicate which one R should play.

▲ 8. A company is considering expanding its warehouses. It designs a two-by-three decision game and a three-by-two decision game as follows:

   a. Row player: decision to expand or leave the same; column player: the future state of business (projected). In this game the future is in control of the subgames.

Future State of Business

		Increased Needs of Storage Capacity	Constant Storage Capacity Needs	Decreased Needs in Storage Capacity
Company	Expansion	1	$-1/2$	-1
	Leave Same	$-1\ 1/4$	0	$-1/4$

   Find the best strategy for the company.

   b. Use the same row and column players as in part **a**, but this time the company is in control of which subgame to play.

Future State of Business

	Increased Need	No Increased Need
Expansion	2	−1
Leave Same	−1	2
Reduce	−2	−1/2

Company

Find the best strategy for the company.

9. A company owns certain plots of land and plans to build stores on the land according to the following games:

a.

Future Business Prospects for the Present Site

	Good	Medium	Bad
Build	3	2	1
Don't Build	−1	0	2

Company

b.

Future Prospects for the Sites

	Good	Bad
Build on Site Now	5	−2
Wait and See	1/3	1/2
Sell Site Now	−5	2

Company

Find the best strategy for the company in each case. ▲

16.3
COMBINED STRATEGIES

In a given matrix game, the column player may decide to *never* play a particular column because at least one other column is better for him in every row (where "better" means *at least as good*). When this happens, the column in question is said to be a *recessive column*, or more commonly, a column that is *dominated* by at least one other column.

Since the column player will never play the dominated column, it might as well not be there. In effect, it is deleted from the game, thereby reducing the game to one of smaller dimensions. This process is called *reduction of the game by dominance*.

Definition **(Dominated column)**
In an m-by-n matrix game (a_{ij}), the kth column,

$$\begin{bmatrix} a_{1k} \\ a_{2k} \\ \vdots \\ a_{mk} \end{bmatrix} \text{ is } \textit{dominated} \text{ if there is some other column } \begin{bmatrix} a_{1j} \\ a_{2j} \\ \vdots \\ a_{mj} \end{bmatrix} \quad (j \neq k),$$

such that $a_{ik} \geq a_{ij}$ for every i, $i = 1, 2, \ldots, m$. In other words, a column is dominated if and only if it is row-by-row greater than or equal to another column. (A larger number is to the column player's disadvantage and is, therefore, one he would *not* prefer.)

Example 16.14 In the matrix game

$$\begin{array}{c} & c_1 & c_2 & c_3 & c_4 & c_5 \\ r_1 & \begin{bmatrix} -3 & -5 & 0 & 5 & -6 \\ -4 & 2 & -3 & -2 & 2 \end{bmatrix}, \\ r_2 & \end{array}$$

the column c_3 is dominated by column c_1. *Regardless* of what row R plays, C would prefer the payoffs in c_1 to those in c_3. (The elements in c_3 are row-by-row greater than those in c_1.) Similarly, c_4 is dominated by c_1, and c_2 is dominated, not by c_1, but by c_5. C prefers c_5's payoffs to those in c_2. We eliminate these three dominated columns and are left with

$$\begin{array}{c} & c_1 & c_5 \\ r_1 & \begin{bmatrix} -3 & -6 \\ -4 & 2 \end{bmatrix}, \\ r_2 & \end{array}$$

which can be solved by the mixed strategy (short method).

Analogously, there are games in which certain rows have such unfavorable payoffs that they would never be played by the row player. These are called *dominated rows*.

Definition (**Dominated row**)
In an m-by-n matrix game (a_{ij}), the kth row $[a_{k1}\ a_{k2}\ \ldots\ a_{kn}]$ is dominated if there is some other row $[a_{j1}\ a_{j2}\ \ldots\ a_{jn}](j \neq k)$, such that $a_{ki} \leq a_{ji}$ for every i, $i = 1, 2, \ldots, n$. In other words, a row is dominated if and only if it is column-by-column less than or equal to another row. Dominated rows can be eliminated, thereby reducing the game by dominance.

Example 16.15 In the matrix game

$$\begin{array}{c} & c_1 & c_2 & c_3 & c_4 \\ r_1 & \begin{bmatrix} -3 & 0 & 1 & 2 \\ 3 & 7 & -4 & 0 \\ 1 & 4 & -6 & -1 \\ -2 & 2 & 1 & 3 \end{bmatrix} = \mathbf{G}, \\ r_2 & \\ r_3 & \\ r_4 & \end{array}$$

the row r_3 is dominated by r_2. (The entries in r_3 are column-by-column less than or equal to those in r_2.) Eliminate r_3. Similarly, r_1 is dominated by r_4. R prefers the r_4 payoffs to those in r_1; so eliminate the dominated row r_1. With the deletion of these two dominated rows (r_1 and r_3), we are left with

$$\begin{array}{c} & c_1 & c_2 & c_3 & c_4 \\ r_2 & \begin{bmatrix} 3 & 7 & -4 & 0 \\ -2 & 2 & 1 & 3 \end{bmatrix} = \mathbf{G}'. \\ r_4 & \end{array}$$

Now, we can eliminate the dominated columns in $\mathbf{G}'$ to reduce the game further. In $\mathbf{G}'$, both c_2 and c_4 are dominated by c_3 (the column player would prefer the payoffs in c_3 to those in either c_2 or c_4). Deleting these columns leaves

16.3 Combined Strategies

$$\begin{array}{c} & c_1 & c_3 \\ r_2 \\ r_4 \end{array} \begin{bmatrix} 3 & -4 \\ -2 & 1 \end{bmatrix} = \mathbf{G}''.$$

Thus, the large four-by-four game **G** was reduced by dominance to the two-by-two game **G″**, which we can easily solve.

Example 16.16 Solve the following games. (Use a combination of reduction by dominance, subgames, and mixed and saddle point strategies.)

a. $\mathbf{G}_1 = \begin{bmatrix} 3 & 7 & 0 \\ -2 & 4 & -2 \\ 8 & 14 & -3 \\ 0 & 2 & 1 \end{bmatrix}$ b. $\mathbf{G}_2 = \begin{bmatrix} 3 & 5 & 6 & 0 & 4 & 2 \\ 1 & 0 & 3 & 4 & -1 & -2 \end{bmatrix}$

Solution a. In game $\mathbf{G}_1$, the second column is dominated by the third column; so eliminate it. This yields the game

$$\mathbf{G}_1' = \begin{bmatrix} 3 & 0 \\ -2 & -2 \\ 8 & -3 \\ 0 & 1 \end{bmatrix}$$

In $\mathbf{G}_1'$, the second row is dominated; so eliminate it, and get

$$\mathbf{G}_1'' = \begin{bmatrix} 3 & 0 \\ 8 & -3 \\ 0 & 1 \end{bmatrix}$$

Now, this game ($\mathbf{G}_1''$) can be solved by the subgame strategy. There are three subgames, which are listed below along with their strategies.

$$\mathbf{H}_1 = \begin{bmatrix} 3 & \boxed{0} \\ 8 & -3 \end{bmatrix}, \qquad \mathbf{H}_2 = \begin{bmatrix} 3 & 0 \\ 0 & 1 \end{bmatrix} \begin{matrix} 1/4 \\ 3/4 \end{matrix}, \qquad \mathbf{H}_3 = \begin{bmatrix} 8 & -3 \\ 0 & 1 \end{bmatrix} \begin{matrix} 1/12 \\ 11/12 \end{matrix}$$

$$\begin{matrix} & 1/4 & 3/4 \end{matrix} \qquad\qquad\qquad \begin{matrix} 4/12 & 8/12 \end{matrix}$$

$$V = 0 \qquad\qquad V = 3/4 \qquad\qquad V = 2/3$$

Therefore, R chooses the subgame $\mathbf{H}_2$. The safe strategy for both sides in the original game is R plays r_1 with probability $1/4$, plays r_2 and r_3 with zero

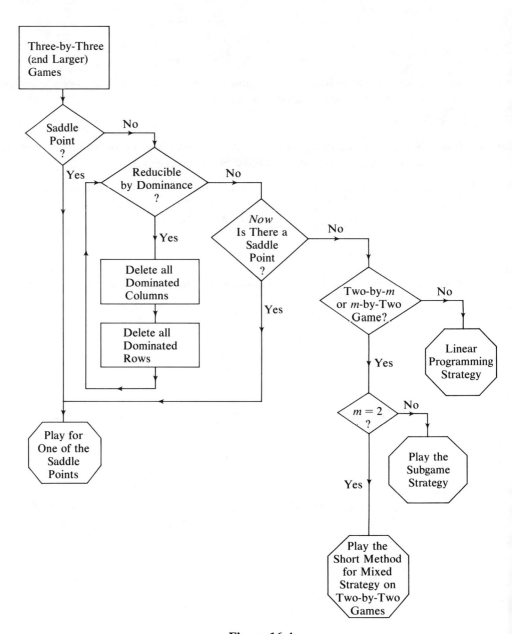

Figure 16.4

16.3 Combined Strategies

probability, and plays r_4 with probability $3/4$. C plays c_1 with probability $1/4$, plays c_2 with probability 0, and plays c_3 with probability $3/4$.

b. In game G_2, the first, second, third, and fifth columns are each dominated by the sixth column. Delete the dominated columns, getting the two-by-two game G_2' shown below along with its strategies:

$$G_2' = \begin{bmatrix} 0 & 2 \\ 4 & -2 \end{bmatrix} \begin{matrix} 3/4 \\ 1/4 \end{matrix}$$

with column probabilities $1/2 \ \ 1/2$.

Up to this point we have discussed four courses of action in solving two-person, zero-sum games. They are:

1. Saddle point for m-by-n games that are strictly determined
2. Mixed strategies for two-by-two nonstrictly determined games
3. Subgame strategies for two-by-n and m-by-two games
4. Reduction by dominance

We have alluded to linear programming, which we will discuss in the next section. The diagram in Figure 16.4 illustrates an interrelationship between the four methods mentioned above and linear programming. The chart shows that if the saddle point and dominance reduction do not produce one of the easily solved games, then linear programming may be employed as a last resort.

EXERCISE 16.3

1. Solve by a combination of strategies:

a. $\begin{bmatrix} -7 & -5 & -4 \\ -2 & 5 & -2 \\ 2 & -4 & 0 \\ -4 & 3 & -5 \end{bmatrix}$ b. $\begin{bmatrix} -4 & 4 & -3 \\ 0 & 2 & 6 \\ 3 & 3 & 1 \end{bmatrix}$

2. Solve by a combination of strategies:

a. $\begin{bmatrix} 3 & 4 & -2 & 0 & -1 \\ 5 & 2 & 1 & -5 & 2 \\ 0 & -1 & -1 & -4 & -3 \\ 10 & 6 & 0 & -2 & 1 \end{bmatrix}$

b. $\begin{bmatrix} 3 & 5 & -6 & 8 & 1 \\ 7 & -2 & 4 & -5 & -6 \end{bmatrix}$

▲ In a buyer's decision problem, a typical question for a buyer is: "Will it be better to buy now or wait for a change in price?" In Problems 3 and 4, find the buyer's best strategy. Unfortunately, the column player controls the subgames.

3. $a > 0$

Future Prices

	Big Price Increase	Small Price Increase	No Increase	Price Decrease
Buy New Refrigerator Now	$2 \cdot a$	a	0	$-a$
Wait to Buy Refrigerator	$-2 \cdot a$	$-a$	0	a

4.

Future Interest Rates

Consumer		Up 3%	Up 1%	Down 2%
	Borrow Money Now	3	1	−2
	Wait to Borrow Money	−3	−1	2

5. Elliot Ness and his federal officers are trying to stop bootleg liquor from coming into Chicago. Ness is the row player and he can choose between inspecting

16.3 Combined Strategies

incoming 1. trucks, 2. airplanes, 3. trains, and 4. ships (on Lake Michigan). The column player is the bootlegger, who can choose from the following methods of transporting the booze into the city: 1. trucks, 2. combinations of airplanes (remote air field) and trucks, 3. trains, and 4. combinations of ships (remote boat dock) and trucks. Suppose the payoff matrix is as shown below. Reduce by dominance and solve.

Bootleggers

		Trucks	Planes and Trucks	Rail	Ships and Trucks
Ness Blockade	Trucks	5	4	−4	3
	Planes	−5	1	−4	−3
	Trains	3	−2	2	−4
	Ships	−6	−5	−5	2

6. Two mutual funds have plans regarding shares of a stock they both hold. The following payoff matrix is given. Reduce by dominance and solve.

Mutual Fund C

		Buy Large Amount	Buy Small Amount	Sell Large Amount	Sell Small Amount
Mutual Fund R	Buy Large Amount	½	−1	3	−10
	Buy Small Amount	2	0	5	½
	Sell Large Amount	6	−4	−½	−1
	Sell Small Amount	7	1	1	0

▲

16.4
LINEAR PROGRAMMING APPLIED TO ZERO-SUM GAMES

Linear programming strategy is really the mixed strategy for games larger than two-by-two. In the linear programming setting, the rows (or colums) produce expected values related to the game value by an inequality. These are the constraint inequalities. The solution for each player is to obtain the probability vector that will make the game value V optimal. The way to carry this out is for C to find the vector that maximizes $1/v$, thereby obtaining the minimum V, and for R to find the vector that minimizes $1/v$, thereby maximizing V. When each of these solution vectors is multiplied by V, they produce the required probability vectors.

We will illustrate the method by an example. Later, a general algorithm will be presented in the form of a flow chart.

Example 16.17 The three-by-three game

$$\mathbf{G} = \begin{bmatrix} 1 & 2 & -1 \\ -2 & 1 & 1 \\ 2 & 0 & 1 \end{bmatrix}$$

has no saddle point, is not reducible by dominance, and is not two-by-n or m-by-two. This leaves linear programming as the method of solution.

STEPS FOR THE COLUMN PLAYER'S LINEAR PROGRAMMING STRATEGY

1. Last resort step. Establish that the other strategies do not solve the game.
2. Translation step. Convert all the entries in the matrix to positive ones by adding a sufficiently large constant to them all. The strategy will not be affected, and the original game can be recovered after all solutions are found by subtracting the added constant.

    Here we add 3 to each entry. (In general, add one more than the absolute value of the *most* negative entry; if there are no negative entries, there will be no need to translate the game.) In this case, after adding three, we get the new game

16.4 Linear Programming Applied to Zero-Sum Games

$$\mathbf{G'} = \begin{array}{c} \\ r_1 \\ r_2 \\ r_3 \end{array} \begin{array}{c} c_1 \ c_2 \ c_3 \\ \left[\begin{array}{ccc} 4 & 5 & 2 \\ 1 & 4 & 4 \\ 5 & 3 & 4 \end{array} \right] \end{array}$$

Since all the entries in $\mathbf{G'}$ are positive, every possible strategy will yield a positive payoff to R; that is, the value V of $\mathbf{G'}$ is positive. Note that if U is the value of the original game $\mathbf{G}$, then $V = U + 3$; therefore, $U = V - 3$. Once we find V, then we can find U.

3. **Probability assignment step.** Regardless of the row played by R, let p_1, p_2, and p_3 be the probabilities with which C will play columns c_1, c_2, and c_3, respectively. Also, note that $p_1 + p_2 + p_3 = 1$.

4. **Expected value step.** The column player's expected value E will be less than or equal to V if he plays the best strategy. In other words, C expects a number *no greater than V, and the smaller the better*.

The three expected value inequalities for all three plays by R are

Row player plays	Expected value for column player
r_1	$4 \cdot p_1 + 5 \cdot p_2 + 2 \cdot p_3 \leq V$
r_2	$1 \cdot p_1 + 4 \cdot p_2 + 4 \cdot p_3 \leq V$
r_3	$5 \cdot p_1 + 3 \cdot p_2 + 4 \cdot p_3 \leq V$

Note that these inequalities are going to evolve into the constraint inequalities, and the equation $p_1 + p_2 + p_3 = 1$ is going to become the objective function.

5. **Conversion step.** Divide all inequalities in Step 4 by V ($V > 0$).

$$4 \frac{p_1}{V} + 5 \frac{p_2}{V} + 2 \frac{p_3}{V} \leq 1$$

$$1 \frac{p_1}{V} + 4 \frac{p_2}{V} + 4 \frac{p_3}{V} \leq 1$$

$$5 \frac{p_1}{V} + 3 \frac{p_2}{V} + 4 \frac{p_3}{V} \leq 1$$

Also, let

$$F = \frac{1}{V} = \frac{p_1}{V} + \frac{p_2}{V} + \frac{p_3}{V}.$$

The objective for C is to find the probabilities p_1, p_2, and p_3 that will provide the minimum value of V. Now, since $F = 1/v$, C seeks values of p_1/v, p_2/v, and p_3/v that will maximize F (the *maximum* of F is the maximum of $1/v$, which occurs at the minimum of V).

Let $y_1 = p_1/v$, $y_2 = p_2/v$, and $y_3 = p_3/v$. Then the linear programming problem is to maximize

$$F = y_1 + y_2 + y_3$$

subject to the constraints

$$4y_1 + 5y_2 + 2y_3 \leq 1$$

$$1y_1 + 4y_2 + 4y_3 \leq 1$$

$$5y_1 + 3y_2 + 4y_3 \leq 1.$$

6. Simplex step. Use the maximizing simplex algorithm from Chapter 9. Let y_4, y_5, and y_6 be the slack variables and start with these variables as the initial basis.*

	C_i	1	1	1	0	0	0		

		y_1	y_2	y_3	y_4	y_5	y_6	y_0	C_j
	y_4	4	5	2	1	0	0	1	0
	y_5	1	4	4	0	1	0	1	0
Pivot row ←	y_6	⑤	3	4	0	0	1	1	0

	z_i	0	0	0	0	0	0	0	
	$C_i - z_i$	1	1	1	0	0	0		

↑
Pivot column

---

*The real importance of linear programming appears in m-by-n games where $m \neq n$. The square matrix game is used here to demonstrate the procedure, but square games can be solved by ordinary simultaneous equation methods (using equalities in Step 5 instead of inequalities). The value of the game would be $1/F$, where F is the *sum of the coordinates in such a solution.*

16.4 Linear Programming Applied to Zero-Sum Games

C_i	1	1	1	0	0	0

		y_1	y_2	y_3	y_4	y_5	y_6	y_0	C_j
Pivot row ←	y_1	1	$3/5$	$4/5$	0	0	$1/5$	$1/5$	1
	y_4	0	ⓘ$13/5$	$-6/5$	1	0	$-4/5$	$1/5$	0
	y_5	0	$17/5$	$16/5$	0	1	$-1/5$	$4/5$	0

	y_1	y_2	y_3	y_4	y_5	y_6	y_0
z_i	1	$3/5$	$4/5$	0	0	$1/5$	$1/5$
$C_i - z_i$	0	$2/5$	$1/5$	0	0	$-1/5$	

↑
Pivot column

	y_1	y_2	y_3	y_4	y_5	y_6	y_0	C_j
y_1	1	0	$14/13$	$-3/13$	0	$5/13$	$2/13$	1
y_2	0	1	$-6/13$	$5/13$	0	$-4/13$	$1/13$	1
y_5	0	0	$62/13$	$-17/13$	1	$11/13$	$7/13$	0

	y_1	y_2	y_3	y_4	y_5	y_6	y_0
z_i	1	1	$8/13$	$2/13$	0	$1/13$	$3/13$
$C_i - z_i$	0	0	$5/13$	$-2/13$	0	$-1/13$	

The final simplex table is

	y_1	y_2	y_3	y_4	y_5	y_6	y_0	C_j
y_1	1	0	0	$2/26$	$-7/31$	$6/13$	$1/31$	1
y_2	0	1	0	$8/31$	$3/31$	$-7/31$	$4/31$	1
y_3	0	0	1	$-17/62$	$13/62$	$11/62$	$7/62$	1

	y_1	y_2	y_3	y_4	y_5	y_6	y_0
z_i	1	1	1	$49/806$	$5/62$	$9/62$	$17/62$
$C_i - z_i$	0	0	0	$-49/806$	$-5/62$	$-9/62$	

Therefore, the maximum value of $F = 1/v = 17/62$. Hence, $V = 62/17$, which is the value of the *translated* game, **G'**. The solution to the linear programming problem is $y_1 = 1/31$, $y_2 = 4/31$, and $y_3 = 7/62$. Now, from the definitions of the y_i's, we get $p_1 = y_1 \cdot V$, $p_2 = y_2 \cdot V$, and $p_3 = y_3 \cdot V$. Substituting the values found above, we get

$$p_1 = \frac{1}{31} \cdot \frac{62}{17} = \frac{2}{17}, \; p_2 = \frac{4}{31} \cdot \frac{62}{17} = \frac{8}{17}, \; p_3 = \frac{7}{62} \cdot \frac{62}{17} = \frac{7}{17}.$$

In other words, the strategy for the original game **G** is

C plays column c_1 with probability $2/17$,
C plays column c_2 with probability $8/17$,
C plays column c_3 with probability $7/17$.

The value of the original game is *3 less than the value of the game* **G'**. Hence, $62/17 - 3 = 11/17 =$ value of **G**.

The linear programming method may also be applied to the problem of finding R's strategy.

STEPS FOR THE ROW PLAYER'S LINEAR PROGRAMMING STRATEGY

1. Last resort step. (Examine the game for saddle points as in the column strategy.)

2. Translation step. (As before, obtaining $V > 0$.)

3. Probability step. Regardless of the column played by C, let q_1, q_2, and q_3 be the probabilities with which R will play rows r_1, r_2, and r_3, respectively.

4. Expected value step. The row player's expected value E will be greater than or equal to V if he plays the best strategy. In other words, R expects a number no less than V. The three expected value inequalities (for all three plays by C) are

Column player plays	Expected value for row player
c_1	$4q_1 + 1q_2 + 5q_3 \geq V$
c_2	$5q_1 + 4q_2 + 3q_3 \geq V$
c_3	$2q_1 + 4q_2 + 4q_3 \geq V$

5. Conversion step. Divide by V ($V > 0$) and replace q_1/v by x_1, q_2/v by x_2, and q_3/v by x_3:

16.4 Linear Programming Applied to Zero-Sum Games

$$4x_1 + 1x_2 + 5x_3 \geq 1$$
$$5x_1 + 4x_2 + 3x_3 \geq 1$$
$$2x_1 + 4x_2 + 4x_3 \geq 1$$

Let

$$F = \frac{1}{V} = x_1 + x_2 + x_3.$$

The objective for R is to find the maximum of V by finding the minimum of $1/v$ (or F) subject to the constraints given above. [Actually, what R really wants is the vector (x_1, x_2, x_3) that furnishes the minimum F; the strategy for R can be computed from this vector.]

6. Simplex step. Introduction of slack variables, x_4, x_5, and x_6, and the use of the simplex algorithm for minimizing the objective yields the solution

$$x_1 = 3/62, \; x_2 = 5/62, \text{ and } x_3 = 9/62.$$

Since $V = 62/17$ for the translated game, then the row strategy is

$$q_1 = x_1 V = 3/17; \; q_2 = x_2 V = 5/17; \; q_3 = x_3 V = 9/17.$$

The value of the original game is $11/17$, as before.

We indicated that linear programming is the method of last resort; the primary reason for this is that the other methods are so much easier.

Strictly speaking, we are not required to exhaust all other methods before using linear programming. On the contrary, linear programming is used to prove that the mixed strategy works in two-by-two games. It is also used to prove that the subgame strategy works in two-by-n games.

The one exception is *never try to use linear programming (or any mixed strategy) for a game that has a saddle point*. Mixed strategies will not uncover the safest strategy in games with saddle points.

Example 16.18 The game

$$\begin{bmatrix} 3 & \boxed{0} \\ 2 & -1 \end{bmatrix}$$

has a minimax solution in the saddle point at (r_1, c_2). The game value is 0, but any attempt to use a mixed strategy will produce four different expected values, none of which is the minimax solution. This is the reason that Theorem 16.1 has in its hypothesis the assumption that the game is not strictly determined.

Finally, we apply linear programming to a special case of the two-by-two game of Theorem 16.1. Let **G** be the two-by-two game

$$\mathbf{G} = \begin{bmatrix} a & b \\ c & d \end{bmatrix}$$

with a, b, c, and d positive, each of a and $d > b$, and each of a and $d > c$. Suppose all entries are positive. Let V be the value of the game; $V > 0$. For any play by R, let p_1 and p_2 be the probabilities for C to play the columns c_1 and c_2.

$$r_1: \quad p_1 a + p_2 b \leq V$$
$$r_2: \quad p_1 c + p_2 d \leq V$$
$$p_1 + p_2 = 1$$

Let $y_1 = p_1/v$, $y_2 = p_2/v$, and let $F = 1/v$. Maximize

$$F = y_1 + y_2 + 0y_3 + 0y_4$$

subject to the constraints

$$ay_1 + by_2 + y_3 + 0y_4 = 1$$
$$cy_1 + dy_2 + 0y_3 + y_4 = 1.$$

The initial simplex table with the slack variables, y_3 and y_4, as the basis is

	C_i	1	1	0	0		

		y_1	y_2	y_3	y_4	y_0	C_j
Pivot row ←	y_3	ⓐ	b	1	0	1	0
	y_4	c	d	0	1	1	0

	z_i	0	0	0	0	0
	$C_i - z_i$	1	1	0	0	

↑
Pivot column

16.4 Linear Programming Applied to Zero-Sum Games

After pivoting in y_1, pivoting out y_3, and solving, we get the table

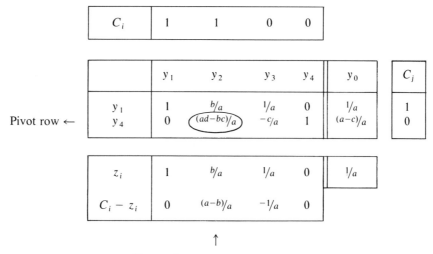

$$V = \frac{ad - bc}{d - b + a - c}, \text{ and } p_1 = y_1 V = \frac{d - b}{ad - bc} \cdot \frac{ad - bc}{d - b + a - c} = \frac{d - b}{d - b + a - c}.$$

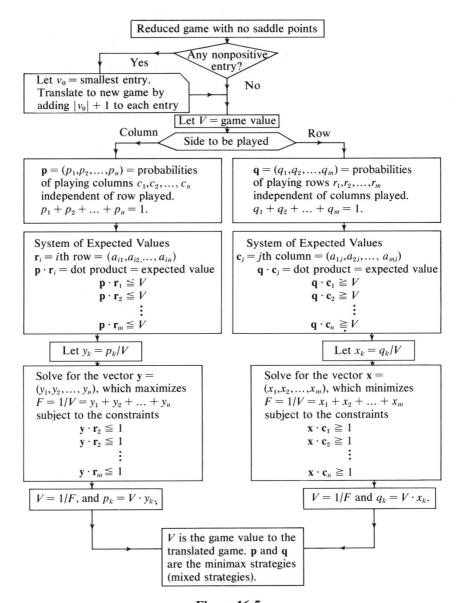

Figure 16.5

16.4 Linear Programming Applied to Zero-Sum Games

Similarly,

$$p_2 = y_2 V = \frac{a-c}{d-b+a-c},$$

and these are the column strategies in the conclusion of Theorem 16.1.

The flow chart in Figure 16.5 gives the general linear programming solution.

EXERCISE 16.4

1. Given the two-by-three game

$$\begin{bmatrix} 2 & 6 & 0 \\ 3 & -1 & 4 \end{bmatrix},$$

   a. Solve by linear programming strategy.

   b. Solve by subgame strategy and compare with part **a**.

2. Given the three-by-two game

$$\begin{bmatrix} 5 & -6 \\ -2 & 4 \\ 3 & -3 \end{bmatrix},$$

   a. Solve by linear programming strategy.

   b. Solve by subgame strategy and compare with part **a**.

3. Solve the following three-by-three game with linear programming. Check your solution by computing the game value for all three rows and all three columns.

$$\begin{bmatrix} 2 & 1 & 3 \\ 1 & 2 & 1 \\ 3 & 1 & 2 \end{bmatrix}$$

ANSWERS TO ODD-NUMBERED PROBLEMS

EXERCISE 1.1 (page 9)

1. a. 18 **b.** 10 **c.** 65 **d.** 15 **e.** 28

3. a. 40 **b.** 10 **c.** 11 **d.** 57 **e.** 27

5. a. 147 **b.** 341

7. a. 39 **b.** 3 **c.** 18

9. a. 122 **b.** 782 **c.** 915 **d.** 793 **e.** 29.7 percent (approximately)

11. a. Eight types: AB+, AB−, A+, A−, B+, B−, O+, and O−.

  b. 12 are AB−, 340 are O+, 60 are O−.

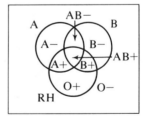

EXERCISE 1.2 (page 22)

1. a. No **b.** Yes **c.** Yes **d.** Yes **e.** Yes **f.** No

513

3.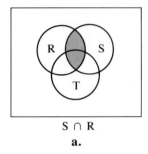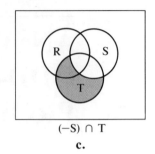

  S ∩ R   T ∪ (S ∩ R)   (−S) ∩ T
    a.   b.   c.

5. **a.** A set of words associated with either colors or moods; **b.** A set of words associated with colors but not with moods; **c.** A set of words associated with moods but not with colors.

7. (Solutions to parts **a, c, e,** and **g** only.) **a.** $-A$ is the set of households with more than two members. No, $x \notin A$ if x has more than 7 members. **c.** No, $y \notin B$. **e.** $A - D$ is the set of households with exactly one or two members but not farm households with less than four members. (*Other verbal descriptions are possible.*) The average income of the households in $A - D$ is $1875.

EXERCISE 1.3 (page 30)

1. **a.** 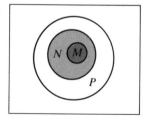   **b.** Yes, $M \subseteq P$.

3. **a.** False   **b.** True   **c.** False   **d.** False   **e.** True   **f.** False

5. If L is a language, then L requires abstract symbols. Let A be the set of languages and B the set of things requiring abstract symbols, then the diagram, with $A \subseteq B$, is: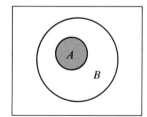

7. a. 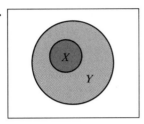 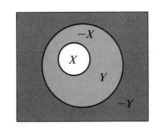 b. Yes, $-Y \subseteq -X$.

c. No

9. a. 8   b. $\{a\}$ $\{b\}$ $\{c\}$ $\{a,b\}$ $\{a,c\}$ $\{b,c\}$

11. 2^n or $2 \cdot 2 \cdot \ldots \cdot 2$ n times.

13. a. Y = the gout set, X = the hyperuricemia set. b. No. c. People with hyperuricemia who do not have gout. d. 54.8 percent (approximately)

EXERCISE 2.1 (page 42)

1. a. $268   b. $394.50

3. a. $^{11}/_{13}$ b. $^{4}/_{9}$ c. $^{51}/_{331}$ d. neither ($^{91}/_{210} = ^{39}/_{90}$)

5. a. True (Axiom 1)   b. True (Axiom 1)   c. True (Axiom 4)

   d. False; for example, let $x = 0$ and $y = \frac{1}{2}$.

7. $x > 2$.

9. $x > 290$; there are 145 solutions.

11. a and e.

13. If $x < x$, then by Axiom 1, $x \neq x$, contradicting Property 1 of equality.

15. $Y_P \leqq 25,300$.

17. a. $2 < 42$.

b. Suppose you can subtract inequalities; then from $10 < 50$ and $3 < 45$, you could get $10 - 3 < 50 - 45$ or $7 < 5$, which is false.

19. By Axiom 2, if $a < b$ and c is any number, then the number $(-c)$ can be added to both sides: $a + (-c) < b + (-c)$. Thus, $a - c < b - c$.

21. Given $a \neq b$, then by Axiom 1 either $a - b = 0$ or $a - b \neq 0$. Suppose $a - b = 0$, then adding b to both sides yields $a = b$, contrary to the given.

23. From Problems 21 and 22, $(a - b)^2 > 0$. Thus, $a^2 - 2ab + b^2 > 0$ and $a^2 + b^2 > 2ab$.

EXERCISE 2.2 (page 52)

1. $$\begin{aligned} x_1 + 3x_2 - x_3 + x_4 + x_5 &= 7. \\ x_1 + 2x_3   + x_6 &= 10. \\ 5x_2 + x_3 + 2x_4 - x_7 &= 3. \end{aligned}$$

3. a. $$\begin{aligned} x_1 + x_2 + x_3 + x_4 &= 100. \\ 300x_1 + 365x_2 + 400x_3 + x_5 &= 33{,}000. \end{aligned}$$

   b. x_4 is the number of unfilled positions in the entire company. x_5 is the unspent dollars available for the payroll.

5. $x_4 = 9$, $x_5 = 755$.

   Hiring options:
   Hire one person in department A and one in department B.
   Hire one person in department A and one in department C.
   Hire two people in department B.
   Other options are possible.

7. Constraint inequalities: Slack variable equations:

$$\begin{aligned} x + y &\leq 100. \\ y &\geq 20. \\ 30x + 45y &\leq 3200. \\ x - 2y &\geq 0. \end{aligned} \qquad \begin{aligned} x + y + s_1 &= 100. \\ y - s_2 &= 20. \\ 30x + 45y + s_3 &= 3200. \\ x - 2y - s_4 &= 0. \end{aligned}$$

Chapter 2

9. a.

	Departments				Total Available Resources	
I	II	III	IV			
$w_1 +$	$w_2 +$	$w_3 +$	w_4	$\leq$	100	(personnel).
$500w_1 +$	$600w_2 +$	$400w_3 +$	$500w_4$	$\leq$	50,000	(floor space).
$5000w_1 +$	$7500w_2 +$	$8500w_3 +$	$9000w_4$	$\leq$	610,000	(budget).

b.

$$w_1 + w_2 + w_3 + w_4 + w_5 = 100.$$
$$500w_1 + 600w_2 + 400w_3 + 500w_4 + w_6 = 50{,}000.$$
$$5000w_1 + 7500w_2 + 8500w_3 + 9000w_4 + w_7 = 610{,}000.$$

w_5 = unfilled positions.
w_6 = unassigned floor space.
w_7 = unexpended budget.

c. If $w_1 = 25$, $w_2 = 21$, $w_3 = 18$, and $w_4 = 18$, then $w_5 = 18$, $w_6 = 8700$, and $w_7 = 12{,}500$. If they hire just one person in either department III or IV, they will use up so much of the budget that they could not hire anyone else. They could hire two people in department I, or one in department I and one in II.

EXERCISE 2.3 (page 58)

1. $\sum_{i=1}^{4} a_i x_i \leq R_1$.

$\sum_{i=1}^{4} b_i x_i \leq R_2$.

$\sum_{i=1}^{4} c_i x_i \leq R_3$.

3. a. $49/20$ or 2.45   **b.** $2514/1716$ or 1.465 (approximately)   **c.** $5369/1800$

5. a. 210   **b.** 185

7. a. 10.42 (approximately)   **b.** 14.17 (approximately)   **c.** 635

9. a. $P = \sum_{j=1}^{6} x_j = 4 + 7 + 10 + 1 + 6 + 3 = 31$; $m = 180/31$ or 5.8 (approximately).

b. $P = 31$; $m = {}^{132}\!/_{31}$ or 4.26 (approximately).

11. a. 9 **b.** 12 **c.** 7 **d.** 108

$x_i \cdot y_i$
2
12
−5
−2
0

13. a. 6160 **b.** 135

15. $\Pi_{i=1}^{4} (4 - 3^i)$.

EXERCISE 3.1 (page 68)

1. 1940: 60¢; 1930: 70¢; 1918: 80¢

3. a. $1.13

 b. Five-year change = +18¢, yearly average change = +3.6¢/year.

5. 370 billion dollars

7. a. About 615 billion dollars **b.** About 11 billion dollars/year

9. a. 40 gallons/minute

 b. Change in velocity = +35 (gallons/minute) per minute.

11. a. 5 gallons **b.** 60 gallons **c.** between 20 and 25 gallons

13. Ordinate is one less than twice the abscissa ($y = 2x - 1$).

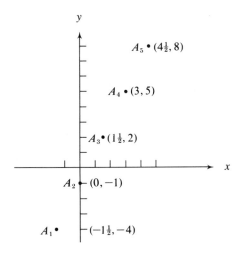

15. All points are 3 units above the x-axis, or all points have ordinate 3, or all points satisfy the equation $y = 3$.

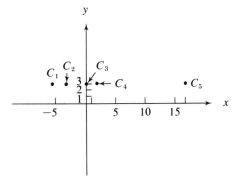

17. a. y-axis   **b.** x-axis

19.

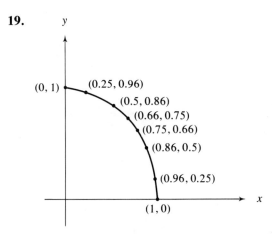

21.

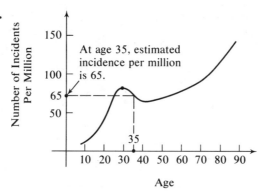

23. a. 0.14   **b.** 0.82   **c.** 0.02

d.

x	z
0	0.00
70	0.02
85	0.16
100	0.50
115	0.84
130	0.98
200	1.00

e.

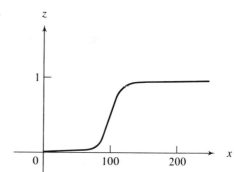

f. If $x = 90$, $z = 0.25$, approximately. Yes, the ordinates of the graph are from the table in **d**.

g. If $x = 105$, $z = 0.55$ approximately.

h. For $x = 120$, $z = 0.95$ approximately.

EXERCISE 3.2 (page 81)

1. b and **d**.

3. a. No   **b.** $(1,1)$ and $(1,-1)$

c. If $x < 0$, then $y^2 < 0$, but no real number squared is negative.

5. a. $x \neq 1$ b. $x \geq -10$ c. $x \geq 0$ and $x \neq 2$ d. $-1 < x < 1$

7. a. 25.92 grams, 15 grams   b. 216.51 grams, 11 grams

c. 9079 grams, 20 pounds

9. a. 0.84 pounds per cubic feet   b. 0.0135 grams per cubic centimeter

11. a. 18   b. $5\frac{1}{2}$

13. a. $L = 240$ b. Then $30/L = 0$, which has no solution in real numbers.

c. As p decreases, Q increases.

EXERCISE 3.3 (page 92)

1. a. 2.23   b. 1.49   c. 4.95   d. 6.07   e. 2.23 (approximately)

3. a. 4   b. 2   c. 6   d. 2   e. $2\frac{1}{4}$ f. $7\frac{1}{4}$

5. a. 0   b. $\frac{2}{3}$ c. $\frac{2}{3}$ d. No

e. $f(x) \to \pm\infty$ as $x \to 2$; that is, f gets indefinitely large (positively or negatively) as x gets close to 2.

f. Domain is the set of all real numbers except 2 and 3.

7. a. $f(0) = 4$, $f(6) = -26$.   b. $\Delta f = -30$.   c. $\Delta f/\Delta x = -30/6 = -5$.

d. $x = \frac{4}{5}$.

9. a. $g(2) = 20$, $g(4) = 80$.   b. $\Delta g = 60$.   c. $\Delta g/\Delta x = 30$.

d. $\Delta g = 5b^2 - 5a^2$, $\Delta x = b - a$.   e. $\Delta g/\Delta x = 5(b + a)$.

11. a. $R(100) = 40{,}000$, $R(150) = 52{,}500$. b. For $x_1 = 100$ and $x_2 = 150$, $\Delta R = 12{,}500$. c. $MR = \Delta R/\Delta x = 250$ for $x_1 = 100$, $x_2 = 150$. d. $R(200)$

= 60,000, $R(250) = 62,500$. **e.** For $x_1 = 200$, $x_2 = 250$, $\Delta R = 2500$. **f.** Between 200 and 250, $MR = 50$. **g.** $R(300) = 60,000$; for $x_1 = 250$, $x_2 = 300$, $\Delta R = -2500$ and $MR = -50$. This means that the revenue is decreasing as x increases (diminishing returns).

13. a. $V(1) = \sqrt{2k/\pi}$ **b.** $V(8) = 3 \cdot \sqrt{k/\pi}$

c. $V(15) = 4 \cdot \sqrt{k/\pi}$, $V(5) = \sqrt{6k/\pi}$, $V(15)/V(5) = 4/\sqrt{6} = 1.63$ approximately.

15. $z(15,80) = 300$
$z(71,12) = 320$
$z(50,50) = 350$ MAXIMUM
$z(40,60) = 340$
$z(35,64) = 332$
$z(22,75) = 313$

17. a. 110   **b.** $-10\,{}^5/_7$

EXERCISE 4.1 (page 104)

1. a.

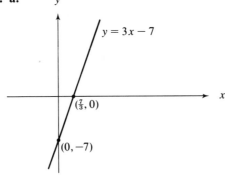

b.

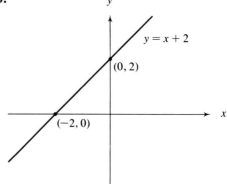

Chapter 4

3. a. 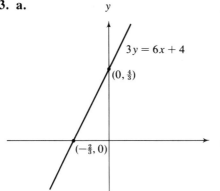   **b.**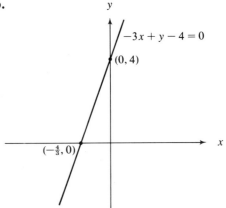

5. a. $c(11{,}000) = 10{,}700$, $c(25{,}000) = 20{,}500$, $c(18{,}000) = 15{,}600$.

 b. $s(11{,}000) = 120$, $s(25{,}000) = 1800$, $s(18{,}000) = 960$.

7. a. $y = -1/7 \cdot x$.   **b.** $(-8000, \, 8000/7)$

9. a. $y = -35/24 \cdot x + 7/2$.   **b.** $(84/59, \, 84/59)$

11. a. $4x - y + 80 = 0$.   **b.** 88 square miles   **c.** 111,360 gallons

13. For x shares of Greyhound and y shares of Xerox: **a.** $y = 450 - 1/4 \cdot x$. **b.** If $x = 1600$, $y = 50$. **c.** All Greyhound, yes!

EXERCISE 4.2 (page 108)

1. $y - 7 = -1/2(x - 3)$.

3. $m = 3/2$.

5. Two points are $(0, 8/7)$ and $(-8/3, 0)$. Alternative points are possible. The slope is $m = 3/7$.

7. Horizontal lines have zero slope.

9. K and L both have slope -3; they are distinct since K contains $(1/2, 0)$, but L does not.

11. a. $11x - 6y - 735 = 0$.   b. 427.5 miles   c. 427,500 miles

EXERCISE 4.3 (page 115)

1. $x = 25/4$, $y = 43/4$.

3. $x = 16$, $y = -9$.

5. a. $0.01x + 0.07y = 300{,}000$.   b. $0.50x + 0.75y = 10{,}000{,}000$.

   c. $x = (190{,}000{,}000)/11$, $y = (20{,}000{,}000)/11$ or $(x,y) = (17{,}200{,}000;\ 1{,}800{,}000)$ approximately.

7. $x = 24$, $y = -27$, $z = -43$.

9. K and $L = (5,-9)$, K and $M = (5,4)$, M and $L = (7/4, 3/4)$.

11. a. 17 men, 7 women   b. 69.2 percent

13. a. $y_B = 12{,}000 + 130 \cdot x$, $y_S = 9000 + 160 \cdot x$, $y_R = 330 \cdot x$.

   b. Revenue on 50 vehicles is $y_R(50) = 16{,}500$. Cost on 50 buses $y_B(50) = 18{,}500$. Cost on 50 streetcars $y_S(50) = 17{,}000$. Buses lose $2000 daily and streetcars lose $500 daily.

   c. Buses break even at $x = 60$, streetcars at $x = 53$ (approximately).

   d. $x = 100$.   e. Streetcars

EXERCISE 5.1 (page 128)

1.

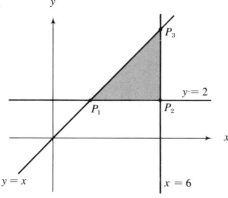

3. For Problem 1, the corner points are $P_1 = (2,2)$; $P_2 = (6,2)$; and $P_3 = (6,6)$.

5.

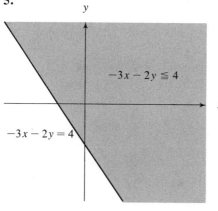
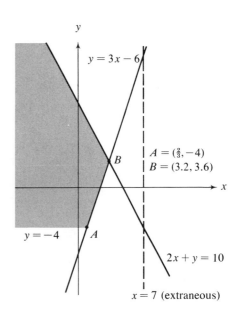

$A = (\frac{2}{3}, -4)$
$B = (3.2, 3.6)$

9. $y \le (1/2)x + 3\,1/4$
$y \ge -(1/4)x + 1$
$y \le -(2/3)x + 3\,1/4$
$y \ge (7/13)x - 3/4$

11. $x \ge 0$
$y \ge 0$
$x + y \le 250$
$-2x + y < 75$

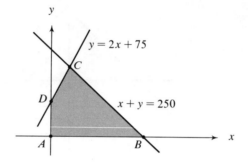

Corner points: $A = (0,0)$; $B = (250,0)$; $C = (175/3, 575/3)$; $D = (0,75)$.

EXERCISE 5.2 (page 135)

1. a.

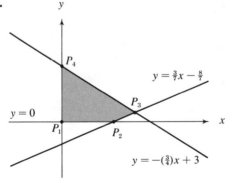

b. Boundary lines: $x = 0$, $y = 0$, $y = -(3/4)x + 3$, and $y = (3/7)x - (8/7)$.

c. Corner points: $P_1 = (0,0)$; $P_2 = (8/3, 0)$; $P_3 = (116/33, 4/11)$; $P_4 = (0,3)$.

d. Maximum $z = 120$ at P_4, minimum $z = 0$ at P_1.

Chapter 5

3. $A = (10,0); z = 65.$
   $B = (20,0); z = 125$ Maximum.
   $C = (10,10); z = 45.$
   $D = (10/3, 20/3); z = 35/3.$

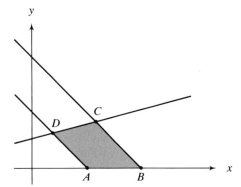

5. **a.** 75 acres of crop A and 100 acres of crop B.

   b. 25 acres of crop A and 150 acres of crop B.

7. Maximum value: $z = 250{,}000$ at corner point $B = (250,0)$.

9. $3\,3/11$ mc of A and $3\,3/11$ mc of B.

EXERCISE 5.3 (page 147)

1. $3x + 4y + s_1 = 10$
   $7x - y + s_2 = 5.$

(I)	$x =$	$y =$	$0,$	$s_1 = 10,$	$s_2 = 5.$	FEASIBLE
(II)	$x =$	$s_1 =$	$0,$	$y = 5/2,$	$s_2 = 15/2.$	FEASIBLE
(III)	$x =$	$s_1 =$	$0,$	$y = -5$		INFEASIBLE
(IV)	$y =$	$s_1 =$	$0,$	$x = 10/3,$	$s_2 = -55/3.$	INFEASIBLE
(V)	$y =$	$s_2 =$	$0,$	$x = 5/7,$	$s_1 = 55/7.$	FEASIBLE
(VI)	$s_1 =$	$s_2 =$	$0,$	$x = 30/31,$	$y = 55/31.$	FEASIBLE

3. $5w + 2x + y + 3z + s_1 = 1.$
   (I) $w = x = y = z = 0,\ s_1 = 1.$
   (II) $w = x = y = s_1 = 0,\ z = 1/3.$
   (III) $w = x = s_1 = z = 0,\ y = 1.$
   (IV) $w = s_1 = y = z = 0,\ x = 1/2.$
   (V) $s_1 = x = y = z = 0,\ w = 1/5.$

5. a.
$$x + y + z + s_1 = 3$$
$$2x + y - z + s_2 = 1$$
$$x + y - s_3 = 2$$
$$-3x + 4y + s_4 = 5$$

b. Solution for the system in which $s_1 = s_2 = s_3 = 0$ is $x = 0$, $y = 2$, $z = 1$, $s_4 = -3$. No, the solution is not feasible.

c. $(2, 3, -2, 11)$, not feasible.

d. If $x = y = s_3 = 0$, then the third equation $x + y - s_3 = 2$ would be meaningless. In such a case there is no solution.

7.

System	x	y	s_1	s_2	s_3	s_4	Feasible?
I	0	0	−4	6	3	3	No
II	0	2	0	4	3	1	Yes
III	0	6	8	0	3	−3	No
IV	0	—	—	—	0	—	Impossible
V	0	3	2	3	3	0	Yes
VI	4	0	0	2	−1	7	No
VII	6	0	2	0	−3	9	No
VIII	3	0	−1	3	0	6	No
IX	−3	0	−7	9	6	0	No
X	8	−2	0	0	−5	13	No
XI	3	1/2	0	5/2	0	11/2	Yes
XII	−2/3	7/3	0	13/3	11/3	0	No
XIII	3	3	5	0	0	3	Yes
XIV	3/2	9/2	13/2	0	3/2	0	Yes
XV	3	6	11	−3	0	0	No

9. a. Constraint inequalities:

$$z \geq 100{,}000$$
$$x - (1/4)y \leq 0$$
$$-x + y - z \geq 500{,}000$$

Slack variable equations:

$$z - s_1 = 100{,}000$$
$$x - (1/4)y + s_2 = 0$$
$$-x + y - z - s_3 = 500{,}000$$

b. $x = s_3 = s_1 = 0$; $y = 600{,}000$; $z = 100{,}000$; $s_2 = 150{,}000$.

EXERCISE 5.4 (page 155)

1. $x_1 + 2x_2 + x_3 = 10$
$-3x_1 + x_2 + x_4 = 2$

Objective function:
$P = 13x_1 + 4x_2 + 0x_3 + 0x_4$.

System	x_1	x_2	x_3	x_4	Feasible?	P
I	0	0	10	2	Yes	0
II	0	5	0	-3	No	—
III	0	2	6	0	Yes	8
IV	10	0	0	32	Yes	130 (Maximum)
V	-2/3	0	32/3	0	No	—
VI	6/7	32/7	0	0	Yes	0

3. Maximum profit $P = 1800$ at $x_1 = 9$, $x_2 = 0$.

5. a. $3x_1 + 2x_2 \leq 20$
$15x_1 + 28x_2 \leq 105$.

b. Maximum $z = {}^{45}\!/_2$ at $x_1 = 0$, $x_2 = {}^{15}\!/_4$.

7. The minimum is at $x_1 = 20$, $x_2 = 0$, $x_3 = 0$, $z = 60$.

9. a. $3x_1 + 5x_2 + x_3 - x_4 + x_5 = 100$
$2x_1 - x_2 + 4x_3 + x_4 + x_6 = 80$.

System	x_1	x_2	x_3	x_4	x_5	x_6	Feasible?
I	0	0	0	0	100	80	Yes
II	0	0	0	-100	0	180	No
III	0	0	0	80	180	0	Yes
IV	0	0	100	0	0	-320	No
V	0	0	20	0	80	0	Yes
VI	0	0	36	-64	0	0	No
VII	0	20	0	0	0	100	Yes
VIII	0	-80	0	0	500	0	No
IX	0	45	0	125	0	0	Yes
X	0	$320/21$	$500/21$	0	0	0	Yes
XI	$100/3$	0	0	0	0	$40/3$	Yes
XII	40	0	0	0	-20	0	No
XIII	36	0	0	8	0	0	Yes
XIV	32	0	4	0	0	0	Yes
XV	$500/13$	$-40/13$	0	0	0	0	No

b. Maximum is $R(0,45,0,125) = 217.5$, the value of R for system IX.

EXERCISE 6.1 (page 165)

1. a. 19   b. -23 c. 23   d. 19

3. a. Each expression is equal to $a_{11}a_{22} - a_{21}a_{12}$.

   b. Each is equal to $a_{11}a_{22} - a_{21}a_{12}$.

5. Any three determinants of the form

$$\begin{vmatrix} a & b \\ c & d \end{vmatrix}$$ in which $ad - bc = 157$ will do. Examples:

$$\begin{vmatrix} 157 & 0 \\ 1 & 1 \end{vmatrix}, \begin{vmatrix} 5 & -7 \\ 1 & 30 \end{vmatrix}, \begin{vmatrix} 11 & 10 \\ -8 & 7 \end{vmatrix}.$$

Chapter 6

7. a. $x = \dfrac{\begin{vmatrix} 7 & 8 \\ 10 & -7 \end{vmatrix}}{\begin{vmatrix} 3 & 8 \\ 5 & -7 \end{vmatrix}} = {-129}/{-61} = 2\,{}^{7}\!/_{61}$, $y = \dfrac{\begin{vmatrix} 3 & 7 \\ 5 & 10 \end{vmatrix}}{-61} = {-5}/{-61} = {}^{5}\!/_{61}$.

b. $x = \dfrac{\begin{vmatrix} 7 & 3 \\ 10 & 5 \end{vmatrix}}{\begin{vmatrix} 8 & 3 \\ -7 & 5 \end{vmatrix}} = {}^{5}\!/_{61}$, $y = {}^{129}\!/_{61} = 2\,{}^{7}\!/_{61}$.

9. $x = \dfrac{\begin{vmatrix} b_1 & a_{12} \\ b_2 & a_{22} \end{vmatrix}}{\begin{vmatrix} a_{11} & a_{12} \\ a_{21} & a_{22} \end{vmatrix}} = \dfrac{b_1 a_{22} - b_2 a_{12}}{a_{11} a_{22} - a_{21} a_{12}}$, $y = \dfrac{\begin{vmatrix} a_{11} & b_1 \\ a_{21} & b_2 \end{vmatrix}}{\begin{vmatrix} a_{11} & a_{12} \\ a_{21} & a_{22} \end{vmatrix}} = \dfrac{a_{11} b_2 - b_1 a_{21}}{a_{11} a_{22} - a_{21} a_{12}}$.

11. $\begin{bmatrix} x \\ y \end{bmatrix} = -({}^{1}\!/_{5}) \begin{bmatrix} -13 \\ -22 \end{bmatrix}$ means $x = {}^{13}\!/_{5}$ and $y = {}^{22}\!/_{5}$; therefore, $x + y = {}^{13}\!/_{5} + {}^{22}\!/_{5} = {}^{35}\!/_{5} = 7$ and $2x - 3y = {}^{26}\!/_{5} - {}^{66}\!/_{5} = -8$.

13. $x = 8$, $y = 2$.

15. $p = {}^{4}\!/_{7}$, $q = {}^{3}\!/_{7}$.

EXERCISE 6.2 (page 174)

1. $-15, -75, 129, 33$.

3. $x = \dfrac{\begin{vmatrix} 0 & 2 & -1 \\ 7 & -1 & 3 \\ -1 & 0 & 5 \end{vmatrix}}{\begin{vmatrix} 3 & 2 & -1 \\ 1 & -1 & 3 \\ 2 & 0 & 5 \end{vmatrix}}, \; y = \dfrac{\begin{vmatrix} 3 & 0 & -1 \\ 1 & 7 & 3 \\ 2 & -1 & 5 \end{vmatrix}}{\begin{vmatrix} 3 & 2 & -1 \\ 1 & -1 & 3 \\ 2 & 0 & 5 \end{vmatrix}},$

$z = \dfrac{\begin{vmatrix} 3 & 2 & 0 \\ 1 & -1 & 7 \\ 2 & 0 & -1 \end{vmatrix}}{\begin{vmatrix} 3 & 2 & -1 \\ 1 & -1 & 3 \\ 2 & 0 & 5 \end{vmatrix}}.$

5. $x = 5, \; y = {-43}/{5}, \; z = {-11}/{5}.$

7. $\begin{bmatrix} x \\ y \\ z \end{bmatrix} = \begin{bmatrix} 5 \\ -43/5 \\ -11/5 \end{bmatrix}.$

9. $a_{11}x_1 + a_{12}x_2 + a_{13}x_3 = b_1$
$a_{21}x_1 + a_{22}x_2 + a_{23}x_3 = b_2$
$a_{31}x_1 + a_{32}x_2 + a_{33}x_3 = b_3.$

11. $x = \pm \sqrt{7/3}.$

13. $x = 160.2$ billion gal/day, $y = 36.6$ billion gal/day, $z = 247.2$ billion gal/day.

EXERCISE 6.3 (page 181)

1. **a.** $(8, 15, 4, 1)$ **b.** $(-34, -60, -5, 13)$ **c.** $(-1/2, 1, 5, 5)$

3. **a.** $\mathbf{p} = (11/3, 20/3, 7/6, -1/2).$ **b.** $\mathbf{q} = (32/5, 9, -13/5, 12/5).$

5. a. 69 **b.** −189 **c.** −834

7. $a \cdot e_1 + b \cdot e_2 = a(1,0) + b(0,1) = (a,0) + (0,b) = (a,b)$.

9. a. $10^5(1.6, 0.7, 2.0, 3.0, 0.8, 1.1, 1.5, 0.6)$ **b.** 80,000

  c. The vector z is a vector whose components represent the numerical change in each occupation between 1968 and 1969.

  d. Secretaries **e.** Total payroll for these jobs

EXERCISE 7.1 (page 195)

1. a. $\begin{bmatrix} 5 & 9 & 8 \\ 13 & 22 & -10 \end{bmatrix}$ **b.** $\begin{bmatrix} 5 & -8 & -11 \\ 14 & 36 & -20 \end{bmatrix}$ **c.** $\begin{bmatrix} 19/2 & 29 & 57/2 \\ 24 & 32 & -12 \end{bmatrix}$

3. a. $C = 1/11 \begin{bmatrix} 18 & 80 & 82 \\ 44 & 40 & -8 \end{bmatrix}$. **b.** $C = \begin{bmatrix} 4 & 31 & 33 \\ 9 & -2 & 6 \end{bmatrix}$.

5. a. $\begin{bmatrix} 34 \\ 63 \\ 18 \end{bmatrix}$ **b.** (3,8,49,11,29)

7. a. $2x + y = 3$ **b.** $x = 8/7, y = 5/7$.
    $3x + 5y = 7$.

9. $\begin{bmatrix} 8/7 \\ 5/7 \end{bmatrix}$.

11. a. $I\mathbf{v} = \begin{bmatrix} 1 & 0 \\ 0 & 1 \end{bmatrix} \begin{bmatrix} x \\ y \end{bmatrix} = \begin{bmatrix} x \cdot 1 + 0 \cdot y \\ 0 \cdot x + 1 \cdot y \end{bmatrix} = \begin{bmatrix} x \\ y \end{bmatrix} = \mathbf{v}$.

  b. $\mathbf{v} - A\mathbf{v} = \begin{bmatrix} x - a_{11}x - a_{12}y \\ y - a_{21}x - a_{22}y \end{bmatrix} = \begin{bmatrix} 1 - a_{11} & -a_{12} \\ -a_{21} & 1 - a_{22} \end{bmatrix} \begin{bmatrix} x \\ y \end{bmatrix} = (\mathbf{I} - \mathbf{A})\mathbf{v}$.

13. a. $Av = \begin{bmatrix} 42 \\ 82 \\ 21 \end{bmatrix}$.

b. 42 = number of lambs born in 1953 to the barren 1952 ewes; 82 = number of lambs born in 1953 to the ewes having a single birth in 1952; 21 = number of lambs born in 1953 to the ewes having twins in 1952.

c. 145   d. (54,64,30)   e. 148

EXERCISE 7.2 (page 205)

1. a. $\begin{bmatrix} -13 & 35 \\ 9 & -34 \\ -6 & 10 \end{bmatrix}$ b. $\begin{bmatrix} 16 & 32 \\ 24 & 48 \end{bmatrix}$ c. $\begin{bmatrix} a & b & c \\ d & e & f \\ g & h & i \end{bmatrix}$

3.
	Revenue	Profit
Store 1	103.50	49.90
Store 2	149.50	75.80
Store 3	142.50	77.40

Interpretation: Each entry is the total revenue or profit (for all items) at each store.

5. $X = \begin{bmatrix} -3 & -1 \\ 5/3 & 3/2 \end{bmatrix}$.

7. $Y = \begin{bmatrix} 7/6 & -5/3 \\ 1/6 & -8/3 \end{bmatrix}$.

9. a. $\begin{bmatrix} 3 & 7 \\ 5 & 2 \end{bmatrix}$ b. $\begin{bmatrix} x & y \\ u & w \end{bmatrix}$ c. Yes

d. $A - AB = \begin{bmatrix} -22 & -34 \\ -36 & -52 \end{bmatrix} = A(I - B)$.   e. Yes

f. $\mathbf{I}\begin{bmatrix}3\\7\end{bmatrix} = \begin{bmatrix}3\\7\end{bmatrix}, \mathbf{I}\begin{bmatrix}u\\v\end{bmatrix} = \begin{bmatrix}u\\v\end{bmatrix}.$

11. a. $D = 8$ b. $\mathbf{A}^{-1} = \begin{bmatrix} 3/8 & -2/8 \\ -5/8 & 6/8 \end{bmatrix}.$

c. $\mathbf{A}^{-1}\mathbf{A} = \begin{bmatrix} (18-10)/8 & (6-6)/8 \\ (-30+30)/8 & (-10+18)/8 \end{bmatrix} = \begin{bmatrix} 1 & 0 \\ 0 & 1 \end{bmatrix}.$

13. Example: If $H = 71$ and $W = 180$, then

$$\mathbf{V} = \begin{bmatrix} 71 \\ 180 \end{bmatrix}, \text{ and the coded vector } \mathbf{U} = \begin{bmatrix} 38 \\ -33 \end{bmatrix}. \text{ Decode by}$$

computing the product $\mathbf{C}^{-1} \cdot \mathbf{U}$.

EXERCISE 7.3 (page 217)

1. $\det(\mathbf{A}) = 31$, and $\mathbf{A}^{-1} = \begin{bmatrix} -2/31 & 3/31 & -21/31 \\ -2/31 & 3/31 & 10/31 \\ 11/31 & -1/31 & 7/31 \end{bmatrix}.$

3. $\mathbf{I} - \mathbf{A} = \begin{bmatrix} 0 & 0 & -3 \\ -4 & -6 & -2 \\ 1 & -1 & 1 \end{bmatrix}$, $\det(\mathbf{I} - \mathbf{A}) = -30$, and $(\mathbf{I} - \mathbf{A})^{-1}$

$= \begin{bmatrix} 8/30 & -3/30 & 18/30 \\ -2/30 & -3/30 & -12/30 \\ -10/30 & 0 & 0 \end{bmatrix}.$

5. $\mathbf{X} = (\mathbf{I} - \mathbf{A})^{-1}\mathbf{Y} = \begin{bmatrix} 77/30 \\ -53/30 \\ -100/30 \end{bmatrix}.$

7. $\mathbf{P} = \begin{bmatrix} 5 & 1 \\ -7 & 3/2 \end{bmatrix}.$

9. a. $X = \begin{bmatrix} 11 & 22 & 2 & 22 & 16 & 5 \\ 26 & 54 & 5 & 52 & 40 & 12 \end{bmatrix}$. b. Yes

EXERCISE 7.4 (page 223)

1. 0.08.

3. 5.33.

5. a. $MX = \begin{bmatrix} 1600 \\ 235 \end{bmatrix}$. b. $X - MX = \begin{bmatrix} 1900 \\ 2265 \end{bmatrix}$.

7. $I - M = \begin{bmatrix} 1 - 0.35 & 0 - 0.15 \\ 0 - 0.06 & 1 - 0.01 \end{bmatrix}$.

9. a. $X = (I - M)^{-1} \begin{bmatrix} 1000 \\ 500 \end{bmatrix} = \begin{bmatrix} 1678 \\ 607 \end{bmatrix}$.

b. $(I - M)X = \begin{bmatrix} 999.7 \\ 500.3 \end{bmatrix} \approx \begin{bmatrix} 1000 \\ 500 \end{bmatrix}$.

EXERCISE 8.1 (page 235)

1.

v	x	y	z	b
x	1	0	0	5/3
y	0	1	0	3
z	0	0	1	8

Chapter 8

3.

v	x	y	z	b
x	1	0	0	20/7
y	0	1	0	37/7
z	0	0	1	−39/7

5. An intermediate step is

The final answer is

$$\begin{bmatrix} x_1 \\ x_2 \\ x_3 \\ x_4 \end{bmatrix} = \begin{bmatrix} 2 \\ -1 \\ 0 \\ 2 \end{bmatrix}.$$

1	0	2	2	6
0	1	4	4	7
0	0	4	2	4
0	0	1	5/2	5

7. a. $1 \cdot c_1 + 0 \cdot c_2 + 7 \cdot c_3 = 3$
   $2 \cdot c_1 + 1 \cdot c_2 + 1 \cdot c_3 = 5$
   $1 \cdot c_1 + 3 \cdot c_2 + 2 \cdot c_3 = 4.$

  b. $c_1 = 37/17$, $c_2 = 9/17$, and $c_3 = 2/17$.

  c. $T = (37p + 9q + 2r)/17.$ At $(1,1,1)$, $T = 2.8$ (approximately).

9. a. $x - 75 = 2y.$

  b. $x + y + z = 550$
  $6x + 8y + 9z = 4025$
  $x - 2y = 75.$

  c.

v	x	y	z	b
x	1	1	1	550
y	6	8	9	4025
z	1	−2	0	75

  d. $x = 275$, $y = 100$, $z = 175.$

EXERCISE 8.2 (page 246)

1.

v	x_1	x_2	x_3	b
x_1	1	0	$-11/20$	$23/10$
x_2	0	1	$-7/20$	$1/10$

that is, $(x_1, x_2, x_3) = (23/10, 1/10, 0)$.

3.

v	x_1	x_2	x_3	b
x_2	$-7/11$	1	0	$-15/11$
x_3	$-20/11$	0	1	$-46/11$

that is $(x_1, x_2, x_3) = (0, -15/11, -46/11)$.

5. a. x_3 and x_4 **b.** No   **c.**

v	x_1	x_2	x_3	x_4	b
x_3	$11/38$	$7/38$	1	0	$13/38$
x_4	$5/19$	$-2/19$	0	1	$-2/38$

Solution: $(x_1, x_2, x_3, x_4) = (0, 0, 13/38, -2/38)$.

d.

v	x_1	x_2	x_3	x_4	b
x_2	$-5/2$	1	0	$-19/2$	$1/2$
x_3	$3/4$	0	1	$7/4$	$1/4$

Solution: $(x_1, x_2, x_3, x_4) = (0, 1/2, 1/4, 0)$.

7. a. The basis column contains x_1, x_3, and x_4, and the coefficient matrix for these three variables is the identity $\begin{bmatrix} 1 & 0 & 0 \\ 0 & 1 & 0 \\ 0 & 0 & 1 \end{bmatrix}$.

b.

v	x_1	x_2	x_3	x_4	x_5	b
x_1	1	0	2	0	17	4
x_2	0	1	$-1/2$	0	$-7/2$	-1
x_4	0	0	$5/2$	1	$31/2$	6

Solution: $(x_1, x_2, x_3, x_4) = (4, -1, 0, 6, 0)$.

9. a. Basis is x_1 and x_5.   Solution: $(x_1, x_5) = (12, 8)$.

v	x_1	x_2	x_3	x_4	x_5	b
x_1	1	$1/9$	$-7/9$	$23/27$	0	12
x_5	0	$1/3$	$2/3$	$-1/9$	1	8

b. New basis is x_1 and x_2.

Solution: $(x_1, x_2) = (9, 24)$. (The value of x_1 is rounded off.)

v	x_1	x_2	x_3	x_4	x_5	b
x_1	1	0	-1	$8/9$	$-1/3$	$28/3$
x_2	0	1	2	$-1/3$	3	24

11. a. An attempt to solve for x_1 and x_2 produces the table

v	x_1	x_2	x_3	x_4	x_5	b
x_1	1	3	6	4	5	10
x_2	0	0	6	-7	5	-9

The second row has zeros for both the x_1 and x_2 coefficients. This means that if the other variables are nonbasic, then $-9 = 0$. Impossible!

b. For x_1 and x_3, we get the table

v	x_1	x_2	x_3	x_4	x_5	b
x_1	1	3	0	11	0	19
x_3	0	0	1	$-7/6$	$5/6$	$-3/2$

EXERCISE 9.1 (page 261)

1. a.

C_j
4
3

The C_j column has the objective function's coefficients for the basic variables.

b.

						z
z_i	4	17	3	1	31	126
$C_i - z_i$	0	-7	0	1	-29	

c. $C_4 - z_4 = 1$ is the largest positive number; therefore, x_4 should be brought into the system for a maximizing problem.

d. $C_5 - z_5 = -29$ is the most negative number; therefore, x_5 should be brought into the system for a minimizing problem.

3. a.

C_j
1
3
0

b.

							z
z_i	10	1	3	8	0	3	202
$C_i - z_i$	5	0	0	4	0	−3	

c. $C_1 - z_1$ is the largest positive number; therefore, x_1 should be brought into the system for a maximizing problem.

d. $C_6 - z_6$ is the most negative; therefore, x_6 should be brought into the system to minimize.

5. a and b.

| C_i | 4 | 10 | 3 | 2 | 2 | | |

V	x_1	x_2	x_3	x_4	x_5	x_0	C_j
x_3	1	5	1	0	9	37	3
x_4	1	2	0	1	4	15	2

						z
z_i	5	19	3	2	35	141
$C_i - z_i$	−1	−9	0	0	−33	

c. There is no positive net potential $C_i - z_i$; therefore, the solution is maximum.

EXERCISE 9.2 (page 272)

1. x_2 row: $b/p = 16/4 = 4$, smallest. x_3 row: $b/p = 9/2 = 4.5$. Therefore, x_2 is the outgoing variable. The new basis is (x_1, x_3, x_4) and the solution for the new system is $x_1 = 5$, $x_3 = 1$, $x_4 = 4$.

3. a. $z = 100$.   b. The best incoming variable is x_4.

   c. The outgoing variable is x_3.

   d. Solution: $(x_1, x_4) = (25/3, 10/3)$, $z = 310/3 = 103\,1/3$. No net potential for this basis is positive.

5. Minimum value of z is 30; it occurs when $(x_1, x_2, x_3, x_4, x_5, x_6)$ is $(0, 15, 0, 20, 6, 0)$.

7. Maximum value of z is 90; it occurs when $(x_1, x_2, x_3, x_4, x_5, x_6)$ is $(0, 18, 0, 46, 0, 32)$.

EXERCISE 9.3 (page 283)

1. $z = 60$; maximum at $x_1 = 10$, $x_2 = 0$, $x_3 = 0$, $x_4 = 0$, $x_5 = 11$.

3. No maximum value of z. The domain is unbounded.

5. $z = 86/3$; minimum at $x_1 = 0$, $x_2 = 46/3$, $x_3 = 40/3$, $x_4 = x_5 = 0$.

7. $z = 15$; maximum at $x_1 = 0$, $x_2 = 3$, $x_3 = 0$, $x_4 = 21$, $x_5 = 4$.

9. $z = 15{,}500$; minimum at $x_3 = 15{,}000$ and $x_5 = 1000$. The other two inequalities are $x_2 + 2x_4 + x_5 \geq 1000$ and $x_3 + x_5 \geq 8000$.

EXERCISE 10.1 (page 297)

1. Five of spades, ten of spades, ace of diamonds, jack of diamonds, and four of diamonds.

3. 54, 14, 94, 98, 47, 48, 12, 85, 95, 78.

5. A draw (or tie).

7. a. $\{X_1\}$ = event of a man with normal vision, $\{X_2\}$ = event of a color-blind man.

b. $\{X_1, X_2\}$ = event of a man, $\{X_1, X_3\}$ = event of a person with normal vision.

9. X_1 = bird uses the sun's position, X_2 = bird uses barometric pressure, X_3 = bird uses polarized light, X_4 = bird uses earth's magnetic field. Example of a union: $\{X_1\} \cup \{X_2\}$ = event that the bird uses either the sun's position or barometric pressure.

EXERCISE 10.2 (page 304)

1. a. $\{X_0\}, \{X_1\}, \{X_2\}, \{X_3\}, \{X_4\}$ **b.** $S = \{X_0, X_1, X_2, X_3, X_4\}$.

   c. $P(\{X_4\}) = 0.08$.

   d. A = the event that at least 101 houses were damaged.

   e. $\{X_2, X_3, X_4\}$ **f.** 0.46 **g.** 0.54 **h.** 32 possible events.

3. a. $P(\{Z\}) = 0.41$. **b.** $P(\{X,Y\}) = 0.59$. **c.** $P(-\{X\}) = 0.65$.

   d. $P(-\{Y\}) = 0.76$. **e.** 8

   f. $P(\emptyset) = 0$, $P(\{X\}) = 0.35$, $P(\{Y\}) = 0.24$, $P(\{Z\}) = 0.41$, $P(\{X,Y\}) = 0.59$, $P(\{X,Z\}) = 0.76$, $P(\{Y,Z\}) = 0.65$, $P(\{X,Y,Z\}) = 1$.

5. C = event that at least three and no more than five phone calls come in, D = event that at least eight phone calls come in.

7. $P(C) = 0.637$, $P(D) = 0.029$.

9. $P(E_0) + P(E_1) + \ldots + P(E_{11}) = 0.000 + 0.001 + \ldots + 0.002 = 1$.

11. $P(B \cap D) = P(\{E_8\}) = 0.023$.

13. Y_1, Y_2, Y_3 covers all possibilities; therefore $P(Y_1, Y_2, Y_3) = 1$.

15. $P(-Y_2) = 0.488$.

17. $P(Y_1) = 0.351$, $P(Y_2) = 0.512$, $P(Y_3) = 0.137$.

19. For example, let $S''' = \{Z_1, Z_2, Z_3, Z_4, Z_5, Z_6, Z_7\}$, where $Z_1 = \{E_0, E_1\}$, $Z_2 = \{E_2\}$, $Z_3 = \{E_3, E_4, E_5\}$, $Z_4 = \{E_6\}$, $Z_5 = \{E_7\}$, $Z_6 = \{E_8\}$, $Z_7 = \{E_9, E_{10}, E_{11}\}$. $P(Z_1) = 0.001$, $P(Z_2) = 0.024$, $P(Z_3) = 0.637$, $P(Z_4) = 0.201$, $P(Z_5) = 0.108$, $P(Z_6) = 0.023$, $P(Z_7) = 0.006$.

EXERCISE 10.3 (page 310)

1. $P(F) = 0.84$.

3. Probability $= {}^{18}\!/_{52}$ (or approximately 0.35).

5. $1\!/_7$.

7. $n(A) = 2,958,930$.

9. 7678.

11. 0.

13.

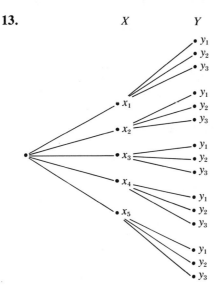

Chapter 10

15. a. $P(\{X_i\}) = 1/8$.   b. The event that exactly two of the coins are heads

c. $P(\{X_2, X_3, X_5\}) = 3/8$.   d. $\{X_4, X_6, X_7, X_8\}$ e. $3/8$ f. $1/2$.

17.

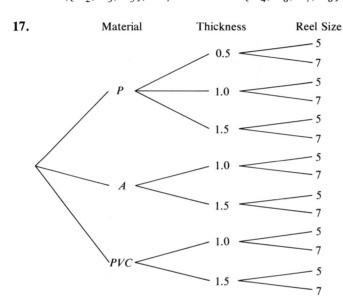

19. 6.

21. $n(S) = 36$, $n(A) = 16$, $n(B) = 12$, $P(A) = 4/9$, $P(B) = 1/3$.

23. $8/9$.

EXERCISE 10.4 (page 319)

1. 720.

3. a. 5040   b. 210.

5. a. 120   b. 250.

7. 220.

9. $(n + k)!/(n!\,k!)$

11. a. 1   b. 6   c. 15   d. 20   e. 15   f. 6   g. 1.

13. There are 28! paintings, and 28! is approximately $(3.05)(10^{29})$, which is approximately 46 million times as large as $(6.6)(10^{21})$.

15. a. $12!/(6!\,6!) = 924$.

　　b. All six are from the eight objects in category I, so there are $8!/(2!\,6!)$ or 28 such samples.

　　c. $[8!/(3!\,5!)] \cdot [4!/(3!\,1!)] = 224$.

17. 11,440.

19. 120.

21. 0.0105.

23. a. 720   b. 0.396.

EXERCISE 11.1 (page 329)

1. a. 0.86   b. 0.125.

3. $P(A \cap B) = 0.004$.

5. $P(A|B) = 0.167$.

7. $P(A|B) = 0.51$, $P(B|A) = 0.33$.

9. $P(A|-B) = 0.52$, $P(-A|B) = 0.49$, $P[(A \cap B)|B] = 0.51$, $P[(A \cap B)|-B] = 0$.

11. $n(A \cup B) = 543$, Yes, $P[A|(A \cup B)] = 0.76$.

13. $P(M \cdot L) = 0.24$, $P(L|M) = 0.30$, $P(L \cup M) = 0.96$, $P[(L \cup M)|-M] = 0.80$.

15. **a.** 0.67   **b.** 0.53.

17. **a.** $1/216$ (or 0.0046)

   b. One answer is $-A$ is the event that a 1 does not appear on the first die.

   c. $P(-A) = P(-B) = P(-C) = 5/6$, so $P[(-A) \cdot (-B) \cdot (-C)] = 125/216$ (or 0.579).

   d. $P[(-A) \cdot B \cdot (-C)] = 0.116$ (approximately).

EXERCISE 11.2 (page 338)

1. **a.** $P(A \cup B) = 8/45$.   **b.** $P(A|B) = 9/10$.   **c.** $P[A|(A \cup B)] = 15/16$.

   d. $P[(A \cup B)|B] = 1$.

3. $P(AB^+) = 0.068$, $P(A^-) = 0.063$, $P(\text{Type O}) = 0.40$.

5. (Answers accurate to three decimal places)   **a.** $P(A \cup B) = 0.483$.

   b. $P[(A \cup B) \cdot B] = P(B) = 0.333$.   **c.** $P[(A \cup B)|B] = 1$.

   d. $P[B|(A \cup B)] = 0.690$.   **e.** $P[(-A) \cdot (-B)] = 0.517$.

   f. $P(-A \cup -B) = 0.75$.

7.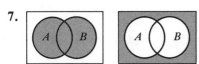

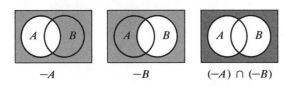

9. Let p = event of pass and f = event of first down, then $P(p) = 0.70$, $P(f|-p) = 0.50$, and $P(f|p) = 0.60$ are given. We seek $P(f)$. Since $f = f(p \cup -p) = fp \cup f(-p)$ is the union of two disjoint sets, then

$$P(f) = P(fp) + P[f(-p)] = P(f|p) \cdot P(p) + P(f|-p) \cdot P(-p)$$
$$= (0.60)(0.70) + (0.50)(0.30) = 0.57.$$

11.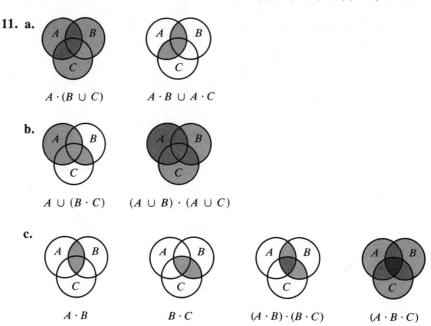

a.

$A \cdot (B \cup C)$ $A \cdot B \cup A \cdot C$

b.

$A \cup (B \cdot C)$ $(A \cup B) \cdot (A \cup C)$

c.

$A \cdot B$ $B \cdot C$ $(A \cdot B) \cdot (B \cdot C)$ $(A \cdot B \cdot C)$

d. Yes

e. $P[A \cdot (B \cup C)] = P(A \cdot B \cup A \cdot C) = P(AB) + P(AC) - P(ABAC)$, etc.

13. Use the formulas in Table 11.1. By #11, $P[-(A \cup B)] = 1 + c - a - b$, and by #9, $P[-(A \cap B)] = 1 - c$. Adding these two equations, we get the result desired.

15. a. No animal can have *only four toes* **and** *four toes plus an imperfect one* on the same foot.

b. $P(X_1 \cup X_2) = 0.90$, Yes, $X_1 \cap X_3 = \emptyset$, $P(X_1 \cup X_3) = 0.87$.

c. $P(X_1 \cup -X_2) = 0.87$.   **d.** $P(-X_1 \cap -X_3) = 0.13$.

e. $P(-X_3) = 0.90$.   **f.** $P(-X_1) = 0.23$.

17. $P(-X_3 | -X_1) = 0.57$.

EXERCISE 11.3 (page 347)

1. **a.** $P(A_1) = 3/10$, $P(A_2) = 2/10$, $P(A_3) = 5/10$.   **b.** $0.3 + 0.2 + 0.5 = 1$.

3. Many answers are possible; one example is $S'' = [C_1, C_2, C_3] = [\{E_1, E_2, E_3\}, \{E_5, E_7, E_8\}, \{E_4, E_6, E_9, E_{10}\}]$

    a. $P(C_1) = 0.3$, $P(C_2) = 0.3$, $P(C_3) = 0.4$.   **b.** $0.3 + 0.3 + 0.4 = 1$.

5. $A_1 \cdot B \cup A_2 \cdot B \cup A_3 \cdot B = \{E_7, E_6, E_3\} = B$, and $A_1 \cdot B \cap A_2 \cdot B = \emptyset$, $A_1 \cdot B \cap A_3 \cdot B = \emptyset$, and $A_2 \cdot B \cap A_3 \cdot B = \emptyset$.

7. **a.** $Y_2 = \{(1,1)\}$, $Y_3 = \{(1,2), (2,1)\}$, $Y_4 = \{(1,3), (2,2), (3,1)\}$, $Y_5 = \{(1,4), (2,3), (3,2), (4,1)\}$, $Y_6 = \{(1,5), (2,4), (3,3), (4,2), (5,1)\}$, $Y_7 = \{(1,6), (2,5), (3,4), (4,3), (5,2), (6,1)\}$, $Y_8 = \{(2,6),(3,5),(4,4),(5,3),(6,2)\}$, $Y_9 = \{(3,6),(4,5),(5,4),(6,3)\}$, $Y_{10} = \{(4,6),(5,5),(6,4)\}$, $Y_{11} = \{(5,6),(6,5)\}$, $Y_{12} = \{(6,6)\}$.

    b.

Y_i	Y_2	Y_3	Y_4	Y_5	Y_6	Y_7	Y_8	Y_9	Y_{10}	Y_{11}	Y_{12}
$P(Y_i)$	1/36	2/36	3/36	4/36	5/36	6/36	5/36	4/36	3/36	2/36	1/36

   c. $S'' = [Z_1, Z_2]$ is a partition of S, since $Z_1 \cup Z_2$ = set of throws with either an odd number on the first die or an even number on the first die = set of all throws possible. Also, $Z_1 \cap Z_2 = \emptyset$, since no dice throw can have its first die be *both* even and odd.

d.

S'\S"	z_1	z_2
Y_2	(1,1)	
Y_3	(1,2)	(2,1)
Y_4	(1,3), (3,1)	(2,2)
Y_5	(1,4), (3,2)	(2,3), (4,1)
Y_6	(1,5), (3,3), (5,1)	(2,4), (4,2)
Y_7	(1,6), (3,4), (5,2)	(2,5), (4,3), (6,1)
Y_8	(3,5), (5,3)	(2,6), (4,4), (6,2)
Y_9	(3,6), (5,4)	(4,5), (6,3)
Y_{10}	(5,5)	(4,6), (6,4)
Y_{11}	(5,6)	(6,5)
Y_{12}		(6,6)

9. $P(A) = P(A|Y_1) \cdot P(Y_1) + P(A|Y_2) \cdot P(Y_2) + P(A|Y_3) \cdot P(Y_3) + P(A|Y_4) \cdot P(Y_4) = 0.3915.$

EXERCISE 12.1 (page 358)

1. a. See the figure below.

b. $P(X = -2) = 0.07$.

c. E_2 and E_5, $P(X = 0) = 0.33 + 0.10 = 0.43$.

d. $P(X = 1 \text{ or } X = 10) = 0.50.$ **e.** $P(X = -2 \text{ or } X = 10) = 0.22.$

f. See the figure below.

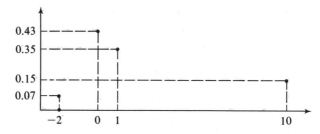

3. a.

	1	2	3	4	5	6
1	1	2	3	4	5	6
2	2	4	6	8	10	12
3	3	6	9	12	15	18
4	4	8	12	16	20	24
5	5	10	15	20	25	30
6	6	12	18	24	30	36

$S = \{1,2,3,4,5,6,8,9,10,12,15,16,18,20,24,25,30,36\}$.

b.

$X = x$	1	2	3	4	5	6	8	9	10
$P(X = x)$	1/36	2/36	2/36	3/36	2/36	4/36	2/36	1/36	2/36

$X = x$	12	15	16	18	20	24	25	30	36
$P(X = x)$	4/36	2/36	1/36	2/36	2/36	2/36	1/36	2/36	1/36

c. $P(X \geq 18) = 5/18$.

5. a. $P(A) = 2/9$, $P(B) = 4/9$, $P(A \cap B) = 5/36$.

b. $P(A|-B) = 3/20$, $P(B \cap -A) = 11/36$.   **c.** $P(-A \cup B) = 11/12$.

7. a. $t = 4$.

b.

$X = x$	-4	-2	0	3
$P(X = x)$	6/15	1/15	8/15	0

c.

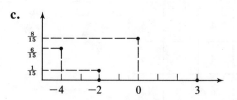

9. a. Range = {0,1,2,3}.

b.

$X = x$	0	1	2	3
$P(X = x)$	1/6	1/2	3/10	1/30

EXERCISE 12.2 (page 366)

1. a. $E(X) = 1.36$.   **b.** $E(X) = 3.32$.

  c. For part **a,** $E(X + 5) = 6.36$, for part **b,** $E(X + 5) = 8.32$.

  d. For part **a,** $E(3X - 1) = 3.08$, for part **b,** $E(3X - 1) = 8.96$.

3. a. $E(X) = \$8.02$.   **b.** Yes   **c.** Lost $56.

  d. No change in player's expected value, but cost to sponsors is reduced.

5. a. {1,2,3,4,5}   **b.** $P(X = 1) = 1/38$.   **c.** LW → $X = 2$.

  d. LLLLW → $X = 5$, and LLLLL → $X = 5$.

  e. $P(X = 3) = P(LLW) = P(L)\,P(L)\,P(W)$.

  f. $E(X) = 1 \cdot P(X = 1) + 2 \cdot P(X = 2) + \ldots + 5 \cdot P(X = 5) = 4.7437$.

EXERCISE 12.3 (page 373)

1. **a.** $P(A) = 0.65$.

   b. $P(E_1|A) = 0.231$, $P(E_2|A) = 0.415$, $P(E_3|A) = 0.092$, $P(E_4|A) = 0.185$, $P(E_5|A) = 0.077$.

   c. $P(A \cap E_1) = 0.15$.   **d.** $P(A \cap -E_2) = 0.38$.

   e. $P(E_3 \cap -A) = 0.24$.   **f.** $P(E_3|-A) = 0.686$.

3. $P(A) = 0.67$, $P(E_1|A) = 0.269$, $P(E_2|A) = 0.269$, $P(E_3|A) = 0.462$.

5. **a.** $P(G) = 1/2$.   **b.** $2/3, 1/3, 0$.

7. **a.** $P(T) = 0.4102$.   **b.** $P(E_2|T) = 0.488$.

   c. $P(E_3|T) = 0.000$ (to four decimal places, it is 0.0004).

9. From $P(A|E_i) = P(A \cap E_i)/P(E_i)$, we get $P(A \cap E_i) = P(A|E_i)P(E_i)$ for all $i = 1, 2,$ or 3.

   b. $P(A) = \Sigma_{i=1}^{3} P(A \cap E_i)$ and $P(A \cap -E_1) = P(A) - P(A \cap E_1)$.

   c. $P(A \cap -E_i) = P(A) - P(A \cap E_i)$ for $i = 1,2,3$.

   d. Yes, $P(A|-E_i) = P(A \cap -E_i)/P(-E_i)$, $i = 1,2,3$.

   e. Yes, $P(-E_i|A) = P(-E_i \cap A)/P(A)$, $i = 1,2,3$.

11. $P(-F) = 0.40$.

13. $P(A \cap -F) = P(A) - P(A \cap F)$ and $P(F \cap -A) = P(F) - P(A \cap F)$ can be subtracted to yield $P(A) = P(F) + P(A \cap -F) - P(F \cap -A) = 0.57$.

15. $P(F|-A) = 0.1628$.

17. $P(-A \cap -F) = 1 + P(A \cap F) - P(A) - P(F)$, and from Problem 13, $P(A) = 0.57$ and $P(A \cap F) = P(A) - P(A \cap -F) = 0.53$, so $P(-A \cap -F) = 0.36$.

19. $P(-F|-A) = 0.837$.

21. $P(A|F) = 0.883$.

EXERCISE 13.1 (page 384)

1. a. $v_1 = \frac{1}{8}$. **b.** $u_3 = 0.26$.

3. $(\frac{5}{21}, \frac{31}{70}, \frac{67}{210})$.

5. $\begin{bmatrix} 5/12 & 7/12 \\ 7/20 & 13/20 \end{bmatrix}$.

7. a. $\mathbf{M}^2 = \mathbf{M}^3 = \mathbf{M}$.

b. $\mathbf{T} = \mathbf{M}^k = \mathbf{M}$ for all positive integers k. Otherwise, let $k =$ the least integer for which $\mathbf{M}^k \neq \mathbf{M}$, then $\mathbf{M}^{k-1} = \mathbf{M}$; that is, $\mathbf{M}^{k-1} \cdot \mathbf{M} = \mathbf{M} \cdot \mathbf{M}$. But this says that $\mathbf{M}^k = \mathbf{M}^2 (=\mathbf{M})$, contrary to hypothesis.

9. a. $\frac{1}{4}$ **b.** $\frac{1}{2}$

c. $\mathbf{B}^2 = \begin{bmatrix} 11/16 & 5/16 \\ 5/8 & 3/8 \end{bmatrix}$.

d. $\mathbf{B}^3 = \begin{bmatrix} 43/64 & 21/64 \\ 21/32 & 11/32 \end{bmatrix}$, $\mathbf{B}^4 = \begin{bmatrix} 171/256 & 85/256 \\ 85/128 & 43/128 \end{bmatrix}$.

e. $\mathbf{T} - \mathbf{B} = \begin{bmatrix} -1/12 & 1/12 \\ 1/6 & -1/6 \end{bmatrix}$, $\mathbf{T} - \mathbf{B}^2 = \begin{bmatrix} -1/48 & 1/48 \\ 1/24 & -1/24 \end{bmatrix}$,

$\mathbf{T} - \mathbf{B}^3 = \begin{bmatrix} -1/192 & 1/192 \\ 1/96 & -1/96 \end{bmatrix}$.

f. $(\frac{2}{3}, \frac{1}{3}) \cdot \begin{bmatrix} 3/4 & 1/4 \\ 1/2 & 1/2 \end{bmatrix} = (\frac{2}{3}, \frac{1}{3})$.

11. a. $\mathbf{N}^2 = \begin{bmatrix} 1 & 0 \\ 0 & 1 \end{bmatrix}$, and for all positive integers k, $\mathbf{N}^k = \begin{bmatrix} 1 & 0 \\ 0 & 1 \end{bmatrix}$. Otherwise, if k is the least positive integer such that $\mathbf{N}^k \neq \begin{bmatrix} 1 & 0 \\ 0 & 1 \end{bmatrix}$, then $\mathbf{N}^{k-1} = \begin{bmatrix} 1 & 0 \\ 0 & 1 \end{bmatrix}$ and $\mathbf{N}^k = \mathbf{N}^{k-1} \cdot \mathbf{N} = \mathbf{N}^2 = \mathbf{N}$, contrary to hypothesis.

b. Any intuitive argument will do. A formal induction answer is

$$\mathbf{A}^k = \begin{bmatrix} 1 & 0 \\ 1 - 1/2^k & 1/2^k \end{bmatrix}, \text{ for all positive integers } k.$$

By direct computation (for $k = 2$), $\mathbf{A}^2 = \begin{bmatrix} 1 & 0 \\ 3/4 & 1/4 \end{bmatrix}$. Suppose the formula does not hold in general, then if h is the least positive integer such that

$$\mathbf{A}^h \neq \begin{bmatrix} 1 & 0 \\ 1 - 1/2^h & 1/2^h \end{bmatrix}, \text{ then } \mathbf{A}^{h-1} = \begin{bmatrix} 1 & 0 \\ 1 - 1/2^{h-1} & 1/2^{h-1} \end{bmatrix},$$

and by direct computation, the product $\mathbf{A} \cdot \mathbf{A}^{h-1}$ is

$$\mathbf{A} \cdot \mathbf{A}^{h-1} = \begin{bmatrix} 1 & 0 \\ 1 - 1/2^h & 1/2^h \end{bmatrix}.$$

But $\mathbf{A} \cdot \mathbf{A}^{h-1}$ is $\mathbf{A}^h$, and the last matrix equation is contrary to the hypothesis. Finally, as k increases indefinitely, $1/2^k$ approaches zero. Therefore, the general expression for $\mathbf{A}^k$ approaches $\begin{bmatrix} 1 & 0 \\ 1 - 0 & 0 \end{bmatrix}$ as k increases indefinitely.

13. From the hint, $ax + b - bx = x$ and $-ax + 1 - b + bx = 1 - x$. In either case $x = b/(b + 1 - a)$. The fixed point is $[b/(b + 1 - a), (1 - a)/(b + 1 - a)]$.

EXERCISE 13.2 (page 394)

1. a.
$$\mathbf{M} = \begin{array}{c} \\ r \\ s \end{array}\begin{array}{c} r \quad\quad s \\ \begin{bmatrix} 0.6 & 0.4 \\ 0.1 & 0.9 \end{bmatrix} \end{array}.$$

b. $M^2 = \begin{bmatrix} 0.40 & 0.60 \\ 0.15 & 0.85 \end{bmatrix}$, $M^3 = \begin{bmatrix} 0.300 & 0.700 \\ 0.175 & 0.825 \end{bmatrix}$.

3. $(0.2625, 0.7375)$.

5. a.

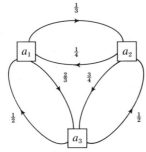

b. $A = \begin{bmatrix} 0 & 1/3 & 2/3 \\ 1/4 & 0 & 3/4 \\ 1/2 & 1/2 & 0 \end{bmatrix}$. **c.** $A^3 = \begin{bmatrix} 15/72 & 19/72 & 38/72 \\ 19/96 & 20/96 & 57/96 \\ 19/48 & 19/48 & 10/48 \end{bmatrix}$.

7. a.

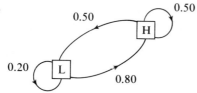

b. $\begin{array}{c} \\ H \\ L \end{array} \begin{array}{c} H \quad L \\ \begin{bmatrix} 0.50 & 0.50 \\ 0.80 & 0.20 \end{bmatrix} \end{array}$.

9. $(0.440, 0.178, 0.252, 0.130)$.

11. a. $(0.2650, 0.2475, 0.2425, 0.2450)$

b. $v \cdot M = (0.95, 0.01, 0.04, 0.00)$.

c. $(v \cdot M) \cdot M = (0.9052, 0.0195, 0.0744, 0.0009)$.

EXERCISE 13.3 (page 406)

1. $A^2 = \begin{bmatrix} 1/2 & 1/2 \\ 1/4 & 3/4 \end{bmatrix}$, no zero entries; therefore, A is regular.

3. C^5 has no zero entries.

5. Fixed point = $(1/5, 2/5, 2/5)$.

7. $T = \begin{bmatrix} 1/3 & 2/3 \\ 1/3 & 2/3 \end{bmatrix}$.

9. a. $v \cdot T = (3/7, 4/7) \cdot \begin{bmatrix} 1/3 & 2/3 \\ 1/3 & 2/3 \end{bmatrix} = (1/3, 2/3)$.

   b. The sequence $v \cdot A, v \cdot A^2, v \cdot A^3, \ldots$ approaches $(1/3, 2/3)$.

11. a. $M = \begin{bmatrix} 0.85 & 0.10 & 0.05 \\ 0.05 & 0.90 & 0.05 \\ 0.10 & 0.10 & 0.80 \end{bmatrix}$.

    b. $M^2 = \begin{bmatrix} 0.7325 & 0.1800 & 0.0875 \\ 0.0925 & 0.8200 & 0.0875 \\ 0.1700 & 0.1800 & 0.6500 \end{bmatrix}$, The probabilities for transitions two cities away from the first.

    c. The probabilities for transitions in cities even farther away.

    d. $\begin{bmatrix} 3/10 & 1/2 & 1/5 \\ 3/10 & 1/2 & 1/5 \\ 3/10 & 1/2 & 1/5 \end{bmatrix}$ e. M_2.

Answers to Odd-Numbered Problems

EXERCISE 14.1 (page 414)

1. a.

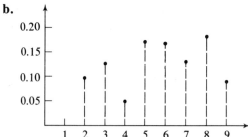

x_i	2	3	4	5	6	7	8	9
f_i	4	5	2	7	6	5	7	3

b.

c. Mean $\bar{x} = (1/39) \Sigma xf = 220/39 = 5.64$. There is no unique mode, but 5 and 8 are the most frequent measurements. For the 39 measurements, the median is the 20th one, which is 6.

d.

Class	[2,4)	[4,6)	[6,8)	[8,10)
Class Frequency	9	9	11	10

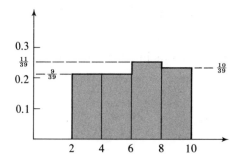

Chapter 14

  e. 6 occurs between the 46th and the 62nd percentile.

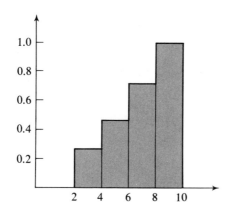

3. a. $\Sigma f = 35$.   b. $\bar{x} = (1/35)\Sigma xf = 74.94$.

  c.

Class	[40,50)	[50,60)	[60,70)	[70,80)	[80,90)	[90,100)
Class Frequency	2	1	6	14	8	4

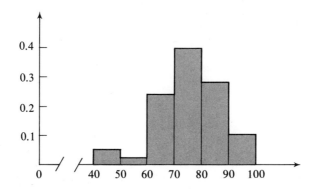

5. Estimated $\bar{x} = 75.57$.

7. The left end point of D (60) is at the 8.6th percentile ($3/35$ of 100) and the right end point (70) is at the 25.7th percentile.

9. From the hours column M and the frequency column f_{ss}, we get the mean $\bar{M}_{ss}$ for suspended solids: $\bar{M}_{ss} = (1/589) \Sigma M \cdot f_{ss} = 1227.7$ (approximately bimonthly).

11. a. Suspended solids; mode: monthly, median: monthly.

    b. *Coliform bacillus*; mode: monthly, median: monthly.

13. Between the 91.2 and the 99.7 percentiles.

15. Answers to parts **a, c, e,** and **g** only.

    a.

x	f	C. f	x	f	C. f	x	f	C. f
0.6	3	3	3.2	1	28	5.6	6	57
0.8	7	10	3.4	1	29	5.8	4	61
1.0	3	13	3.6	2	31	6.0	3	64
1.2	1	14	3.8	2	33	6.4	2	66
1.4	2	16	4.0	5	38	6.6	1	67
1.6	3	19	4.2	3	41	7.2	1	68
2.2	2	21	4.4	1	42	7.4	1	69
2.4	1	22	4.6	1	43	7.6	1	70
2.6	2	24	4.8	3	46	7.8	1	71
2.8	2	26	5.0	1	47			
3.0	1	27	5.2	4	51			

    Note that x is in units of one percent ($x = 1.0$ means $x = 1$ percent). Also, only those rates included in the original chart are used (for example, 1.8 was not in the original chart, so 1.8 is not listed here; its frequency is zero.) C. f = cumulative frequency.

    c. $\bar{x} = (1/71) \Sigma xf = 267.2/71 = 3.76$ percent.

e.

Class	Frequency	C. f
[0.6, 1.4)	14	14
[1.4, 2.2)	5	19
[2.2, 3.0)	7	26
[3.0, 3.8)	5	31
[3.8, 4.6)	11	42
[4.6, 5.4)	9	51
[5.4, 6.2)	13	64
[6.2, 7.0)	3	67
[7.0, 7.8)	3	70
[7.8, 8.6)	1	71

g. The C. f column in part **a** is the cumulative frequency column. 3.0 is at $27/71$ of 100 percent (the 38th percentile). 3.7 is between the 43.7 and the 46.5 percentiles. 5.6 is between the 71.8 and the 80.3 percentiles.

EXERCISE 14.2 (page 427)

1. a. Mean $\bar{x} = 4.25$, variance $= {}^{193.5}\!/_{8} = 24.1875$, standard deviation $= 4.918$.

  b. 11 is at the 87.5 percentile.

3. $\Sigma f = 17$, $\bar{x} = {}^{142}\!/_{17} = 8.35$, variance $= {}^{193.88}\!/_{17} = 11.40$, standard deviation $= 3.38$.

5. a. $n = \Sigma f = 43$; $m =$ midpoints 35, 45, 55, 65, 75, 85, 95; $\bar{x} =$ mean $= (1/43) \Sigma mf = {}^{3175}\!/_{43} = 73.83$; variance $= (1/43) \Sigma (m - \bar{x})^2 f = {}^{9241.86}\!/_{43} = 214.93$; standard deviation $= 14.66$.

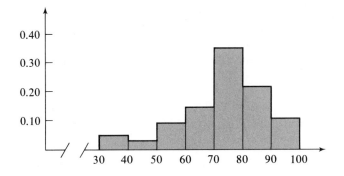

c. Between 30.2 and 65.1 percentiles.

7. a. $n = \Sigma f = 100$, $\bar{x} = {}^{656}/_{100} = 6.56$.

   b. variance $= 11.25$, standard deviation $= 3.35$.

   c. Between 93 and 96 percentiles.

   d. Between 89 and 93 percentiles.

9. a. ${}^{14990000}/_{88000} = 170.34 =$ mean.

   b. variance $= 22.61$; standard deviation $= 4.755$.

11. Use the probability P as the relative frequency f/n. The formulas are mean $= \Sigma x \cdot P(X = x)$, variance $= \Sigma (x - \bar{x})^2 P(X = x)$ or $\Sigma x^2 P(X = x) - \bar{x}^2$. In decimals, $\bar{x} = E(X) = 0 + (0.0988) + 2(0.2963) + 3(0.3951) + 4(0.1975) = 2.6667$; variance $= 0.8886$; standard deviation $= 0.9427$.

EXERCISE 14.3 (page 436)

1. a. Let $Y =$ number of successes in four trials, $P(Y \geq 3) = p^4 + 4p^3 q = 0.0272$.

   b. $X =$ number of successes in five trials, $P(X = 2) = [5!/(2!)(3!)](p^2 q^3) = 0.2048$.

   c. Let $Z =$ number of successes in 100 trials, $\bar{Z} = 100(0.20) = 20$; variance$(Z) = 16$; standard deviation $= 4$.

3. $(p + q)^5 = p^5 + 5p^4 q + 10p^3 q^2 + 10p^2 q^3 + 5pq^4 + q^5$.

5. Success means a hit, $p = 0.35$, let $X =$ the number of successes in 5 times at bat; $P(X = 0) = (0.65)^5 = 0.116$ (five failures).

7. At least two hits $= P(X \geq 2)$: $P(X \geq 2) = (0.35)^5 + 5(0.35)^4 (0.65) + 10(0.35)^3 (0.65)^2 + 10(0.35)^2 (0.65)^3 = 0.572$.

9. $P(X \geq 2) = 0.9988$.

11. The mean is the expected value $\bar{x} = \Sigma xP(X = x)$. Here it is

$$\bar{y} = \sum (n - k) \cdot \frac{n!}{(n-k)! \cdot k!} \cdot p^{n-k}(1-p)^k \qquad \text{Definition of mean.}$$

$$= \sum \frac{n!}{(n-k-1)! \cdot k!} \cdot p^{n-k}(1-p)^k \qquad \frac{n-k}{(n-k)!} = \frac{1}{(n-k-1)!}.$$

$$= np \cdot \sum \frac{(n-1)!}{(n-1-k)! \cdot k!} \cdot p^{n-1-k}(1-p)^k \qquad \text{Factoring out } np.$$

$$= np \cdot (p + (1-p))^{n-1} = np \cdot 1 = np.$$

13. Mean = $(170)(0.85) \approx 144$, standard deviation = 4.655.

15. $(p + q)^9 = p^9 + 9p^8 q + 36p^7 q^2 + 84p^6 q^3 + 126p^5 q^4 + 126p^4 q^5 + 84p^3 q^6 + 36p^2 q^7 + 9pq^8 + q^9$.

17. **a.** $p = 0.8 =$ probability of success (failure probability $= q = 0.2$). Probability of four or fewer successes = the terms of $(p + q)^9$ corresponding to 4, 3, 2, 1, and 0 successes. The sum of these terms is $126p^4 q^5 + 84p^3 q^6 + 36p^2 q^7 + 9pq^8 + q^9 = 0.01958$. **b.** Yes.

EXERCISE 14.4 (page 442)

1. **a.** $z = -1.66$ corresponds to 4.95th percentile.

   b. $z = 0.5$ corresponds to 69.5th percentile.

   c. $z = 1.583$ corresponds to 94th percentile.

   d. $z = -2.75$ corresponds to 0.3th percentile.

3. For the given value $\mu = 37.5$ and $\sigma = 1.1$, 45.5 is more than seven standard deviations greater than the mean and 55 is more than 15 standard deviations greater than the mean. If the organ weight satisfies normal distributions, then the weights are too rare to occur by chance.

5. **a.** $z = -1.6$ implies $x = 13.6$.

b. $z = 3.1$ implies $x = 32.4$ for this given mean and standard deviation.

7. a. 48.8   **b.** 50.4   **c.** 57.2   **d.** 45.4.

9. City X: 99.9 percent, City Y: 50 percent.

11. City X: 89.6 percent, City Y: 7 percent.

13. 78.06 percentile.

15. A: 51.3, B: 47.6, C: 48.9, D: 52.6.

17. a. Far above the 99.9 percentile.

b. There is a smaller standard deviation for the individual.

EXERCISE 15.1 (page 455)

1. a. 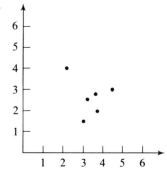   **b.** $y = 3.6957 - 0.314x$.

c. $y = 2.4397$ when $x = 4$.

d. $r = -0.2755$.

3. a. $y = 1.5 + 0.3 x$.   **b.** $y = 3$ at $x = 5$.   **c.** $r = 0.25$.

d. There is no good reason to think so.

5. a. $y = 4.062 + 0.906x$,   **b.** $r = 0.803$.

EXERCISE 15.2 (page 466)

1. Oral report.

3. a. $b_1 = 0.375$, $b_2 = 0.000$, $a = 2.125$, $Y = 2.125 + 0.375X_1$.

   b. $r_{12} = 0.738$, $r_{Y1.2} = 0.707$, $r_{Y2.1} = 0.522$.

   c. Use b_1 and b_2 of part **a**, which are $b_{Y1.2}$ and $b_{Y2.1}$, respectively. Now the β coefficients are $\beta_{V1.2} = b_{Y1.2}(S_1/S_Y) = 0.707$ and $\beta_{V2.1} = b_{Y2.1}(S_2/S_Y) = 0$. Here S_1, S_2, and S_Y are the standard deviations for X_1, X_2, and Y, respectively. Also if $b_{21} = \Sigma x_2 x_1 / \Sigma x_1^2$, then $\beta_{21} = b_{21}(S_1/S_2) = 0.738$.

   d.

   $U_1 = (X_1 - \bar{X}_1)/S_1,\ U_2 = (X_2 - \bar{X}_2)/S_2,\ V = (Y - \bar{Y})/S_Y$.

5. a. $b_1 = 0.049$, $b_2 = 0.221$, $a = 4.74$.

   b. $r_{12} = -0.6$, $r_{Y1.2} = 0.03$, $r_{Y2.1} = 0.10$.

   c. $\beta_{V1.2} = 0.14$, $\beta_{V2.1} = 0.18$.

EXERCISE 16.1 (page 482)

1. a. Value: 5, saddle point: (r_1, c_1).

   b. Value: -2, saddle point: (r_1, c_1).

   c. Value: -2, saddle point: (r_1, c_3).

3. **a.** $\begin{bmatrix} \boxed{-3} & 4 \\ -6 & -2 \end{bmatrix}$ **b.** $\begin{bmatrix} -4 & -2 & 1 & 5 & 9 \\ \boxed{-2} & -1 & 0 & 4 & 8 \end{bmatrix}$ **c.** $\begin{bmatrix} \boxed{30} & 40 \\ -45 & 48 \end{bmatrix}$

5. **a.** Saddle point: row 1, column 2. (Pollution control, state J_2.)

   b. $\begin{bmatrix} 6 & 3 & 4 \\ 1 & -5 & 7 \end{bmatrix}$

   No, it cannot return to the no-control decision.

EXERCISE 16.2 (page 492)

1. **a.** R plays r_1 with probability $7/8$ and r_2 with probability $1/8$.
   C plays c_1 with probability $5/8$ and c_2 with probability $3/8$.

   b. R plays r_1 with probability $14/23$ and r_2 with probability $9/23$.
   C plays c_1 with probability $20/69$ and c_2 with probability $49/69$.

   c. R plays each row with probability $1/2$, C plays each column with probability $1/2$.

3. **a.** $V = 21/8$. **b.** $V = 1/23$. **c.** $V = 0.49$.

5. $n!/(n-2)!2! = n \cdot (n-1)/2$.

7. Subgame I   Subgame II   Subgame III
   $\begin{bmatrix} 6 & -1 \\ -2 & 4 \end{bmatrix}$ $\begin{bmatrix} 6 & \boxed{-1} \\ 1/2 & -1 \end{bmatrix}$ $\begin{bmatrix} -2 & 4 \\ 1/2 & -1 \end{bmatrix}$
   Value = $22/13$ Value = -1 Value = 0

   R should play Subgame I. The strategy for Subgame I is to play r_1 with probability $6/13$, r_2 with probability $7/13$, and r_3 with probability 0.

9. **a.** R plays r_1 with probability $2/3$ and r_2 with probability $1/3$. His game value $V = 4/3$, independent of C's moves.

   b. R plays r_1 with probability $1/43$, r_2 with probability $42/43$, and r_3 with probability 0.

EXERCISE 16.3 (page 499)

1. a. r_1 is dominated; remove it. In the resulting game, r_4 is dominated; remove it. In the resulting game, the column c_1 is dominated; remove it. The final result is the two-by-two game

$$\begin{bmatrix} 5 & -2 \\ -4 & 0 \end{bmatrix},$$

in which R plays the strategy $(4/11, 7/11)$ with game value $-8/11$. The strategy for the original game is to play r_1, r_2, r_3, and r_4 with strategies 0, $4/11$, $7/11$, and 0, respectively. C's strategies are 0, $2/11$, and $9/11$.

b. Column c_2 is dominated; remove it. In the resulting game, r_1 is dominated; remove it. The resulting two-by-two game is

$$\begin{bmatrix} 0 & 6 \\ 3 & 1 \end{bmatrix}.$$

The strategy for the original game is $R: (0, 1/4, 3/4)$ and $C: (5/8, 0, 3/8)$. The game value is $V = 9/4$.

3. C can play several different strategies with a zero game value. If R assumes C is an intelligent player, then R should play r_1 and r_2 with probability $1/2$ each.

5. The game reduces to

$$\begin{array}{c} & c_3 \quad c_4 \\ r_1 \\ r_3 \end{array} \begin{bmatrix} -4 & 3 \\ 2 & -4 \end{bmatrix}$$

with a game value of $-10/13$. R's strategy is to play the rows r_1, r_2, r_3, and r_4 of the original game with probabilities $6/13$, 0, $7/13$, and 0, respectively. C plays the columns with probabilities 0, 0, $7/13$, and $6/13$.

EXERCISE 16.4 (page 511)

1. a. Follow the chart given in Figure 16.5. Translate by adding 2 to each entry, thereby creating the new game

$$\mathbf{G}' = \begin{bmatrix} 4 & 8 & 2 \\ 5 & 1 & 6 \end{bmatrix}.$$

Let V = value of $\mathbf{G}'$. We will find the linear programming strategy for playing the columns. Let p_1, p_2, and p_3 be the probabilities ($p_1 + p_2 + p_3 = 1$) for playing the columns regardless of the rows being played. Then,

$$4p_1 + 8p_2 + 2p_3 \leq V$$
$$5p_1 + p_2 + 6p_3 \leq V.$$

Let $y_k = p_k/V$; now, the problem is to maximize $F = 1/V = y_1 + y_2 + y_3$ subject to

$$4y_1 + 8y_2 + 2y_3 \leq 1$$
$$5y_1 + y_2 + 6y_3 \leq 1.$$

Introduce slack variables y_4 and y_5 and assume the basis is y_4, y_5. Then the initial simplex table is

	C_i	1	1	1	0	0		
		y_1	y_2	y_3	y_4	y_5	y_0	C
y_4		4	8	2	1	0	1	0
y_5		5	1	6	0	1	1	0
z		0	0	0	0	0	0	
$C - z$		1	1	1	0	0		

The final table is

	y_1	y_2	y_3	y_4	y_5	y_0	C
y_2	7/23	1	0	6/46	-1/29	4/46	1
y_3	18/23	0	1	-1/46	4/23	7/46	1
z	25/23	1	1	5/46	3/46	11/46	
$C - z$	-2/23	0	0	-5/46	-3/46		

Maximum of $1/V$

Chapter 16

The value of game $\mathbf{G}'$ is $V = {}^{46}/_{11}$. The table shows the solution to be $y_1 = 0$, $y_2 = {}^{4}/_{46}$, and $y_3 = {}^{7}/_{46}$. By the definition of y_k, we have $p_k = y_k \cdot V$; so, $p_1 = 0 \cdot ({}^{46}/_{11}) = 0$, $p_2 = ({}^{4}/_{46})({}^{46}/_{11}) = {}^{4}/_{11}$, and $p_3 = ({}^{7}/_{46})({}^{46}/_{11}) = {}^{7}/_{11}$. The value of the original game is $V - 2 = {}^{24}/_{11}$. C's strategy is to play the original game with strategies $(0, {}^{4}/_{11}, {}^{7}/_{11})$. For the row player, let q_1 and q_2 be the strategies with $q_1 + q_2 = 1$; also let $x_k = q_k/v$. The objective to be minimized is $F = 1/v = x_1 + x_2$ subject to the constraints

$$4x_1 + 5x_2 \geq 1$$
$$8x_1 + x_2 \geq 1$$
$$2x_1 + 6x_2 \geq 1.$$

Again the linear programming solution produces the minimum of ${}^{11}/_{46}$ for $1/v$ for $x_1 = {}^{5}/_{46}$, $x_2 = {}^{6}/_{46}$, and x_3 (a slack variable) $= {}^{4}/_{46}$. But since $x_k = q_k \cdot V$, then $q_1 = ({}^{5}/_{46})({}^{46}/_{11}) = {}^{5}/_{11}$ and $q_2 = {}^{6}/_{11}$.

b. Use of the subgame method produces the same strategies as the linear programming method.

3. For the column player, let p_1, p_2, and p_3 be the probabilities for playing the columns. $V =$ game value. Let $y_k = p_k/v$, then maximize $F = 1/v = y_1 + y_2 + y_3$, subject to

$$2y_1 + y_2 + 3y_3 \leq 1$$
$$y_1 + 2y_2 + y_3 \leq 1$$
$$3y_1 + y_2 + 2y_3 \leq 1.$$

The solution is $V = {}^{8}/_{5}$ at $y_1 = {}^{1}/_{8}$, $y_2 = {}^{3}/_{8}$, $y_3 = {}^{1}/_{8}$, and $p_1 = {}^{1}/_{5}$, $p_2 = {}^{3}/_{5}$, and $p_3 = {}^{1}/_{5}$. For R the probabilities are the same.

INDEX

A

Abscissa of a point, 66
Absorbing Markov chain, 405
Absorbing state, 404
Abuse of notation, 303
Acceptance level, 435
Addition
 of matrices, 187
 of vectors, 177
Algorithm, 132
Arithmetic mean, 410
Artificial variables, 275
 maximizing problems, 276
 minimizing problems, 280
Axioms of order, 39

B

Basic variables, 144
Basis, 249
Bayes' theorem, 368–77
 conditional probabilities, 369
Bell-shaped curve, 438ff
Best-fitting line, 446–47
Binomial
 distribution, 430–36
 experiment, 430
Boundary lines, 120
Break-even analysis, 112

C

Ceiling analogy (objective function), 131
Celsius temperature, 103
Class, 412ff

Class (*continued*)
 boundaries, 412
 frequency, 412
 interval, 412
Coefficient
 correlation, 452, 458ff
 least squares line, 449
 path, 463
 regression, 452
 weights in constraints, 46
Column maxima (game theory), 475
Column vector, 164
Combinations, 313
 n things k at a time, 316
Combined probabilities (Table 11.1), 337
Complement, 14
Components of a vector, 164
Conditional probabilities, 323–29
Conjointly exhaustive subsets, 343
Constant sequence, 58
Constraint inequality, 45
Convex, 130
Coordinates, 66
Corner points, 131
Correlation, 445–55
 concept, 445
 high, 445
 linear, 452
Correlation coefficients, 452ff
 formula, 452, 453
 partial, 461
Counting
 combinations, 316

Counting (*continued*)
 elements of sets, 2
 multiplicative principle, 308
 permutations, 315
Cramer's rule, 163
 three-by-three systems, 171
Cross partition, 345
Cumulative frequency, 414

D

Data, 409
Data variability, 419–27
Decision theory, 471
 (*See also* Games)
Delta notation, 89
De Morgan's laws, 335
Dependent variable, 80
Desire to win postulate (game theory), 472
Determinants, 160ff
 used in solving systems of equations, 163, 167
 third-order, 167
 two-by-two, 160
 value of, 160, 168
Deterministic model, 291
Deviation from the mean, 419
Deviation of a point from a line, 446
Dice, 312
Difference $A - B$ of two sets, 18
Distribution
 binomial, 430–36
 frequency, 410
 graph, 355ff

570

Index

Distribution (*continued*)
  normal, 439
  induced by the probability function, 354
  of a random variable, 354
Distributive law, scalar multiplication of vectors, 181
Domain of a function, 78
Dominated column (Game theory), 495
Dominated row, 496

E

Elements
  of a determinant, 161
  of a matrix, 185
Elimination, 226
Empty set, 21
Equality properties, 40
Equally likely events, 306
Equations
  linear, 97ff
  systems of, 109
Equiprobable space, 306
Events, 295
  independent, 327
  probability of, 301ff
  random, 294
  simple, 295
Expansion by minors, 169
Expectation, 365
Expected value(s)
  definition, 362
  laws, 365-66
  statistics, 410
Experiment, 294
  outcomes, 295
  sample space for, 301
Extreme point, 133

F

Factorial, 313
Fahrenheit temperature, 103
Failure (outcome of a binomial experiment), 431
Fair game, expected value of, 365
Feasible solution, 133
Finite probability, 302ff
  function, 302
  model, 303
Fixed point, 386, 399
Floor plan analogy (constraint region), 131

Frequency distribution, 410
Function
  concept, 78
  domain of, 78
  graph of, 78
  linear, 97
  notation, 86
  objective, 133
  probability, 302
  random variable, 353
  range of, 78
  two or more variables, 90

G

Gambler's ruin, 405
Game(s)
  matrix form, 473
  nonstrictly determined, 487
  payoff matrix, 473
  plan, 480
  postulates, 471, 472
  strategies, 475ff
  strictly determined, 476
  theory, 471–511
  two-person, 471
  value of, 478, 486, 502
  zero-sum, 473, 475
Gauss-Jordan elimination, 225
  solution of general linear systems, 234
  two-by-two and three-by-three systems, 226–33
Geometric linear programming, 132
Geometry of inequalities, 120
Graphs
  area under a graph, 63
  of a function, 78
  path analysis, 463
  probability distribution, 352
  rate of change, 63
  transition diagrams, 389
Gross product (matrix model), 220
Grouped data, 425ff
Growth rates, 88, 102

H

Hierarchical system of equations in a causal model, 84, 465
Histogram, 413

I

Identity matrix, 205, 212, 242
Incoming variable, 249
Independent events, 327
Independent variable, 80
Inequalities, 35
  axioms, 39
  constraint, 45
  symbols, 36
  theorems, 41
Infeasible solution, 125
Initial feasible solution, 274ff
Initial vector, 392
Input-output model, 219ff
Intelligent player postulate (Game theory), 472
Intersection of sets, 17
Invariance in game theory
  multiplication, 479
  translation, 479
Inverse matrix, multiplicative, 209

J

Join of sets, 19

L

Laws
  De Morgan's, 335
  of expectation, 365
  of vectors, 180
Least squares line, 448ff
  formulas, 449
Leontief, 219ff
  economic model, 219
  input-output matrix, 219
Linear, 98
Linear equations, 98
  equivalent systems, 111
  systems of, 109
Linear functions, 97
Linear inequalities, 120
  plane regions, 120–28
Linear programming, 130ff, 249ff
  artificial variables, 275
  basic variables, 144
  basis, 249
  game theory, 502
  geometric algorithm, 132
  history, 286
  incoming variable, 249
  objective function, 133
  opportunity cost, 252

Linear programming (*continued*)
  outgoing variable, 264
  pivoting, 274
  potential net contribution, 252
  replacement quotient, 265
  replacement rule, 265
  simplex algorithm, 287-88
  slack variable algorithm, 153
Linear regression, 445-55
Line equation, 97
  graph of, 98
  point-slope form, 106
  two-point form, 101
Lines
  intersecting, 112
  area between, 113
Logical implications and subsets, 29

M

Marginal analysis, 88
Markov process, 387
Matrix, 185ff
  addition, 187
  elements, 185
  equality, 186
  identity, 205
  inverse (multiplicative), 209
  multiplication
    by a matrix, 199
    by a scalar, 187
    by a vector, 191, 194
  singular, 216
  steady-state, 402
  stochastic, 382
  transition, 390
Matrix games, 471ff
Maximizing solution, 133
Mean, 410
  deviations from the mean, 419
Median, 412
Minimax strategy for games, 476
Minimizing solution, 133
Mixed constraints, 282
Mixed strategies, 485ff
Mnemonic device, 162, 167
Mode, 412
Multiple regression, 458-66
Multiplication invariance in game theory, 479
Multiplicative principle in counting, 308

N

Net production (matrix economic model), 221
Nonbasic variables, 144
Nonoverlapping sets, 4
Normal distribution, 438-42
  z-scores, 439-40
Normally distributed, 439
N-simplex, 287
Number of elements in a set, 2

O

Objective function, 133
Opportunity cost, 252
Ordered pair, 65
Ordinate of a point, 66
Origin of a coordinate system, 66
Outcomes of an experiment, 295
Outgoing variable, 264

P

Parallel lines, 108
Partition(s), 341-47
  cross, 345
  definition, 343
  theorem, 346
Path
  analysis, 465-66
  coefficients, 463
  diagram, 463
  model, 464
Payoff matrix, 473
Percentile, 412
Permutations, 313
  n things k at a time, 315
Pivoting, 274
Postulates (Game theory), 471-72
  desire to win, 472
  intelligent player, 472
  well-defined utility, 472
Potential net contribution, 252
Primitive term, 1
Probability, 291-322
  of combined events, 337
  conditional, 325
  De Morgan's laws, 335
  equiprobable, 306
  events, 295
  experiment, 294
  finite model, 303
  function, 302

Probability (*continued*)
  relative frequency, 299
  sample space, 301
  of simultaneous events, 324
  strategy in games, 486ff
  total (union of events), 332, 334
  vectors, 379

R

Random events, 294
Random numbers, 292
Random variable(s), 351-58
  concept, 351ff
  definition, 353
  distribution, 354
  probability function, 354
Range of a function, 78
Rate of change, 63, 89, 101
Reduction by dominance, 495-96
Regions of the plane, 120
  boundary line, 120
Regression
  coefficient, 452
  linear, 452
  multiple, 458-59
  partial, 461ff
Regular stochastic matrix, 398
Rejection level, 435
Relation, 73ff
  expressed by equations and inequalities, 74
  function, 78
  inverse, 76
  as a set of ordered pairs, 73
  between sets of numbers, 73
Relative frequency, 299
  distribution, 410
Replacement quotient, 265

S

Saddle point, 476
Sample point, 295
Sample space, 301
Scalar product of vectors, 179
Scatter diagram, 451
Sequence, 58
Set
  builders (braces), 16
  complement, 14
  empty, 21
  intersection, 17
  membership, 14
  primitive term, 1

Index

Set (*continued*)
  singleton, 16
  solution, 39
  subsets, 27
  union, 19
  universe, 30
  Venn diagrams, 1
Simple events, 295
Simplex algorithm, 287
Simultaneous events, 323
Simultaneous linear equations, 109ff
  methods of solution, 110
Singleton set, 16
Singular matrix, 216
Slack variable
  applications to management, 47–51
  geometric meaning, 140
  linear programming algorithm, 153–54
  objective function, 151ff
  theorem, 47
  use in finding corner points, 142
Slope of a line, 106
Solution set, 39
Squared deviation from the mean, 421
Squared error, 446
  sum of, 446ff
Standard deviation, 422
Standardized scores, 440
States, 387–94
  absorbing, 404
  Markov process, 387ff
  transitions, 390
Steady-state matrix, 402
Stochastic matrix
  concept, 381
  definition, 382
  regular, 398
Strategies
  flow diagram, 477, 499
  linear programming, 502–10
  reduction by dominance, 495–96

Strategies (*continued*)
  saddle point, 476
  subgames, 490
Strictly determined games, 476
Subgames, 490
Subscripts, 45, 66
Subset, 27
Subtraction, $A - B$ of two sets, 18
Success (outcome in a binomial experiment), 431
Summation notation, 56
Systems of equations, 109–14
  Cramer's rule, 163, 171
  Gauss-Jordan elimination, 226–33
  hierarchical, 84, 465
  slack variables, 48, 138ff
  (*See also* Linear programming)

T

Tableau, 255
Tabulation of the problem, 249
Terminal matrix, 402
Total probabilities, 332–38
  disjoint sets, 332
  overlapping sets, 334
  table of formulas, 337
Transitions, 387–94
  diagram, 389
  matrix, 390
  states, 387ff
  trees, 389
  two-step, 388ff
Transpose of a matrix, 211
Tree diagram, 309
  transition, 389ff
Trials in a binomial experiment, 430ff

U

Union of sets, 19
Universe, 30
  of discourse, 30

V

Value of a determinant, 160, 168
Value of a function, 86
Value of a game, 478, 486, 502
Variable(s), 35
  artificial, 275
  basic, 144
  dependent, 80
  independent, 80
  nonbasic, 144
  slack, 47
Variance, 421
Vectors, 176ff
  addition, 177
  column, 164
  component of, 164
  equality, 177
  laws, 180
  notation, 164
  probability, 379
  scalar product, 179
Venn diagrams, 1ff
  complement of a set, 14
  counting problems, 2
  intersection of sets, 17
  probabilities, 328, 336
  subsets, 27
  union of sets, 19

W

Well-defined sets, 1
Well-defined utility postulate in game theory, 472

Z

z-Scores, 439
Zero-sum games, 473

INDEX TO APPLIED PROBLEMS AND EXAMPLES

Adrenal gland size related to stress, 442
Advertising vs sales, 456
Age of TV viewers, 438
Age of U.S. voters, 440, 442
Agricultural economy, 459
Air pollution, 193
American economy, 219ff
Animal weight estimated from length, 83
Anthropology
   intergenerational birthplace movement, 397
   nasal bone length, 444
   primitive society, 206
Aptitude tests, 443
Automobiles
   accidents, 326, 330
   electric, 108
   exceeding certain speeds, 72
   repair service, 299

Baseball
   base runners, 311
   batting averages, 436
   batting order, 314
Basketball scores, 90
Biology
   animal weight, 83ff
   biochemical process, 175
   biological membranes, 176
   cell identification by light scattering, 457
   genotypes, 392ff
   insect invasion, 115
   lung, heart, and intestine energy supply, 94
   meristic variability, 340
   mutation in cell division, 362, 364
   sampling
      hereditary traits, 375
      organ disorder, 431, 434
   sense receptors, 321
Birth of lambs, 197
Birth rate in North America, 180
Black lung payment program, 429
Blood chemistry, 445
Blood classification, 13
Boxing match, radio report, 376
Brand loyalty (*See also* Consumer loyalty), 198
Break-even point, 112ff
British Forestry commission, 63ff
Brokerage house, 2
Business
   advertising vs sales, 456
   break-even point, 112
   brokerage house, 2
   chain of discount stores, 157
   consumer decision game, 500
   consumer survey, 11, 339
   door-to-door salesman, 485
   farm income, 95, 106
   fast food stores, 11
   future business prospects game, 493ff
   GNP, 68ff
   insurance company, 361, 364
   marginal costs, 94
   marginal revenue, 88ff, 94
   mutual fund game, 501
   pocket calculator sales, 247ff
   prime commercial paper rates, 418
   public judgment game, 483

   rental service stores, 4
   retail stores, 206
   sampling for quality control, 357ff, 360, 431, 433, 437
   stockbroker's commission, 35
   stockholder's horoscope, 331
   stock portfolio, 106

Car pools and bridge tolls, 116
Cell identification by light scattering, 457
Chess
   Karpov's first move, 332
   matrix game against Petrosian, 478ff
   possible outcomes, 297
   tournament, 367
   world champions, 16
City climates, 443ff
Coin tossing, 303, 309, 311, 320, 331
College selection game, 480ff
Color blind individuals, 298
Color words and mood words, 23
Communication
   information theory, 407ff
   international flag code, 317
   signal horn, 320
   telephone switchboard, 305ff
   television sets, 438
Community council, 318
Constraints (*See* Management)
Consumer decision game, 500
Consumer loyalty, 388ff
Consumer price index, 68
Consumer survey, 11, 339
Consumption and income, 104
Course grade, 361, 364
Crowding problem, 59

Applied Problems and Examples

Decaying bark vs salmon eggs, 109
Defective items, 357ff, 360, 431, 433, 437
Dice throws, 311ff, 331, 333, 334, 356, 359
Dinosaurs, 23
Disease incidence by age, 72
Door-to-door salesman, 485
Durable goods, 177

Ecology
  air polution, 193
  ecosystems, 156
  industrial consumption of natural resources, 53
  oil slick, 105
  pollution-free car, 108
  tree planting, 63ff
  water quality control, 416ff
  water supply and use in the U.S., 175
Economics
  agricultural, 459
  durable goods, 177
  employment distribution, 202ff
  input-output systems, 156
  Leontief's economic models, 219ff
  ore production, 178, 186
  savings and income, 104
  simple economy, 222
Electric car, 108
Elementary school students' self-evaluation, 457
Eliot Ness game, 500
Employee test scores, 369ff
Employment, 182, 202
Energy supplied to lung, heart, and intestine, 94
Extrasensory perception, 321ff

False alarms, 377
Farm income, 95, 106
Farm management, 50ff, 136
Fast food stores, 11
Fisherman's probabilities, 336
Flush toilets, 25ff
Football player, 339
Football watchers, 456
Friday the thirteenth, 310
Fruit farmers, 11
Future business prospects game, 493ff

Gambling
  card selection, 293ff, 297, 326, 352
  coin tossing, 303, 309, 311, 320, 331
  dice throws, 311ff, 331, 333, 356
  gambler's ruin, 405ff
  roulette wheel, 367ff
  simulation of a poker hand, 293ff, 297
Genotypes, 392ff
Gout and hyperuricemia, 32
Gross National Product, 68
Growing conditions, 349

Health sciences
  adrenal gland size related to stress, 442
  alcohol vs marihuana, 61ff
  blood chemistry, 445
  blood classification, 13, 338
  color blind individuals, 23
  disease incidence by age, 72
  drug effectiveness, 437
  gout and hyperuricemia, 32
  ill effects of smoking, 298
  immunology, 468
  nursing, 43
  obstetrical nurse, 294ff, 300ff
  radio-isotope tracers, 137
  warts, 304
  wound healing rates, 105
Homing pigeons, 298
Hospital management, 137
Hotel services, 183
Household conveniences, 25ff, 324
Housing development, 314, 316
Hypnosis to treat warts, 304

Ill effects of smoking, 298
Immunology, 468
Industrial consumption of natural resources, 53
Information theory, 407ff
In-out game, 14
Input-output systems (*See also* Leontief's), 156
Insect invasion, 115
Insurance company, 361, 364
Intergenerational birthplace movement, 397
Intergenerational occupation movement, 396
International flag code, 317

Jury trial, 372ff

Labor and payroll constraints, 52
Learning models, 138, 236ff
Leontief's economic models, 219ff
Lloyd's index of mean crowding, 59

Magnetic recording tape, 311
Major subjects in college game, 481
Management
  city manager budget, 119
  as consumers of resources, 44-50, 91, 119, 137, 151ff
  farm management, 50ff, 136ff
  hiring options, 49ff, 54
  hospital management, 137
  housing development, 314, 316
  labor and payroll constraints, 52
  milkman's route, 95, 134ff
  paper, lumber, and chemical industry, 53
  personnel hiring, 49ff, 130, 137
  petroleum refinery, 150, 157
  police management, 54
  shipping problem, 53, 150
Marginal cost, 94
Marginal revenue, 88ff, 94
Meristic variability, 340ff
Meteorology
  meteorological conditions, 301ff, 304
  tornado damage, 295ff, 304
  weather, 333, 335
Milkman's route, 95, 134ff
Mutations in cell division, 362, 364
Mutual fund game, 501

Nasal bone length, 444
Negotiation matrix, 385ff
North American birthrate, 179
Nursing, 43

Obstetrical nurse, 294ff, 300
Occupational changes between generations, 396
Oil slick, 105
Order of candidates on a ballot, 319ff
Ore production, 178, 186

Personality and disease, 327ff
Personnel hiring, 49ff, 130, 137
Petroleum refinery, 150, 157
Playing card selection, 293, 326, 352
Pocket calculator sales, 247
Poker hand, 293ff, 297
Police department management, 54
Political science
  order of candidates on a ballot, 319ff
  precinct vote, 379, 383
  U.S. House of Representatives, 40, 43, 346
  U.S. Senate, 349
  U.S. voters' ages, 440ff
  voting habits, 395
Pollution free car, 108
Postal service, 437
Poverty households, 24
Precinct vote, 379, 383
Prime commercial rates, 418
Primitive society, 206
Prison admissions, 429
Psychology
  aptitude tests, 443
  color words and mood words, 23
  elementary school students' self-evaluation, 457
  employee test scores, 369ff
  extrasensory perception, 321ff
  learning models, 138, 236ff
  personality and disease, 327ff
Public judgment game, 483
Public transit systems, 116, 117

Radio-isotope tracers, 137
Rental service stores, 4ff
Retail stores, 206
Roulette wheel, 367ff
Royal family, 361, 363

Sampling (biology), 375, 431, 434ff
Sampling (manufacturing), 357ff, 360, 431, 437
Savings and income, 104
Seating capacity, 443
Secretaries and executives, 110ff
Shipping problem, 53, 150
Signal flags, 317ff
Signal horn, 320
Simple economy matrix, 222ff

Simulation of a poker hand, 293ff, 297
Sociology
  false alarms, 376ff
  household conveniences, 25ff
  in-out game, 14
  intergenerational birthplace movement, 397
  intergenerational socio-economic changes, 396ff
  jury trial, 372ff
  Lloyd's index of mean crowding, 59
  poverty households, 24
  primitive society, 206
  prison admissions, 429
  student attitudes, 12ff
  war deaths, 189
Spaceship war game, 483ff
Stockbroker's commission, 35ff
Stockholder's horoscope, 331
Stock portfolio, 106
Stockroom inventory, 395
Storage problem, 238
Swine
  mammae, 351
  litter size, 428

Telephone switchboard, 305ff
Television sets, 438
Television story writer, 308ff
Temperature conversion, 103ff
Tornado damage, 295ff, 304
Transportation
  car pools and bridge tolls, 116
  public transit systems, 117
  shipping problem, 53, 150
  traffic flow, 85
Tree planting, 63ff
Trim loss problem, 285ff

U.S. House of Representatives, 40, 43, 346
U.S. Senate, 349
U.S. voters' ages, 440ff

Voting habits, 395

Walnuts, 297
War deaths, 189
Warehouse needs, 493ff
Warehouse storage, 238
Warts, 304
Water flow through a pipe, 70

Water quality control, 416ff
Water supply and use in the U.S., 175
Weather (*See also* Meteorology), 333, 335
Wine consumption, 409
Wound healing rate, 105